# Algebra: Essential Mathematics

# Algebra: Essential Mathematics

Editor: Logan Phillips

STATES
ACADEMIC PRESS
www.statesacademicpress.com

Published by States Academic Press
109 South 5th Street,
Brooklyn, NY 11249, USA
www.statesacademicpress.com

Algebra: Essential Mathematics
Edited by Logan Phillips



International Standard Book Number: 978-1-63989-039-2 (Hardback)



**Cataloging-in-Publication Data**

Algebra : essential mathematics / edited by Logan Phillips.
p. cm.
Includes bibliographical references and index.
ISBN 978-1-63989-039-2
1. Algebra. 2. Mathematics. 3. Mathematical analysis. I. Phillips, Logan.
QA155 .A44 2022
512--dc23

# Contents

# Preface

Algebra is the study of mathematical symbols and the rules of manipulating these symbols. It is essential in almost all of mathematics. Algebra encompasses solving elementary equations and the study of abstractions such as groups, rings, and fields. Elementary algebra comprises the basic parts of algebra whereas the more abstract aspects fall under abstract algebra or modern algebra. Elementary algebra is required for any study of mathematics, science, or engineering, as well as medicine and economics. Abstract algebra is used in advanced mathematics, and is studied primarily by professional mathematicians. This book is an essential guide for both academicians and those who wish to pursue this discipline further. The topics included herein on algebra are of utmost significance and bound to provide incredible insights to readers. This book aims to equip students and experts with the advanced topics and upcoming concepts in this area.

After months of intensive research and writing, this book is the end result of all who devoted their time and efforts in the initiation and progress of this book. It will surely be a source of reference in enhancing the required knowledge of the new developments in the area. During the course of developing this book, certain measures such as accuracy, authenticity and research focused analytical studies were given preference in order to produce a comprehensive book in the area of study.

This book would not have been possible without the efforts of the authors and the publisher. I extend my sincere thanks to them. Secondly, I express my gratitude to my family and well-wishers. And most importantly, I thank my students for constantly expressing their willingness and curiosity in enhancing their knowledge in the field, which encourages me to take up further research projects for the advancement of the area.

**Editor**

# 1

# FREE MODIFIED ROTA-BAXTER ALGEBRAS AND HOPF ALGEBRAS

Xigou Zhang, Xing Gao and Li Guo

Communicated by Christian Lomp

ABSTRACT. The notion of a modified Rota-Baxter algebra comes from the combination of those of a Rota-Baxter algebra and a modified Yang-Baxter equation. In this paper, we first construct free modified Rota-Baxter algebras. We then equip a free modified Rota-Baxter algebra with a bialgebra structure by a cocycle construction. Under the assumption that the generating algebra is a connected bialgebra, we further equip the free modified Rota-Baxter algebra with a Hopf algebra structure.

**Mathematics Subject Classification (2010)**: 16T99, 16W99, 16S10
**Keywords**: Modified Rota-Baxter algebra, Rota-Baxter algebra, Hopf algebra, bracketed word, cocycle

## 1. Introduction

This paper studies free objects in the category of modified Rota-Baxter algebras, a concept coming from the combination of a Rota-Baxter algebra and a modified Yang-Baxter equation. It also equips the free objects with bialgebra and Hopf algebra structures.

For a fixed constant $\lambda$, a *Rota-Baxter operator* of weight $\lambda$ is a linear operator $P$ on an associative algebra $R$ that satisfies the *Rota-Baxter equation*:

$$P(x)P(y) = P(P(x)y) + P(xP(y)) + \lambda P(xy), \quad \forall x, y \in R. \tag{1}$$

An associative algebra $R$ equipped with a Rota-Baxter operator is called a *Rota-Baxter algebra*, a notion originated from the probability study of G. Baxter [8] in 1960. Later it attracted the attention of well-known mathematicians such as Atkinson, Cartier and Rota [2,10,26]. After some years of dormancy, its study experienced a quite remarkable renascence since late 1990s, with many applications in mathematics and physics [1,3,14,17,19,21,22,24,25]. In particular, it appeared as one of the fundamental algebraic structures in the profound work of Connes and Kreimer on renormalization of quantum field theory [11]. See [16] for further details and references.

The concept of the classical Yang-Baxter equation arose from the study of inverse scattering theory and is also related to Schouten bracket in differential geometry. Further it can be regarded as the classical limit of the quantum Yang-Baxter equation, named after C. N. Yang and R. Baxter. In the 1980s, Semonov-Tian-Shansky [27] found that, under suitable conditions, the operator form of the classical Yang-Baxter equation is precisely the Rota-Baxter identity (1) (of weight 0) on a Lie algebra. As a modified form of the operator form of the classical Yang-Baxter equation, he also introduced in that paper the *modified classical Yang-Baxter equation:*

$$[P(x), P(y)] = P[P(x), y] + P[x, P(y)] - [x, y], \tag{2}$$

later found applications in the study of generalized Lax pairs and affine geometry on Lie groups [4,9,20]. As the associative analogue of Eq. (2), the equation

$$P(x)P(y) = P(P(x)y) + P(xP(y)) - xy. \tag{3}$$

is called the *modified associative Yang-Baxter equation*, which has been applied to the study of extended $\mathcal{O}$-operators, associative Yang-Baxter equations, infinitesimal bialgebras and dendriform algebras [6,7,12].

In the spirit of the aforementioned Yang-Baxter equation to Rota-Baxter operator connection, a linear operator $P$ satisfying Eq. (3) is called a *modified Rota-Baxter operator* and an associative algebra $R$ equipped with a modified Rota-Baxter operator is called a *modified Rota-Baxter algebra.*

Integrating the notions of the Rota-Baxter algebra and modified Rota-Baxter algebra, the concept of a modified Rota-Baxter algebra with a weight was introduced in [6] as a special case of extended $\mathcal{O}$-operators in connection with the extended associative Yang-Baxter equation. The latter motivated their study in the Lie algebra context [5]. In [29], free commutative modified Rota-Baxter algebras were constructed by means of a modified quasi-shuffle product and modified stuffle product, in analogy to the case of free commutative Rota-Baxter algebras [10,17].

Considering the close relationship between the modified Rota-Baxter (associative) algebras and the modified Yang-Baxter equation for Lie algebras, it is especially interesting to consider noncommutative modified Rota-Baxter algebras. This is the subject of study of this paper, focusing on the construction of the free objects and the Hopf algebra structures on the free objects. More precisely, in Section 2, we obtain an explicit construction of the free modified Rota-Baxter algebra on an algebra, by giving a natural basis of the algebra and the corresponding multiplication table. In Section 3, we further provide a bialgebra and then a Hopf algebra structure on the free modified Rota-Baxter algebra.

**Notations.** For the rest of this paper, unless otherwise specified, algebras are *associative unitary* algebras over a commutative unitary ring $\mathbf{k}$.

## 2. Free modified Rota-Baxter algebras

In this section we construct free modified Rota-Baxter algebras. We give the construction in Section 2.1, leading to the main Theorem 2.6 of this section. The proof of the theorem is completed in Section 2.2.

### 2.1. The general construction of the free modified Rota-Baxter algebras.

We begin with the general definition of modified Rota-Baxter algebras.

**Definition 2.1.** Let $R$ be a $\mathbf{k}$-algebra and $\kappa \in \mathbf{k}$. A linear map $P : R \to R$ is called a *modified Rota-Baxter operator* of weight $\kappa$ if $P$ satisfies the operator identity

$$P(u)P(v) = P(uP(v)) + P(P(u)v) + \kappa uv, \quad \text{for all } u, v \in R. \tag{4}$$

Then the pair $(R, P)$ or simply $R$ is called a *modified Rota-Baxter algebra* of weight $\kappa$.

Together with the algebra homomorphisms between the algebras that preserves the linear operators, the class of modified Rota-Baxter algebras of weight $\kappa$ forms a category. We refer the reader to [29] and the references therein for basic properties of modified Rota-Baxter algebras and focus our attention to the construction of free modified Rota-Baxter algebras. We first give the definition.

**Definition 2.2.** Let $A$ be a $\mathbf{k}$-algebra. A *free modified Rota-Baxter algebra on $A$* is a modified Rota-Baxter algebra $(F(A), P_A)$ together with an algebra homomorphism $j : A \longrightarrow F(A)$ with the property that, for any given modified Rota-Baxter algebra $(R, P)$ and algebra homomorphism $f : A \longrightarrow R$, there is a unique homomorphism $\bar{f} : F(A) \longrightarrow R$ of modified Rota-Baxter algebras such that $\bar{f}j = f$.

Note that taking $A$ to be the free algebra $\mathbf{k}\langle Y\rangle$ on a set $Y$, we obtain the free modified Rota-Baxter algebra on the set $Y$. Let $A$ be a $\mathbf{k}$-algebra with a $\mathbf{k}$-basis $X$. We first display a $\mathbf{k}$-basis $\mathfrak{X}_\infty$ of free modified Rota-Baxter algebras in terms of bracketed words from the alphabet set $X$.

**Remark 2.3.** *The set* $\mathfrak{X}_\infty$ *is called the set of* Rota-Baxter words *that was applied to construct free Rota-Baxter algebras* [14]. *Enumeration properties and generating functions of Rota-Baxter words were obtained in* [18] *to which we refer the reader for further details.*

Let $\lfloor$ and $\rfloor$ be two different symbols not in $X$, called brackets, and let $X' := X \cup \{\lfloor, \rfloor\}$. Denote by $M(X')$ the free monoid generated by $X'$.

**Definition 2.4.** ([13,16]) Let $Y, Z$ be two subsets of $M(X')$. Define the *alternating product* of $Y$ and $Z$ to be

$$\Lambda(Y,Z) = \Big(\bigsqcup_{r\geq 1}(Y\lfloor Z\rfloor)^r\Big)\bigsqcup\Big(\bigsqcup_{r\geq 0}(Y\lfloor Z\rfloor)^r Y\Big)\bigsqcup\Big(\bigsqcup_{r\geq 1}(\lfloor Z\rfloor Y)^r\Big)\bigsqcup\Big(\bigsqcup_{r\geq 0}(\lfloor Z\rfloor Y)^r\lfloor Z\rfloor\Big).$$

Here $\sqcup$ stands for disjoint union.

For example, $y_1\lfloor z_1\rfloor y_2$, $\lfloor z_1\rfloor y_1\lfloor z_2\rfloor$, $y_1, y_2 \in Y$, $z_1, z_2 \in Z$, are elements in $\Lambda(Y,Z)$. But $\lfloor z_1\rfloor\lfloor z_2\rfloor$ are not in $\Lambda(Y,Z)$.

We construct a sequence $\mathfrak{X}_n$ of subsets of $M(X')$ by the following recursion on $n \geq 0$. For the initial step, we define $\mathfrak{X}_0 := X \cup \{1\}$. For the inductive step, we define

$$\mathfrak{X}_{n+1} := \Lambda(X, \mathfrak{X}_n) \cup \{1\} \text{ for } n \geq 1.$$

For example, for $x_1, x_2, x_3 \in X$, the elements $\lfloor x_1\rfloor x_2\lfloor x_3\rfloor$, $x_1\lfloor x_2\lfloor x_3\rfloor\rfloor$ and $\lfloor\lfloor x_1\rfloor x_2\lfloor\lfloor x_3\rfloor\rfloor\rfloor$ are all in $\mathfrak{X}_3$, the first two are in $\mathfrak{X}_2$ and the first one is in $\mathfrak{X}_1$.

From the definition we have $\mathfrak{X}_1 \supseteq \mathfrak{X}_0$. Assuming $\mathfrak{X}_n \supseteq \mathfrak{X}_{n-1}$, we get

$$\mathfrak{X}_{n+1} = \Lambda(X, \mathfrak{X}_n) \supseteq \Lambda(X, \mathfrak{X}_{n-1}) = \mathfrak{X}_n.$$

Thus we can define

$$\mathfrak{X}_\infty := \varinjlim \mathfrak{X}_n = \bigcup_{n\geq 0} \mathfrak{X}_n.$$

For $\mathfrak{x} \in \mathfrak{X}_\infty$, we define the *depth* $\operatorname{dep}(\mathfrak{x})$ of $\mathfrak{x}$ to be

$$\operatorname{dep}(\mathfrak{x}) := \min\{n \mid \mathfrak{x} \in \mathfrak{X}_n\}.$$

Further, every $\mathfrak{x} \in \mathfrak{X}_\infty \setminus \{1\}$ has a unique *standard decomposition:*

$$\mathfrak{x} = \mathfrak{x}_1 \cdots \mathfrak{x}_b, \tag{5}$$

where $\mathfrak{x}_i$, $1 \leq i \leq b$, are alternatively in $X$ or in $\lfloor \mathfrak{X}_\infty \rfloor$. We call $b$ to be the *breadth* of $\mathfrak{x}$, denoted by $\operatorname{bre}(\mathfrak{x})$. We define the *head* $h(\mathfrak{x})$ of $\mathfrak{x}$ to be 0 (resp. 1) if $\mathfrak{x}_1$ is in $X$ (resp. in $\lfloor \mathfrak{X}_\infty \rfloor$). Similarly define the *tail* $t(\mathfrak{x})$ of $\mathfrak{x}$ to be 0 (resp. 1) if $\mathfrak{x}_b$ is in $X$ (resp. in $\lfloor \mathfrak{X}_\infty \rfloor$).

Fix a $\kappa \in \mathbf{k}$. We will equip the free $\mathbf{k}$-module

$$F_\kappa(A) = \mathbf{k}\,\mathfrak{X}_\infty = \bigoplus_{\mathfrak{x}\in\mathfrak{X}_\infty} \mathbf{k}\mathfrak{x} \tag{6}$$

with a multiplication $\diamond := \diamond_\kappa$. This is accomplished by defining $\mathfrak{x}\diamond\mathfrak{x}' \in F_\kappa(A)$ for basis elements $\mathfrak{x},\mathfrak{x}' \in \mathfrak{X}_\infty$ and then extending bilinearly. Roughly speaking, the product of $\mathfrak{x}$ and $\mathfrak{x}'$ is defined to be the concatenation whenever $t(\mathfrak{x}) \neq h(\mathfrak{x}')$. When $t(\mathfrak{x}) = h(\mathfrak{x}')$, the product is defined by the product in $A$ or by the modified Rota-Baxter identity in Eq. (4).

To be precise, we use induction on the sum $n := \operatorname{dep}(\mathfrak{x}) + \operatorname{dep}(\mathfrak{x}') \geq 0$ to define $\mathfrak{x}\diamond\mathfrak{x}'$. For the initial step of $n = 0$, $\mathfrak{x},\mathfrak{x}'$ are in $X$ and so are in $A$. Then we define

$$\mathfrak{x}\diamond\mathfrak{x}' := \mathfrak{x}\cdot\mathfrak{x}' \in A \subseteq F_\kappa(A).$$

Here $\cdot$ is the product in $A$.

For the inductive step, let $k \geq 0$ be given and assume that $\mathfrak{x}\diamond\mathfrak{x}'$ have been defined for all $\mathfrak{x},\mathfrak{x}' \in \mathfrak{X}_\infty$ with $n = \operatorname{dep}(\mathfrak{x}) + \operatorname{dep}(\mathfrak{x}') \leq k$. Then consider $\mathfrak{x},\mathfrak{x}' \in \mathfrak{X}_\infty$ with $n = \operatorname{dep}(\mathfrak{x}) + \operatorname{dep}(\mathfrak{x}') = k+1$. First treat the case when $\operatorname{bre}(\mathfrak{x}) = \operatorname{bre}(\mathfrak{x}') = 1$. Then $\mathfrak{x}$ and $\mathfrak{x}'$ are in $X$ or $\lfloor \mathfrak{X}_\infty \rfloor$. Since $n = k+1 \geq 1$, $\mathfrak{x}$ and $\mathfrak{x}'$ cannot be both in $X$. We accordingly define

$$\mathfrak{x}\diamond\mathfrak{x}' := \begin{cases} \mathfrak{x}\mathfrak{x}', & \text{if } \mathfrak{x}\in X \text{ and } \mathfrak{x}' \in \lfloor \mathfrak{X}_\infty \rfloor, \\ \mathfrak{x}\mathfrak{x}', & \text{if } \mathfrak{x}\in \lfloor \mathfrak{X}_\infty \rfloor \text{ and } \mathfrak{x}' \in X, \\ \lfloor \lfloor \bar{\mathfrak{x}} \rfloor \diamond \bar{\mathfrak{x}}' \rfloor + \lfloor \bar{\mathfrak{x}} \diamond \lfloor \bar{\mathfrak{x}}' \rfloor \rfloor + \kappa \bar{\mathfrak{x}} \diamond \bar{\mathfrak{x}}', & \text{if } \mathfrak{x} = \lfloor \bar{\mathfrak{x}} \rfloor \text{ and } \mathfrak{x}' = \lfloor \bar{\mathfrak{x}}' \rfloor \in \lfloor \mathfrak{X}_\infty \rfloor. \end{cases} \tag{7}$$

Here the product in the first and second case are by concatenation and in the third case is by the induction hypothesis since for the three products on the right hand side we have

$$\begin{aligned} \operatorname{dep}(\lfloor \bar{\mathfrak{x}} \rfloor) + \operatorname{dep}(\bar{\mathfrak{x}}') &= \operatorname{dep}(\lfloor \bar{\mathfrak{x}} \rfloor) + \operatorname{dep}(\lfloor \bar{\mathfrak{x}}' \rfloor) - 1 = \operatorname{dep}(\mathfrak{x}) + \operatorname{dep}(\mathfrak{x}') - 1 = k, \\ \operatorname{dep}(\bar{\mathfrak{x}}) + \operatorname{dep}(\lfloor \bar{\mathfrak{x}}' \rfloor) &= \operatorname{dep}(\lfloor \bar{\mathfrak{x}} \rfloor) + \operatorname{dep}(\lfloor \bar{\mathfrak{x}}' \rfloor) - 1 = \operatorname{dep}(\mathfrak{x}) + \operatorname{dep}(\mathfrak{x}') - 1 = k, \\ \operatorname{dep}(\bar{\mathfrak{x}}) + \operatorname{dep}(\bar{\mathfrak{x}}') &= \operatorname{dep}(\lfloor \bar{\mathfrak{x}} \rfloor) - 1 + \operatorname{dep}(\lfloor \bar{\mathfrak{x}}' \rfloor) - 1 = \operatorname{dep}(\mathfrak{x}) + \operatorname{dep}(\mathfrak{x}') - 2 = k - 1. \end{aligned}$$

We next treat the case when $\operatorname{bre}(\mathfrak{x}) > 1$ or $\operatorname{bre}(\mathfrak{x}') > 1$. Let $\mathfrak{x} = \mathfrak{x}_1 \cdots \mathfrak{x}_b$ and $\mathfrak{x}' = \mathfrak{x}'_1 \cdots \mathfrak{x}'_{b'}$ be the standard decompositions from Eq. (5). We then define

$$\mathfrak{x}\diamond\mathfrak{x}' = \mathfrak{x}_1 \cdots \mathfrak{x}_{b-1}(\mathfrak{x}_b \diamond \mathfrak{x}'_1)\,\mathfrak{x}'_2 \cdots \mathfrak{x}'_{b'} \tag{8}$$

where $\mathfrak{x}_b \diamond \mathfrak{x}'_1$ is defined by Eq. (7) and the rest is given by concatenation. Extending $\diamond$ bilinearly, we obtain a binary operation

$$\diamond : F_\kappa(A) \otimes F_\kappa(A) \to F_\kappa(A).$$

This completes the definition of $\diamond$.

**Lemma 2.5.** *Let $\mathfrak{x}, \mathfrak{x}' \in \mathfrak{X}_\infty$.*

(a) $h(\mathfrak{x}) = h(\mathfrak{x} \diamond \mathfrak{x}')$ *and* $t(\mathfrak{x}') = t(\mathfrak{x} \diamond \mathfrak{x}')$.

(b) *If* $t(\mathfrak{x}) \neq h(\mathfrak{x}')$, *then* $\mathfrak{x} \diamond \mathfrak{x}' = \mathfrak{x}\mathfrak{x}'$ *(concatenation).*

(c) *If* $t(\mathfrak{x}) \neq h(\mathfrak{x}')$, *then for any* $\mathfrak{x}'' \in \mathfrak{X}_\infty$,

$$(\mathfrak{x}\mathfrak{x}') \diamond \mathfrak{x}'' = \mathfrak{x}(\mathfrak{x}' \diamond \mathfrak{x}'') \ \text{ and } \ \mathfrak{x}'' \diamond (\mathfrak{x}\mathfrak{x}') = (\mathfrak{x}'' \diamond \mathfrak{x})\mathfrak{x}'.$$

**Proof.** Items (a) and (b) follow from the definition of $\diamond$. The proof of Item (c) is the same as [16, Lemma 4.4.5]. □

We next define a linear operator

$$P_A : F_\kappa(A) \to F_\kappa(A), \ \mathfrak{x} \mapsto \lfloor \mathfrak{x} \rfloor.$$

*In the rest of the paper,* we will use the infix notation $\lfloor \mathfrak{x} \rfloor$ interchangeably with $P_A(\mathfrak{x})$ for any $\mathfrak{x} \in F_\kappa(A)$. Let

$$j_X : X \hookrightarrow \mathfrak{X}_\infty \hookrightarrow F_\kappa(A)$$

be the natural injection which extends to an algebra injection

$$j_A : A \to F_\kappa(A).$$

Now we state our first main result, to be proved in the next subsection.

**Theorem 2.6.** *Let $A$ be a* **k***-algebra with a* **k***-basis $X$ and $\kappa \in \mathbf{k}$ be given.*

(a) *The pair $(F_\kappa(A), \diamond)$ is an algebra.*

(b) *The triple $(F_\kappa(A), \diamond, P_A)$ is a modified Rota-Baxter algebra of weight $\kappa$.*

(c) *The triple $(F_\kappa(A), \diamond, P_A)$ together with the embedding $j_A$ is the free modified Rota-Baxter algebra of weight $\kappa$ on the algebra $A$.*

### 2.2. The proof of Theorem 2.6.

**Proof.** (a) It is enough to verify the associativity for basis elements:

$$(\mathfrak{x}' \diamond \mathfrak{x}'') \diamond \mathfrak{x}''' = \mathfrak{x}' \diamond (\mathfrak{x}'' \diamond \mathfrak{x}'''), \ \text{ for all } \mathfrak{x}', \mathfrak{x}'', \mathfrak{x}''' \in \mathfrak{X}_\infty. \tag{9}$$

We carry out the verification by induction on the sum of the depths

$$n := \operatorname{dep}(\mathfrak{x}') + \operatorname{dep}(\mathfrak{x}'') + \operatorname{dep}(\mathfrak{x}''') \geq 0.$$

If $n = 0$, then

$$\operatorname{dep}(\mathfrak{x}') = \operatorname{dep}(\mathfrak{x}'') = \operatorname{dep}(\mathfrak{x}''') = 0$$

and so $\mathfrak{x}', \mathfrak{x}'', \mathfrak{x}''' \in X$. In this case the product $\diamond$ is given by the product in $A$ and so is associative.

Assume that Eq. (9) holds for $n \leq k$ for any given $k \geq 0$ and consider $\mathfrak{x}', \mathfrak{x}'', \mathfrak{x}''' \in \mathfrak{X}_\infty$ with

$$n = \operatorname{dep}(\mathfrak{x}') + \operatorname{dep}(\mathfrak{x}'') + \operatorname{dep}(\mathfrak{x}''') = k + 1 \geq 1.$$

If $t(\mathfrak{x}') \neq h(\mathfrak{x}'')$, then by Lemma 2.5,

$$(\mathfrak{x}' \diamond \mathfrak{x}'') \diamond \mathfrak{x}''' = (\mathfrak{x}'\mathfrak{x}'') \diamond \mathfrak{x}''' = \mathfrak{x}'(\mathfrak{x}'' \diamond \mathfrak{x}''') = \mathfrak{x}' \diamond (\mathfrak{x}'' \diamond \mathfrak{x}''').$$

A similar argument holds when $t(\mathfrak{x}'') \neq h(\mathfrak{x}''')$. Thus we only need to verify the associativity when $t(\mathfrak{x}') = h(\mathfrak{x}'')$ and $t(\mathfrak{x}'') = h(\mathfrak{x}''')$. We next reduce the proof to the breadths of the words and depart to show a lemma.

**Lemma 2.7.** *If Eq. (9) holds for all $\mathfrak{x}', \mathfrak{x}''$ and $\mathfrak{x}'''$ in $\mathfrak{X}_\infty$ of breadth one, then it holds for all $\mathfrak{x}', \mathfrak{x}''$ and $\mathfrak{x}'''$ in $\mathfrak{X}_\infty$.*

**Proof.** We use induction on the sum of breadths $m := \operatorname{bre}(\mathfrak{x}') + \operatorname{bre}(\mathfrak{x}'') + \operatorname{bre}(\mathfrak{x}''') \geq 3$. The case when $m = 3$ is the assumption of the lemma. Assume the associativity holds for $m \leq j$ for some $j \geq 3$ and take $\mathfrak{x}', \mathfrak{x}'', \mathfrak{x}''' \in \mathfrak{X}_\infty$ with $m = j + 1 \geq 4$. So at least one of $\mathfrak{x}', \mathfrak{x}'', \mathfrak{x}'''$ has breadth greater than or equal to 2.

First assume that $\operatorname{bre}(\mathfrak{x}') \geq 2$. Then we may write

$$\mathfrak{x}' = \mathfrak{x}'_1\mathfrak{x}'_2, \quad \text{where } \mathfrak{x}'_1,\, \mathfrak{x}'_2 \in \mathfrak{X}_\infty \text{ and } t(\mathfrak{x}'_1) \neq h(\mathfrak{x}'_2).$$

By Lemma 2.5, we obtain

$$(\mathfrak{x}' \diamond \mathfrak{x}'') \diamond \mathfrak{x}''' = ((\mathfrak{x}'_1\mathfrak{x}'_2) \diamond \mathfrak{x}'') \diamond \mathfrak{x}''' = (\mathfrak{x}'_1(\mathfrak{x}'_2 \diamond \mathfrak{x}'')) \diamond \mathfrak{x}''' = \mathfrak{x}'_1((\mathfrak{x}'_2 \diamond \mathfrak{x}'') \diamond \mathfrak{x}''').$$

Similarly,

$$\mathfrak{x}' \diamond (\mathfrak{x}'' \diamond \mathfrak{x}''') = (\mathfrak{x}'_1\mathfrak{x}'_2) \diamond (\mathfrak{x}'' \diamond \mathfrak{x}''') = \mathfrak{x}'_1(\mathfrak{x}'_2 \diamond (\mathfrak{x}'' \diamond \mathfrak{x}''')).$$

Thus

$$(\mathfrak{x}' \diamond \mathfrak{x}'') \diamond \mathfrak{x}''' = \mathfrak{x}' \diamond (\mathfrak{x}'' \diamond \mathfrak{x}''')$$

whenever

$$(\mathfrak{x}'_2 \diamond \mathfrak{x}'') \diamond \mathfrak{x}''' = \mathfrak{x}'_2 \diamond (\mathfrak{x}'' \diamond \mathfrak{x}'''),$$

which follows from the induction hypothesis. A similar proof works if $\operatorname{bre}(\mathfrak{x}''') \geq 2$.

Finally if $\operatorname{bre}(\mathfrak{x}'') \geq 2$, we may write

$$\mathfrak{x}'' = \mathfrak{x}''_1\mathfrak{x}''_2 \text{ where } \mathfrak{x}''_1,\, \mathfrak{x}''_2 \in \mathfrak{X}_\infty \text{ and } t(\mathfrak{x}''_1) \neq h(\mathfrak{x}''_2).$$

By Lemma 2.5 again, we get

$$(\mathfrak{x}'\diamond\mathfrak{x}'')\diamond\mathfrak{x}''' = (\mathfrak{x}'\diamond(\mathfrak{x}_1''\mathfrak{x}_2''))\diamond\mathfrak{x}''' = ((\mathfrak{x}'\diamond\mathfrak{x}_1'')\mathfrak{x}_2'')\diamond\mathfrak{x}''' = (\mathfrak{x}'\diamond\mathfrak{x}_1'')(\mathfrak{x}_2''\diamond\mathfrak{x}''').$$

In the same way, we have

$$(\mathfrak{x}'\diamond\mathfrak{x}_1'')(\mathfrak{x}_2''\diamond\mathfrak{x}''') = \mathfrak{x}'\diamond(\mathfrak{x}''\diamond\mathfrak{x}''').$$

This proves the associativity. □

In summary, the proof of the associativity has been reduced to the special case when $\mathfrak{x}',\mathfrak{x}'',\mathfrak{x}'''\in\mathfrak{X}_\infty$ are chosen so that

(a) $n=\operatorname{dep}(\mathfrak{x}')+\operatorname{dep}(\mathfrak{x}'')+\operatorname{dep}(\mathfrak{x}''')=k+1\geq 1$ with the assumption that the associativity holds when $n\leq k$.

(b) the elements have breadth one and

(c) $t(\mathfrak{x}')=h(\mathfrak{x}'')$ and $t(\mathfrak{x}'')=h(\mathfrak{x}''')$.

By Item (b), the head and tail of each of the elements are the same. Therefore by Item (c), either all the three elements are in $X$ or they are all in $\lfloor\mathfrak{X}_\infty\rfloor$. If all of $\mathfrak{x}',\mathfrak{x}'',\mathfrak{x}'''$ are in $X$, then as already shown, the associativity follows from the associativity in $A$. So it remains to consider the case when $\mathfrak{x}',\mathfrak{x}'',\mathfrak{x}'''$ are all in $\lfloor\mathfrak{X}_\infty\rfloor$. Then we may write

$$\mathfrak{x}'=\lfloor\bar{\mathfrak{x}}'\rfloor,\mathfrak{x}''=\lfloor\bar{\mathfrak{x}}''\rfloor,\mathfrak{x}'''=\lfloor\bar{\mathfrak{x}}'''\rfloor \text{ with } \bar{\mathfrak{x}}',\bar{\mathfrak{x}}'',\bar{\mathfrak{x}}'''\in\mathfrak{X}_\infty.$$

Applying Eq. (7) and bilinearity of the product $\diamond$, we get

$$\begin{aligned}
(\mathfrak{x}'\diamond\mathfrak{x}'')\diamond\mathfrak{x}''' &= \big(\lfloor\lfloor\bar{\mathfrak{x}}'\rfloor\diamond\bar{\mathfrak{x}}''\rfloor+\lfloor\bar{\mathfrak{x}}'\diamond\lfloor\bar{\mathfrak{x}}''\rfloor\rfloor+\kappa\bar{\mathfrak{x}}'\diamond\bar{\mathfrak{x}}''\big)\diamond\lfloor\bar{\mathfrak{x}}'''\rfloor\\
&= \lfloor\lfloor\bar{\mathfrak{x}}'\rfloor\diamond\bar{\mathfrak{x}}''\rfloor\diamond\lfloor\bar{\mathfrak{x}}'''\rfloor+\lfloor\bar{\mathfrak{x}}'\diamond\lfloor\bar{\mathfrak{x}}''\rfloor\rfloor\diamond\lfloor\bar{\mathfrak{x}}'''\rfloor+\kappa(\bar{\mathfrak{x}}'\diamond\bar{\mathfrak{x}}'')\diamond\lfloor\bar{\mathfrak{x}}'''\rfloor\\
&= \lfloor\lfloor\lfloor\bar{\mathfrak{x}}'\rfloor\diamond\bar{\mathfrak{x}}''\rfloor\diamond\bar{\mathfrak{x}}'''\rfloor+\lfloor\big(\lfloor\bar{\mathfrak{x}}'\rfloor\diamond\bar{\mathfrak{x}}''\big)\diamond\lfloor\bar{\mathfrak{x}}'''\rfloor\rfloor+\kappa\big(\lfloor\bar{\mathfrak{x}}'\rfloor\diamond\bar{\mathfrak{x}}''\big)\diamond\bar{\mathfrak{x}}'''\\
&\quad+\lfloor\lfloor\bar{\mathfrak{x}}'\diamond\lfloor\bar{\mathfrak{x}}''\rfloor\rfloor\diamond\bar{\mathfrak{x}}'''\rfloor+\lfloor\big(\bar{\mathfrak{x}}'\diamond\lfloor\bar{\mathfrak{x}}''\rfloor\big)\diamond\lfloor\bar{\mathfrak{x}}'''\rfloor\rfloor+\kappa\big(\bar{\mathfrak{x}}'\diamond\lfloor\bar{\mathfrak{x}}''\rfloor\big)\diamond\bar{\mathfrak{x}}'''\\
&\quad+\kappa(\bar{\mathfrak{x}}'\diamond\bar{\mathfrak{x}}'')\diamond\lfloor\bar{\mathfrak{x}}'''\rfloor\\
&= \lfloor\lfloor\lfloor\bar{\mathfrak{x}}'\rfloor\diamond\bar{\mathfrak{x}}''\rfloor\diamond\bar{\mathfrak{x}}'''\rfloor+\lfloor\big(\lfloor\bar{\mathfrak{x}}'\rfloor\diamond\bar{\mathfrak{x}}''\big)\diamond\lfloor\bar{\mathfrak{x}}'''\rfloor\rfloor+\kappa\big(\lfloor\bar{\mathfrak{x}}'\rfloor\diamond\bar{\mathfrak{x}}''\big)\diamond\bar{\mathfrak{x}}'''\\
&\quad+\lfloor\lfloor\bar{\mathfrak{x}}'\diamond\lfloor\bar{\mathfrak{x}}''\rfloor\rfloor\diamond\bar{\mathfrak{x}}'''\rfloor\rfloor+\lfloor\bar{\mathfrak{x}}'\diamond\lfloor\lfloor\bar{\mathfrak{x}}''\rfloor\diamond\bar{\mathfrak{x}}'''\rfloor\rfloor+\lfloor\bar{\mathfrak{x}}'\diamond\lfloor\bar{\mathfrak{x}}''\diamond\lfloor\bar{\mathfrak{x}}'''\rfloor\rfloor\\
&\quad+\kappa\lfloor\bar{\mathfrak{x}}'\diamond\bar{\mathfrak{x}}''\diamond\bar{\mathfrak{x}}'''\rfloor+\kappa\big(\bar{\mathfrak{x}}'\diamond\lfloor\bar{\mathfrak{x}}''\rfloor\big)\diamond\bar{\mathfrak{x}}'''+\kappa(\bar{\mathfrak{x}}'\diamond\bar{\mathfrak{x}}'')\diamond\lfloor\bar{\mathfrak{x}}'''\rfloor.
\end{aligned}$$

Similarly we obtain

$$\begin{aligned}
\mathfrak{x}'\diamond\big(\mathfrak{x}''\diamond\mathfrak{x}'''\big) &= \lfloor\bar{\mathfrak{x}}'\rfloor\diamond\big(\lfloor\lfloor\bar{\mathfrak{x}}''\rfloor\diamond\bar{\mathfrak{x}}'''\rfloor+\lfloor\bar{\mathfrak{x}}''\diamond\lfloor\bar{\mathfrak{x}}'''\rfloor\rfloor+\kappa\bar{\mathfrak{x}}''\diamond\bar{\mathfrak{x}}'''\big)\\
&= \lfloor\bar{\mathfrak{x}}'\rfloor\diamond\lfloor\lfloor\bar{\mathfrak{x}}''\rfloor\diamond\bar{\mathfrak{x}}'''\rfloor+\lfloor\bar{\mathfrak{x}}'\rfloor\diamond\lfloor\bar{\mathfrak{x}}''\diamond\lfloor\bar{\mathfrak{x}}'''\rfloor\rfloor+\kappa\lfloor\bar{\mathfrak{x}}'\rfloor\diamond\big(\bar{\mathfrak{x}}''\diamond\bar{\mathfrak{x}}'''\big)
\end{aligned}$$

$$
\begin{aligned}
&= \lfloor\lfloor\bar{\mathfrak{x}}'\rfloor \diamond (\lfloor\bar{\mathfrak{x}}''\rfloor \diamond \bar{\mathfrak{x}}''')\rfloor + \lfloor\bar{\mathfrak{x}}' \diamond \lfloor\lfloor\bar{\mathfrak{x}}''\rfloor \diamond \bar{\mathfrak{x}}'''\rfloor\rfloor + \kappa\bar{\mathfrak{x}}' \diamond (\lfloor\bar{\mathfrak{x}}''\rfloor \diamond \bar{\mathfrak{x}}''') \\
&\quad + \lfloor\lfloor\bar{\mathfrak{x}}'\rfloor \diamond (\bar{\mathfrak{x}}'' \diamond \lfloor\bar{\mathfrak{x}}'''\rfloor)\rfloor + \lfloor\bar{\mathfrak{x}}' \diamond \lfloor\bar{\mathfrak{x}}'' \diamond \lfloor\bar{\mathfrak{x}}'''\rfloor\rfloor\rfloor + \kappa\bar{\mathfrak{x}}' \diamond (\bar{\mathfrak{x}}'' \diamond \lfloor\bar{\mathfrak{x}}'''\rfloor) \\
&\quad + \kappa\lfloor\bar{\mathfrak{x}}'\rfloor \diamond (\bar{\mathfrak{x}}'' \diamond \bar{\mathfrak{x}}''') \\
&= \lfloor\lfloor\lfloor\bar{\mathfrak{x}}'\rfloor \diamond \bar{\mathfrak{x}}''\rfloor \diamond \bar{\mathfrak{x}}'''\rfloor + \lfloor\lfloor\bar{\mathfrak{x}}' \diamond \lfloor\bar{\mathfrak{x}}''\rfloor\rfloor \diamond \bar{\mathfrak{x}}'''\rfloor \\
&\quad + \kappa\lfloor\bar{\mathfrak{x}}' \diamond \bar{\mathfrak{x}}'' \diamond \bar{\mathfrak{x}}'''\rfloor + \lfloor\bar{\mathfrak{x}}' \diamond \lfloor\lfloor\bar{\mathfrak{x}}''\rfloor \diamond \bar{\mathfrak{x}}'''\rfloor\rfloor + \kappa\bar{\mathfrak{x}}' \diamond (\lfloor\bar{\mathfrak{x}}''\rfloor \diamond \bar{\mathfrak{x}}''') \\
&\quad + \lfloor\lfloor\bar{\mathfrak{x}}'\rfloor \diamond (\bar{\mathfrak{x}}'' \diamond \lfloor\bar{\mathfrak{x}}'''\rfloor)\rfloor + \lfloor\bar{\mathfrak{x}}' \diamond \lfloor\bar{\mathfrak{x}}'' \diamond \lfloor\bar{\mathfrak{x}}'''\rfloor\rfloor\rfloor + \kappa\bar{\mathfrak{x}}' \diamond (\bar{\mathfrak{x}}'' \diamond \lfloor\bar{\mathfrak{x}}'''\rfloor) \\
&\quad + \kappa\lfloor\bar{\mathfrak{x}}'\rfloor \diamond (\bar{\mathfrak{x}}'' \diamond \bar{\mathfrak{x}}''').
\end{aligned}
$$

Now by the induction hypothesis, the $i$-th term in the expansion of $(\mathfrak{x}' \diamond \mathfrak{x}'') \diamond \mathfrak{x}'''$ coincides with the $\sigma(i)$-th term in the expansion of $\mathfrak{x}' \diamond (\mathfrak{x}'' \diamond \mathfrak{x}''')$. Here $\sigma \in \Sigma_9$ is the permutation given by

$$
\sigma = \begin{pmatrix} 1 & 2 & 3 & 4 & 5 & 6 & 7 & 8 & 9 \\ 1 & 6 & 9 & 2 & 4 & 7 & 3 & 5 & 8 \end{pmatrix}.
$$

This completes the proof of Theorem 2.6 (a).

(b). The proof follows from the definition $P_A(\mathfrak{x}) = \lfloor\mathfrak{x}\rfloor$ and Eq. (7).

(c). Let $(M, *, P)$ be a modified Rota-Baxter algebra with multiplication $*$ and let $f : A \to M$ be a $\mathbf{k}$-algebra homomorphism. We will construct a $\mathbf{k}$-linear map $\bar{f} : F_\kappa(A) \to M$ by defining $\bar{f}(\mathfrak{x})$ for $\mathfrak{x} \in \mathfrak{X}_\infty$. We achieve this by defining $\bar{f}(\mathfrak{x})$ for $\mathfrak{x} \in \mathfrak{X}_n$, $n \geq 0$, inductively on $n$. For $\mathfrak{x} \in \mathfrak{X}_0 := X$, define $\bar{f}(\mathfrak{x}) = f(\mathfrak{x})$. Then $j\bar{f} = f$ is satisfied. Suppose $\bar{f}(\mathfrak{x})$ has been defined for $\mathfrak{x} \in \mathfrak{X}_n$ and consider $\mathfrak{x}$ in $\mathfrak{X}_{n+1}$ which is, by definition,

$$
\begin{aligned}
\Lambda(X, \mathfrak{X}_n) &= \Big(\bigsqcup_{r\geq 1}(X\lfloor\mathfrak{X}_n\rfloor)^r\Big) \bigsqcup \Big(\bigsqcup_{r\geq 0}(X\lfloor\mathfrak{X}_n\rfloor)^r X\Big) \\
&\quad \bigsqcup \Big(\bigsqcup_{r\geq 0}\lfloor\mathfrak{X}_n\rfloor(X\lfloor\mathfrak{X}_n\rfloor)^r\Big) \bigsqcup \Big(\bigsqcup_{r\geq 0}\lfloor\mathfrak{X}_n\rfloor(X\lfloor\mathfrak{X}_n\rfloor)^r X\Big).
\end{aligned}
$$

Let $\mathfrak{x}$ be in the first union component $\bigsqcup_{r\geq 1}(X\lfloor\mathfrak{X}_n\rfloor)^r$ above. Then

$$
\mathfrak{x} = \prod_{i=1}^{r}(\mathfrak{x}_{2i-1}\lfloor\mathfrak{x}_{2i}\rfloor)
$$

for $\mathfrak{x}_{2i-1} \in X$ and $\mathfrak{x}_{2i} \in \mathfrak{X}_n$, $1 \leq i \leq r$. By the construction of the multiplication $\diamond$ and the modified Rota-Baxter operator $P_A$, we have

$$
\mathfrak{x} = \diamond_{i=1}^{r}(\mathfrak{x}_{2i-1} \diamond \lfloor\mathfrak{x}_{2i}\rfloor) = \diamond_{i=1}^{r}(\mathfrak{x}_{2i-1} \diamond P_A(\mathfrak{x}_{2i})).
$$

Define

$$
\bar{f}(\mathfrak{x}) = *_{i=1}^{r}\big(\bar{f}(\mathfrak{x}_{2i-1}) * P_A\big(\bar{f}(\mathfrak{x}_{2i})\big)\big). \tag{10}
$$

where the right hand side is well-defined by the induction hypothesis. Similarly define $\bar{f}(\mathfrak{x})$ if $\mathfrak{x}$ is in the other union components. For any $\mathfrak{x} \in \mathfrak{X}_\infty$, we have $P_A(\mathfrak{x}) = \lfloor \mathfrak{x} \rfloor \in \mathfrak{X}_\infty$, and by the definition of $\bar{f}$ in (Eq. (10)), we have

$$\bar{f}(\lfloor \mathfrak{x} \rfloor) = P(\bar{f}(\mathfrak{x})). \tag{11}$$

So $\bar{f}$ commutes with the modified Rota-Baxter operators. Combining this equation with Eq. (10) we see that if $\mathfrak{x} = \mathfrak{x}_1 \cdots \mathfrak{x}_b$ is the standard decomposition of $\mathfrak{x}$, then

$$\bar{f}(\mathfrak{x}) = \bar{f}(\mathfrak{x}_1) * \cdots * \bar{f}(\mathfrak{x}_b).$$

Note that this is the only possible way to define $\bar{f}(\mathfrak{x})$ in order for $\bar{f}$ to be a modified Rota-Baxter algebra homomorphism extending $f$. It remains to prove that the map $\bar{f}$ defined in Eq. (10) is indeed an algebra homomorphism. For this we only need to check the multiplicity

$$\bar{f}(\mathfrak{x} \diamond \mathfrak{x}') = \bar{f}(\mathfrak{x}) * \bar{f}(\mathfrak{x}') \tag{12}$$

for all $\mathfrak{x}, \mathfrak{x}' \in \mathfrak{X}_\infty$. For this we use induction on the sum of depths $n := \mathrm{bre}(\mathfrak{x}) + \mathrm{bre}(\mathfrak{x}')$. Then $n \geq 0$. When $n = 0$, we have $\mathfrak{x}, \mathfrak{x}' \in X$. Then Eq. (12) follows from the multiplicity of $f$. Assume the multiplicity holds for $\mathfrak{x}, \mathfrak{x}' \in \mathfrak{X}_\infty$ with $n \geq k$ and take $\mathfrak{x}, \mathfrak{x}' \in \mathfrak{X}_\infty$ with $n = k+1$. Let $\mathfrak{x} = \mathfrak{x}_1 \cdots \mathfrak{x}_b$ and $\mathfrak{x}' = \mathfrak{x}'_1 \cdots \mathfrak{x}'_{b'}$ be the standard decompositions. Since $n = k+1 \geq 1$, at least one of $\mathfrak{x}_b$ and $\mathfrak{x}'_{b'}$ is in $\lfloor \mathfrak{X}_\infty \rfloor$. Then by Eq. (7) we have

$$\bar{f}(\mathfrak{x}_b \diamond \mathfrak{x}'_1) = \begin{cases} \bar{f}(\mathfrak{x}_b \mathfrak{x}'_1), & \text{if } \mathfrak{x}_b \in X, \mathfrak{x}'_1 \in \lfloor \mathfrak{X}_\infty \rfloor, \\ \bar{f}(\mathfrak{x}_b \mathfrak{x}'_1), & \text{if } \mathfrak{x}_b \in \lfloor \mathfrak{X}_\infty \rfloor, \mathfrak{x}'_1 \in X, \\ \bar{f}(\lfloor \lfloor \bar{\mathfrak{x}}_b \rfloor \diamond \bar{\mathfrak{x}}'_1 \rfloor + \lfloor \bar{\mathfrak{x}}_b \diamond \lfloor \bar{\mathfrak{x}}'_1 \rfloor \rfloor + \kappa \bar{\mathfrak{x}}_b \diamond \bar{\mathfrak{x}}'_1), & \text{if } \mathfrak{x}_b = \lfloor \bar{\mathfrak{x}}_b \rfloor, \mathfrak{x}'_1 = \lfloor \bar{\mathfrak{x}}'_1 \rfloor \in \lfloor \mathfrak{X}_\infty \rfloor. \end{cases}$$

In the first two cases, the right hand side is $\bar{f}(\mathfrak{x}_b) * \bar{f}(\mathfrak{x}'_1)$ by the definition of $\bar{f}$. In the third case, applying Eq. (11), the induction hypothesis and the modified Rota-Baxter relation of the operator $P$ on $M$, we have

$$\begin{aligned} & \bar{f}(\lfloor \lfloor \bar{\mathfrak{x}}_b \rfloor \diamond \bar{\mathfrak{x}}'_1 \rfloor + \lfloor \bar{\mathfrak{x}}_b \diamond \lfloor \bar{\mathfrak{x}}'_1 \rfloor \rfloor + \kappa \bar{\mathfrak{x}}_b \diamond \bar{\mathfrak{x}}'_1) \\ = \; & \bar{f}(\lfloor \lfloor \bar{\mathfrak{x}}_b \rfloor \diamond \bar{\mathfrak{x}}'_1 \rfloor) + \bar{f}(\lfloor \bar{\mathfrak{x}}_b \diamond \lfloor \bar{\mathfrak{x}}'_1 \rfloor \rfloor) + \kappa \bar{f}(\bar{\mathfrak{x}}_b \diamond \bar{\mathfrak{x}}'_1) \\ = \; & P(\bar{f}(\lfloor \bar{\mathfrak{x}}_b \rfloor \diamond \bar{\mathfrak{x}}'_1)) + P(\bar{f}(\bar{\mathfrak{x}}_b \diamond \lfloor \bar{\mathfrak{x}}'_1 \rfloor)) + \kappa \bar{f}(\bar{\mathfrak{x}}_b \diamond \bar{\mathfrak{x}}'_1) \\ = \; & P(\bar{f}(\lfloor \bar{\mathfrak{x}}_b \rfloor) * \bar{f}(\bar{\mathfrak{x}}'_1)) + P(\bar{f}(\bar{\mathfrak{x}}_b) * \bar{f}(\lfloor \bar{\mathfrak{x}}'_1 \rfloor)) + \kappa \bar{f}(\bar{\mathfrak{x}}_b) * \bar{f}(\bar{\mathfrak{x}}'_1) \\ = \; & P(P(\bar{f}(\bar{\mathfrak{x}}_b)) * \bar{f}(\bar{\mathfrak{x}}'_1)) + P(\bar{f}(\bar{\mathfrak{x}}_b) * P(\bar{f}(\bar{\mathfrak{x}}'_1))) + \kappa(\bar{f}(\bar{\mathfrak{x}}_b) * \bar{f}(\bar{\mathfrak{x}}'_1)) \\ = \; & P(\bar{f}(\bar{\mathfrak{x}}_b)) * P(\bar{f}(\bar{\mathfrak{x}}'_1)) \end{aligned}$$

$$\begin{aligned} &= \bar{f}(\lfloor \bar{\mathfrak{x}}_b \rfloor) * \bar{f}(\lfloor \bar{\mathfrak{x}}'_1 \rfloor) \\ &= \bar{f}(\mathfrak{x}_b) * \bar{f}(\mathfrak{x}'_1). \end{aligned}$$

Therefore $\bar{f}(\mathfrak{x}_b \diamond \mathfrak{x}'_1) = \bar{f}(\mathfrak{x}_b) * \bar{f}(\mathfrak{x}'_1)$. Then

$$\begin{aligned} \bar{f}(\mathfrak{x} \diamond \mathfrak{x}') &= \bar{f}(\mathfrak{x}_1 \cdots \mathfrak{x}_{b-1}(\mathfrak{x}_b \diamond \mathfrak{x}'_1)\mathfrak{x}'_2 \cdots \mathfrak{x}'_{b'}) \\ &= \bar{f}(\mathfrak{x}_1) * \cdots * \bar{f}(\mathfrak{x}_{b-1}) * \bar{f}(\mathfrak{x}_b \diamond \mathfrak{x}'_1) * \bar{f}(\mathfrak{x}'_2) \cdots \bar{f}(\mathfrak{x}'_{b'}) \\ &= \bar{f}(\mathfrak{x}_1) * \cdots * \bar{f}(\mathfrak{x}_{b-1}) * \bar{f}(\mathfrak{x}_b) * \bar{f}(\mathfrak{x}'_1) * \bar{f}(\mathfrak{x}'_2) \cdots \bar{f}(\mathfrak{x}'_{b'}) \\ &= \bar{f}(\mathfrak{x}) * \bar{f}(\mathfrak{x}'), \end{aligned}$$

as required.

This completes the proof of Theorem 2.6 □

## 3. The Hopf algebra structure on free modified Rota-Baxter algebras

In this section, starting with the assumption that $A$ is a bialgebra with its coproduct $\Delta_A$ and its counit $\varepsilon_A$, we provide a bialgebraic and then a Hopf algebraic structure on the free modified Rota-Baxter algebras $F_\kappa(A)$ obtained in Section 2, when $\kappa = -\lambda^2$. It would be interesting to see how to extend this construction to other weights $\kappa$. For Hopf algebra structures on free Rota-Baxter algebras, see [15,28] for Hopf algebra structures on free Rota-Baxter algebras.

**3.1. The bialgebraic structure.** We now build on results from previous subsections to obtain a bialgebra structure on $F_{-\lambda^2}(A)$. We first record some lemmas for a preparation.

**Lemma 3.1.** *Let $\lambda$ be a given element of $\mathbf{k}$.*

(a) *The linear map $-\lambda\mathrm{id} : \mathbf{k} \to \mathbf{k}$ is a modified Rota-Baxter operator of weight $-\lambda^2$ on $\mathbf{k}$.*

(b) *There exists a unique modified Rota-Baxter algebra morphism $\varepsilon_{\mathrm{M}} : F_{-\lambda^2}(A) \to \mathbf{k}$ such that*

$$\varepsilon_{\mathrm{M}} \circ j_A = \varepsilon_A \text{ and } \varepsilon_{\mathrm{M}} \circ P_A = -\lambda\mathrm{id} \circ \varepsilon_{\mathrm{M}}. \tag{13}$$

**Proof.** (a) It follows from

$$\begin{aligned} (-\lambda\mathrm{id})(a)(-\lambda\mathrm{id})(b) &= \lambda^2 ab = \lambda^2 ab + \lambda^2 ab - \lambda^2 ab \\ &= (-\lambda\mathrm{id})(a(-\lambda\mathrm{id})(b)) + (-\lambda\mathrm{id})((-\lambda\mathrm{id})(a)b) - \lambda^2 ab. \end{aligned}$$

(b) By Item (a), $(\mathbf{k}, -\lambda \mathrm{id})$ is a modified Rota-Baxter algebra of weight $-\lambda^2$. Then the remainder follows from Theorem 2.6 (c). □

Note that $P_A$ is a modified Rota-Baxter operator on $F_{-\lambda^2}(A)$; however $P_A \otimes P_A$ is not a modified Rota-Baxter operator on $F_{-\lambda^2}(A) \otimes F_{-\lambda^2}(A)$. The following result constructs a modified Rota-Baxter operator on $F_{-\lambda^2}(A) \otimes F_{-\lambda^2}(A)$.

**Lemma 3.2.** *Let $\lambda$ be a given element of $\mathbf{k}$. Define the linear map*

$$Q : F_{-\lambda^2}(A) \otimes F_{-\lambda^2}(A) \to F_{-\lambda^2}(A) \otimes F_{-\lambda^2}(A)$$

*by taking*

$$Q(\mathfrak{x} \otimes \mathfrak{x}') := (P_A(\mathfrak{x}) + \lambda \mathfrak{x}) \otimes \varepsilon_M(\mathfrak{x}')1 + \mathfrak{x} \otimes P_A(\mathfrak{x}') \quad \text{for } \mathfrak{x}, \mathfrak{x}' \in F_{-\lambda^2}(A). \tag{14}$$

*Then $Q$ is a modified Rota-Baxter operator of weight $-\lambda^2$ on $F_{-\lambda^2}(A) \otimes F_{-\lambda^2}(A)$.*

**Proof.** Let $\mathfrak{x}_1, \mathfrak{x}_2, \mathfrak{x}_1', \mathfrak{x}_2' \in F_{-\lambda^2}(A)$. On the one hand,

$$\begin{aligned}
&Q(\mathfrak{x}_1 \otimes \mathfrak{x}_1') \diamond' Q(\mathfrak{x}_2 \otimes \mathfrak{x}_2') \\
&= \Big((P_A(\mathfrak{x}_1) + \lambda \mathfrak{x}_1) \otimes \varepsilon_M(\mathfrak{x}_1')1 + \mathfrak{x}_1 \otimes P_A(\mathfrak{x}_1')\Big) \diamond' \Big((P_A(\mathfrak{x}_2) + \lambda \mathfrak{x}_2) \otimes \varepsilon_M(\mathfrak{x}_2')1 + \mathfrak{x}_2 \otimes P_A(\mathfrak{x}_2')\Big) \\
&= \Big((P_A(\mathfrak{x}_1) + \lambda \mathfrak{x}_1) \diamond (P_A(\mathfrak{x}_2) + \lambda \mathfrak{x}_2)\Big) \otimes \varepsilon_M(\mathfrak{x}_1')\varepsilon_M(\mathfrak{x}_2')1 + \Big((P_A(\mathfrak{x}_1) + \lambda \mathfrak{x}_1) \diamond \mathfrak{x}_2\Big) \otimes \varepsilon_M(\mathfrak{x}_1')P_A(\mathfrak{x}_2') \\
&\quad + \Big(\mathfrak{x}_1 \diamond (P_A(\mathfrak{x}_2) + \lambda \mathfrak{x}_2)\Big) \otimes P_A(\mathfrak{x}_1')\varepsilon_M(\mathfrak{x}_2') + (\mathfrak{x}_1 \diamond \mathfrak{x}_2) \otimes \Big(P_A(\mathfrak{x}_1') \diamond P_A(\mathfrak{x}_2')\Big) \\
&= \Big(P_A(\mathfrak{x}_1) \diamond P_A(\mathfrak{x}_2) + \lambda P_A(\mathfrak{x}_1) \diamond \mathfrak{x}_2 + \lambda \mathfrak{x}_1 \diamond P_A(\mathfrak{x}_2) + \lambda^2 \mathfrak{x}_1 \diamond \mathfrak{x}_2\Big) \otimes \varepsilon_M(\mathfrak{x}_1')\varepsilon_M(\mathfrak{x}_2')1 \\
&\quad + \Big(P_A(\mathfrak{x}_1) \diamond \mathfrak{x}_2\Big) \otimes \varepsilon_M(\mathfrak{x}_1')P_A(\mathfrak{x}_2') + \lambda(\mathfrak{x}_1 \diamond \mathfrak{x}_2) \otimes \varepsilon_M(\mathfrak{x}_1')P_A(\mathfrak{x}_2') + \Big(\mathfrak{x}_1 \diamond P_A(\mathfrak{x}_2)\Big) \otimes P_A(\mathfrak{x}_1')\varepsilon_M(\mathfrak{x}_2') \\
&\quad + \lambda(\mathfrak{x}_1 \diamond \mathfrak{x}_2) \otimes P_A(\mathfrak{x}_1')\varepsilon_M(\mathfrak{x}_2') + (\mathfrak{x}_1 \diamond \mathfrak{x}_2) \otimes \Big(P_A(\mathfrak{x}_1') \diamond P_A(\mathfrak{x}_2')\Big) \\
&= \Big(P_A(\mathfrak{x}_1 \diamond P_A(\mathfrak{x}_2)) + P_A(P_A(\mathfrak{x}_1) \diamond \mathfrak{x}_2) + \lambda P_A(\mathfrak{x}_1) \diamond \mathfrak{x}_2 + \lambda \mathfrak{x}_1 \diamond P_A(\mathfrak{x}_2)\Big) \otimes \varepsilon_M(\mathfrak{x}_1')\varepsilon_M(\mathfrak{x}_2')1 \\
&\quad + (P_A(\mathfrak{x}_1) \diamond \mathfrak{x}_2) \otimes \varepsilon_M(\mathfrak{x}_1')P_A(\mathfrak{x}_2') + \lambda(\mathfrak{x}_1 \diamond \mathfrak{x}_2) \otimes \varepsilon_M(\mathfrak{x}_1')P_A(\mathfrak{x}_2') + (\mathfrak{x}_1 \diamond P_A(\mathfrak{x}_2)) \otimes P_A(\mathfrak{x}_1')\varepsilon_M(\mathfrak{x}_2') \\
&\quad + \lambda(\mathfrak{x}_1 \diamond \mathfrak{x}_2) \otimes P_A(\mathfrak{x}_1')\varepsilon_M(\mathfrak{x}_2') + (\mathfrak{x}_1 \diamond \mathfrak{x}_2) \otimes \Big(P_A(\mathfrak{x}_1 \diamond P_A(\mathfrak{x}_2)) + P_A(P_A(\mathfrak{x}_1) \diamond \mathfrak{x}_2) - \lambda^2 \mathfrak{x}_1' \diamond \mathfrak{x}_2'\Big).
\end{aligned}$$

(by Theorem (2.6) (b))

On the other hand,

$$\begin{aligned}
&Q\Big((\mathfrak{x}_1 \otimes \mathfrak{x}_1') \diamond' Q(\mathfrak{x}_2 \otimes \mathfrak{x}_2')\Big) + Q\Big(Q(\mathfrak{x}_1 \otimes \mathfrak{x}_1') \diamond' (\mathfrak{x}_2 \otimes \mathfrak{x}_2')\Big) - \lambda^2\Big((\mathfrak{x}_1 \otimes \mathfrak{x}_1') \diamond' (\mathfrak{x}_2 \otimes \mathfrak{x}_2')\Big) \\
&= Q\Big((\mathfrak{x}_1 \otimes \mathfrak{x}_1') \diamond' \Big((P_A(\mathfrak{x}_2) + \lambda \mathfrak{x}_2) \otimes \varepsilon_M(\mathfrak{x}_2')1 + \mathfrak{x}_2 \otimes P_A(\mathfrak{x}_2')\Big)\Big) \\
&\quad + Q\Big(\Big((P_A(\mathfrak{x}_1) + \lambda \mathfrak{x}_1) \otimes \varepsilon_M(\mathfrak{x}_1')1 + \mathfrak{x}_1 \otimes P_A(\mathfrak{x}_1')\Big) \diamond' (\mathfrak{x}_2 \otimes \mathfrak{x}_2')\Big) - \lambda^2(\mathfrak{x}_1 \diamond \mathfrak{x}_2) \otimes (\mathfrak{x}_1' \diamond \mathfrak{x}_2') \\
&= Q\Big((\mathfrak{x}_1 \diamond P_A(\mathfrak{x}_2)) \otimes \mathfrak{x}_1'\varepsilon_M(\mathfrak{x}_2') + \lambda(\mathfrak{x}_1 \diamond \mathfrak{x}_2) \otimes \mathfrak{x}_1'\varepsilon_M(\mathfrak{x}_2') + (\mathfrak{x}_1 \diamond \mathfrak{x}_2) \otimes (\mathfrak{x}_1' \diamond P_A(\mathfrak{x}_2'))\Big)
\end{aligned}$$

$$
\begin{aligned}
&+Q\Big(\big(P_A(\mathfrak{x}_1)\diamond\mathfrak{x}_2\big)\otimes\varepsilon_{\mathrm{M}}(\mathfrak{x}_1')\mathfrak{x}_2'+\lambda(\mathfrak{x}_1\diamond\mathfrak{x}_2)\otimes\varepsilon_{\mathrm{M}}(\mathfrak{x}_1')\mathfrak{x}_2'+(\mathfrak{x}_1\diamond\mathfrak{x}_2)\otimes\big(P_A(\mathfrak{x}_1')\diamond\mathfrak{x}_2'\big)\Big)\\
&-\lambda^2(\mathfrak{x}_1\diamond\mathfrak{x}_2)\otimes(\mathfrak{x}_1'\diamond\mathfrak{x}_2')\\
=&\Big(P_A\big(\mathfrak{x}_1\diamond P_A(\mathfrak{x}_2)\big)+\lambda\mathfrak{x}_1\diamond P_A(\mathfrak{x}_2)\Big)\otimes\varepsilon_{\mathrm{M}}\big(\mathfrak{x}_1'\varepsilon_{\mathrm{M}}(\mathfrak{x}_2')\big)1+\big(\mathfrak{x}_1\diamond P_A(\mathfrak{x}_2)\big)\otimes P_A\big(\mathfrak{x}_1'\varepsilon_{\mathrm{M}}(\mathfrak{x}_2')\big)\\
&+\lambda\Big(P_A(\mathfrak{x}_1\diamond\mathfrak{x}_2)+\lambda\mathfrak{x}_1\diamond\mathfrak{x}_2\Big)\otimes\varepsilon_{\mathrm{M}}\big(\mathfrak{x}_1'\varepsilon_{\mathrm{M}}(\mathfrak{x}_2')\big)1+\lambda(\mathfrak{x}_1\diamond\mathfrak{x}_2)\otimes P_A\big(\mathfrak{x}_1'\varepsilon_{\mathrm{M}}(\mathfrak{x}_2')\big)\\
&+\Big(P_A(\mathfrak{x}_1\diamond\mathfrak{x}_2)+\lambda\mathfrak{x}_1\diamond\mathfrak{x}_2\Big)\otimes\varepsilon_{\mathrm{M}}\big(\mathfrak{x}_1'P_A(\mathfrak{x}_2')\big)1+(\mathfrak{x}_1\diamond\mathfrak{x}_2)\otimes P_A\big(\mathfrak{x}_1'P_A(\mathfrak{x}_2')\big)\\
&+\Big(P_A\big(P_A(\mathfrak{x}_1)\diamond\mathfrak{x}_2\big)+\lambda P_A(\mathfrak{x}_1)\diamond\mathfrak{x}_2\Big)\otimes\varepsilon_{\mathrm{M}}\big(\varepsilon_{\mathrm{M}}(\mathfrak{x}_1')\mathfrak{x}_2'\big)1+\big(P_A(\mathfrak{x}_1)\diamond\mathfrak{x}_2\big)\otimes P_A\big(\varepsilon_{\mathrm{M}}(\mathfrak{x}_1')\mathfrak{x}_2'\big)\\
&+\lambda\Big(P_A(\mathfrak{x}_1\diamond\mathfrak{x}_2)+\lambda\mathfrak{x}_1\diamond\mathfrak{x}_2\Big)\otimes\varepsilon_{\mathrm{M}}\big(\varepsilon_{\mathrm{M}}(\mathfrak{x}_1')\mathfrak{x}_2'\big)1+\lambda(\mathfrak{x}_1\diamond\mathfrak{x}_2)\otimes P_A\big(\varepsilon_{\mathrm{M}}(\mathfrak{x}_1')\mathfrak{x}_2'\big)\\
&+\Big(P_A(\mathfrak{x}_1\diamond\mathfrak{x}_2)+\lambda\mathfrak{x}_1\diamond\mathfrak{x}_2\Big)\otimes\varepsilon_{\mathrm{M}}\big(P_A(\mathfrak{x}_1')\mathfrak{x}_2'\big)+(\mathfrak{x}_1\diamond\mathfrak{x}_2)\otimes P_A\big(P_A(\mathfrak{x}_1')\mathfrak{x}_2'\big)-\lambda^2(\mathfrak{x}_1\diamond\mathfrak{x}_2)\otimes(\mathfrak{x}_1'\diamond\mathfrak{x}_2')\\
=&\Big(P_A\big(\mathfrak{x}_1\diamond P_A(\mathfrak{x}_2)\big)+\lambda\mathfrak{x}_1\diamond P_A(\mathfrak{x}_2)\Big)\otimes\varepsilon_{\mathrm{M}}(\mathfrak{x}_1')\varepsilon_{\mathrm{M}}(\mathfrak{x}_2')1+\big(\mathfrak{x}_1\diamond P_A(\mathfrak{x}_2)\big)\otimes P_A(\mathfrak{x}_1')\varepsilon_{\mathrm{M}}(\mathfrak{x}_2')\\
&+\lambda\Big(P_A(\mathfrak{x}_1\diamond\mathfrak{x}_2)+\lambda\mathfrak{x}_1\diamond\mathfrak{x}_2\Big)\otimes\varepsilon_{\mathrm{M}}(\mathfrak{x}_1')\varepsilon_{\mathrm{M}}(\mathfrak{x}_2')1+\lambda(\mathfrak{x}_1\diamond\mathfrak{x}_2)\otimes P_A(\mathfrak{x}_1')\varepsilon_{\mathrm{M}}(\mathfrak{x}_2')\\
&+\Big(P_A(\mathfrak{x}_1\diamond\mathfrak{x}_2)+\lambda\mathfrak{x}_1\diamond\mathfrak{x}_2\Big)\otimes\varepsilon_{\mathrm{M}}(\mathfrak{x}_1')\varepsilon_{\mathrm{M}}(P_A(\mathfrak{x}_2'))1+(\mathfrak{x}_1\diamond\mathfrak{x}_2)\otimes P_A\big(\mathfrak{x}_1'P_A(\mathfrak{x}_2')\big)\\
&+\Big(P_A\big(P_A(\mathfrak{x}_1)\diamond\mathfrak{x}_2\big)+\lambda P_A(\mathfrak{x}_1)\diamond\mathfrak{x}_2\Big)\otimes\varepsilon_{\mathrm{M}}(\mathfrak{x}_1')\varepsilon_{\mathrm{M}}(\mathfrak{x}_2')1+\big(P_A(\mathfrak{x}_1)\diamond\mathfrak{x}_2\big)\otimes\varepsilon_{\mathrm{M}}(\mathfrak{x}_1')P_A(\mathfrak{x}_2')\\
&+\lambda\Big(P_A(\mathfrak{x}_1\diamond\mathfrak{x}_2)+\lambda\mathfrak{x}_1\diamond\mathfrak{x}_2\Big)\otimes\varepsilon_{\mathrm{M}}(\mathfrak{x}_1')\varepsilon_{\mathrm{M}}(\mathfrak{x}_2')1+\lambda(\mathfrak{x}_1\diamond\mathfrak{x}_2)\otimes\varepsilon_{\mathrm{M}}(\mathfrak{x}_1')P_A(\mathfrak{x}_2')\\
&+\Big(P_A(\mathfrak{x}_1\diamond\mathfrak{x}_2)+\lambda\mathfrak{x}_1\diamond\mathfrak{x}_2\Big)\otimes\varepsilon_{\mathrm{M}}(P_A(\mathfrak{x}_1'))\varepsilon_{\mathrm{M}}(\mathfrak{x}_2')1+(\mathfrak{x}_1\diamond\mathfrak{x}_2)\otimes P_A\big(P_A(\mathfrak{x}_1')\mathfrak{x}_2'\big)\\
&-\lambda^2(\mathfrak{x}_1\diamond\mathfrak{x}_2)\otimes(\mathfrak{x}_1'\diamond\mathfrak{x}_2')\qquad\text{(by $\varepsilon_{\mathrm{M}}$ is $\mathbf{k}$-linear and $\varepsilon_{\mathrm{M}}$ is a homomorphism)}\\
=&\Big(P_A\big(\mathfrak{x}_1\diamond P_A(\mathfrak{x}_2)\big)+\lambda\mathfrak{x}_1\diamond P_A(\mathfrak{x}_2)\Big)\otimes\varepsilon_{\mathrm{M}}(\mathfrak{x}_1')\varepsilon_{\mathrm{M}}(\mathfrak{x}_2')1+\big(\mathfrak{x}_1\diamond P_A(\mathfrak{x}_2)\big)\otimes P_A(\mathfrak{x}_1')\varepsilon_{\mathrm{M}}(\mathfrak{x}_2')\\
&+\lambda\Big(P_A(\mathfrak{x}_1\diamond\mathfrak{x}_2)+\lambda\mathfrak{x}_1\diamond\mathfrak{x}_2\Big)\otimes\varepsilon_{\mathrm{M}}(\mathfrak{x}_1')\varepsilon_{\mathrm{M}}(\mathfrak{x}_2')1+\lambda(\mathfrak{x}_1\diamond\mathfrak{x}_2)\otimes P_A(\mathfrak{x}_1')\varepsilon_{\mathrm{M}}(\mathfrak{x}_2')\\
&-\lambda\Big(P_A(\mathfrak{x}_1\diamond\mathfrak{x}_2)+\lambda\mathfrak{x}_1\diamond\mathfrak{x}_2\Big)\otimes\varepsilon_{\mathrm{M}}(\mathfrak{x}_1')\varepsilon_{\mathrm{M}}(\mathfrak{x}_2')1+(\mathfrak{x}_1\diamond\mathfrak{x}_2)\otimes P_A\big(\mathfrak{x}_1'P_A(\mathfrak{x}_2')\big)\\
&+\Big(P_A\big(P_A(\mathfrak{x}_1)\diamond\mathfrak{x}_2\big)+\lambda P_A(\mathfrak{x}_1)\diamond\mathfrak{x}_2\Big)\otimes\varepsilon_{\mathrm{M}}(\mathfrak{x}_1')\varepsilon_{\mathrm{M}}(\mathfrak{x}_2')1+\big(P_A(\mathfrak{x}_1)\diamond\mathfrak{x}_2\big)\otimes\varepsilon_{\mathrm{M}}(\mathfrak{x}_1')P_A(\mathfrak{x}_2')\\
&+\lambda\Big(P_A(\mathfrak{x}_1\diamond\mathfrak{x}_2)+\lambda\mathfrak{x}_1\diamond\mathfrak{x}_2\Big)\otimes\varepsilon_{\mathrm{M}}(\mathfrak{x}_1')\varepsilon_{\mathrm{M}}(\mathfrak{x}_2')1+\lambda(\mathfrak{x}_1\diamond\mathfrak{x}_2)\otimes\varepsilon_{\mathrm{M}}(\mathfrak{x}_1')P_A(\mathfrak{x}_2')\\
&-\lambda\Big(P_A(\mathfrak{x}_1\diamond\mathfrak{x}_2)+\lambda\mathfrak{x}_1\diamond\mathfrak{x}_2\Big)\otimes\varepsilon_{\mathrm{M}}(\mathfrak{x}_1')\varepsilon_{\mathrm{M}}(\mathfrak{x}_2')1+(\mathfrak{x}_1\diamond\mathfrak{x}_2)\otimes P_A\big(P_A(\mathfrak{x}_1')\mathfrak{x}_2'\big)\\
&-\lambda^2(\mathfrak{x}_1\diamond\mathfrak{x}_2)\otimes(\mathfrak{x}_1'\diamond\mathfrak{x}_2')\qquad\text{(Using Eq. (13) in the fifth and eleventh terms)}
\end{aligned}
$$

$$
\begin{aligned}
&= \Big(P_A(\mathfrak{x}_1 \diamond P_A(\mathfrak{x}_2)) + \lambda \mathfrak{x}_1 \diamond P_A(\mathfrak{x}_2) + P_A(P_A(\mathfrak{x}_1)\diamond \mathfrak{x}_2) + \lambda P_A(\mathfrak{x}_1)\diamond \mathfrak{x}_2\Big) \otimes \varepsilon_{\mathrm{M}}(\mathfrak{x}_1')\varepsilon_{\mathrm{M}}(\mathfrak{x}_2')1 \\
&\quad + (\mathfrak{x}_1 \diamond P_A(\mathfrak{x}_2)) \otimes P_A(\mathfrak{x}_1')\varepsilon_{\mathrm{M}}(\mathfrak{x}_2') + \lambda(\mathfrak{x}_1 \diamond \mathfrak{x}_2) \otimes P_A(\mathfrak{x}_1')\varepsilon_{\mathrm{M}}(\mathfrak{x}_2') + (\mathfrak{x}_1 \diamond \mathfrak{x}_2) \otimes P_A(\mathfrak{x}_1' P_A(\mathfrak{x}_2')) \\
&\quad + (P_A(\mathfrak{x}_1)\diamond \mathfrak{x}_2) \otimes \varepsilon_{\mathrm{M}}(\mathfrak{x}_1')P_A(\mathfrak{x}_2') + \lambda(\mathfrak{x}_1 \diamond \mathfrak{x}_2) \otimes \varepsilon_{\mathrm{M}}(\mathfrak{x}_1')P_A(\mathfrak{x}_2') + (\mathfrak{x}_1 \diamond \mathfrak{x}_2) \otimes P_A(P_A(\mathfrak{x}_1')\mathfrak{x}_2') \\
&\quad - \lambda^2(\mathfrak{x}_1 \diamond \mathfrak{x}_2) \otimes (\mathfrak{x}_1' \diamond \mathfrak{x}_2').
\end{aligned}
$$

This completes the proof. □

With a similar argument, we can obtain

**Lemma 3.3.** *Let $\lambda$ be a given element of* $\mathbf{k}$. *Define the linear map*

$$\tilde{Q} : F_{-\lambda^2}(A) \otimes F_{-\lambda^2}(A) \otimes F_{-\lambda^2}(A) \to F_{-\lambda^2}(A) \otimes F_{-\lambda^2}(A) \otimes F_{-\lambda^2}(A)$$

*by taking*

$$
\begin{aligned}
\tilde{Q}(\mathfrak{x} \otimes \mathfrak{x}' \otimes \mathfrak{x}'') :=& (P_A(\mathfrak{x}) + \lambda \mathfrak{x}) \otimes \varepsilon_{\mathrm{M}}(\mathfrak{x}')1 \otimes \varepsilon_{\mathrm{M}}(\mathfrak{x}'')1 + \mathfrak{x} \otimes (P_A(\mathfrak{x}') + \lambda \mathfrak{x}') \otimes \varepsilon_{\mathrm{M}}(\mathfrak{x}'')1 \\
&+ \mathfrak{x} \otimes \mathfrak{x}' \otimes P_A(\mathfrak{x}'') \quad \text{for } \mathfrak{x}, \mathfrak{x}' \in F_{-\lambda^2}(A).
\end{aligned} \tag{15}
$$

*Then $\tilde{Q}$ is a modified Rota-Baxter operator of weight $-\lambda^2$ on $F_{-\lambda^2}(A) \otimes F_{-\lambda^2}(A) \otimes F_{-\lambda^2}(A)$.*

Now we are ready for our main result of this subsection. Recall $\varepsilon_{\mathrm{M}} : F_{-\lambda^2}(A) \to \mathbf{k}$ is an algebra homomorphism given in Lemma 3.1. Let $j_A : A \to F_\kappa(A)$ be the natural embedding. By Theorem 2.6 (c) and Lemma 3.2, there is a (unique) modified Rota-Baxter algebra morphism

$$\Delta_M : F_{-\lambda^2}(A) \to F_{-\lambda^2}(A) \otimes F_{-\lambda^2}(A)$$

such that $\Delta_M \circ j_A = \Delta_A$.

**Theorem 3.4.** *Let $A$ be a bialgebra and $\lambda \in \mathbf{k}$. Then the quintuple $(F_{-\lambda^2}(A), \diamond, 1, \Delta_{\mathrm{M}}, \varepsilon_{\mathrm{M}})$ is a bialgebra.*

**Proof.** It suffices to prove the counity of $\varepsilon_{\mathrm{M}}$ and coassociativity of $\Delta_{\mathrm{M}}$. For the former, denote by

$$\phi := (\varepsilon_{\mathrm{M}} \otimes \mathrm{id})\Delta_M : F_{-\lambda^2}(A) \to F_{-\lambda^2}(A)$$

Then $\phi$ is an algebra homomorphism, since $\varepsilon_{\mathrm{M}}$ and $\Delta_M$ are algebra homomorphisms. Further it is a modified Rota-Baxter algebra morphism. Indeed, for any $\mathfrak{x} \in F_{-\lambda^2}(A)$,

$$\phi \circ P_A(\mathfrak{x}) = \big((\varepsilon_{\mathrm{M}} \otimes \mathrm{id})\Delta_M\big)P_A(\mathfrak{x}) = (\varepsilon_{\mathrm{M}} \otimes \mathrm{id})(\Delta_M P_A)(\mathfrak{x})$$

$$
\begin{aligned}
&=(\varepsilon_{\mathrm{M}}\otimes \mathrm{id})(Q\Delta_M)(\mathfrak{x}) \quad (\text{by } \Delta_M \text{ being an modified Rota-Baxter algebra morphism})\\
&=(\varepsilon_{\mathrm{M}}\otimes \mathrm{id})Q\Big(\sum_{(\mathfrak{x})}\mathfrak{x}_{(1)}\otimes \mathfrak{x}_{(2)}\Big) \quad (\text{by Sweedler's notation})\\
&=\sum_{(\mathfrak{x})}(\varepsilon_{\mathrm{M}}\otimes \mathrm{id})\Big((P_A(\mathfrak{x}_{(1)})+\lambda\mathfrak{x}_{(1)})\otimes \varepsilon_{\mathrm{M}}(\mathfrak{x}_{(2)})1+\mathfrak{x}_{(1)}\otimes P_A(\mathfrak{x}_{(2)})\Big) \quad (\text{by Eq. (14)})\\
&=\sum_{(\mathfrak{x})}\Big(\varepsilon_{\mathrm{M}}(P_A(\mathfrak{x}_{(1)}))\otimes \varepsilon_{\mathrm{M}}(\mathfrak{x}_{(2)})1+\lambda\varepsilon_{\mathrm{M}}(\mathfrak{x}_{(1)})\otimes \varepsilon_{\mathrm{M}}(\mathfrak{x}_{(2)})1+\varepsilon_{\mathrm{M}}(\mathfrak{x}_{(1)})\otimes P_A(\mathfrak{x}_{(2)})\Big)\\
&=\sum_{(\mathfrak{x})}\Big(-\lambda\varepsilon_{\mathrm{M}}(\mathfrak{x}_{(1)})\otimes \varepsilon_{\mathrm{M}}(\mathfrak{x}_{(2)})1+\lambda\varepsilon_{\mathrm{M}}(\mathfrak{x}_{(1)})\otimes \varepsilon_{\mathrm{M}}(\mathfrak{x}_{(2)})1+\varepsilon_{\mathrm{M}}(\mathfrak{x}_{(1)})\otimes P_A(\mathfrak{x}_{(2)})\Big)\\
&\qquad\qquad\qquad\qquad\qquad\qquad\qquad\qquad\qquad\qquad (\text{by Eq. (13)})\\
&=\sum_{(\mathfrak{x})}\varepsilon_{\mathrm{M}}(\mathfrak{x}_{(1)})\otimes P_A(\mathfrak{x}_{(2)})=\sum_{(\mathfrak{x})}P_A(\varepsilon_{\mathrm{M}}(\mathfrak{x}_{(1)})\mathfrak{x}_{(2)})=P_A((\varepsilon_{\mathrm{M}}\otimes \mathrm{id})\Delta_M(\mathfrak{x}))\\
&=P_A\circ\phi(\mathfrak{x}).
\end{aligned}
$$

By unicity in the universal property of $F_{-\lambda^2}(A)$, we have

$$(\varepsilon_{\mathrm{M}}\otimes \mathrm{id})\Delta_M=\phi=\mathrm{id}_{F_{-\lambda^2}(A)}$$

and so $\varepsilon_{\mathrm{M}}$ is a left counit. By symmetry, we can prove $\varepsilon_{\mathrm{M}}$ is also a right counit.

Moreover, both $(\Delta_M\otimes\mathrm{id})\Delta_M$ and $(\mathrm{id}\otimes\Delta_M)\Delta_M$ are modified Rota-Baxter algebra morphisms from $F_{-\lambda^2}(A)$ to $F_{-\lambda^2}(A)\otimes F_{-\lambda^2}(A)\otimes F_{-\lambda^2}(A)$, which is equipped with the modified Rota-Baxter operator $\tilde{Q}$ of weight $-\lambda^2$ given in Lemma 3.3. As they coincide on $A$

$$(\Delta_M\otimes \mathrm{id})\Delta_M|_A=(\Delta_A\otimes \mathrm{id})\Delta_A=(\mathrm{id}\otimes\Delta_A)\Delta_A=(\mathrm{id}\otimes\Delta_M)\Delta_M|_A,$$

they are equal and so $\Delta_M$ is coassociative. Here $\Delta_A$ is the coproduct on $A$. Thus the quintuple $(F_{-\lambda^2}(A),\diamond,1,\Delta_{\mathrm{M}},\varepsilon_{\mathrm{M}})$ is a bialgebra. □

**Remark 3.5.** *For any $\mathfrak{x}\in F_{-\lambda^2}(A)$, we have*

$$
\begin{aligned}
\Delta_M\circ P_A(\mathfrak{x}) &=Q\circ\Delta_M(\mathfrak{x}) \quad (\textit{by } \Delta_M \textit{ being a modified Rota-Baxter algebra morphism})\\
&=Q\Big(\sum_{(\mathfrak{x})}\mathfrak{x}_{(1)}\otimes \mathfrak{x}_{(2)}\Big) \quad (\textit{by Sweedler's notation})\\
&=\sum_{(\mathfrak{x})}\Big((P_A(\mathfrak{x}_{(1)})+\lambda\mathfrak{x}_{(1)})\otimes \varepsilon_{\mathrm{M}}(\mathfrak{x}_{(2)})1+\mathfrak{x}_{(1)}\otimes P_A(\mathfrak{x}_{(2)})\Big) \quad (\textit{by Eq. (14)})\\
&=\sum_{(\mathfrak{x})}\Big(P_A(\mathfrak{x}_{(1)})\otimes \varepsilon_{\mathrm{M}}(\mathfrak{x}_{(2)})1+\lambda\mathfrak{x}_{(1)}\otimes \varepsilon_{\mathrm{M}}(\mathfrak{x}_{(2)})1+\mathfrak{x}_{(1)}\otimes P_A(\mathfrak{x}_{(2)})\Big)
\end{aligned}
$$

$$=\sum_{(\mathfrak{x})}\Big(P_A(\mathfrak{x}_{(1)}\varepsilon_{\mathrm{M}}(\mathfrak{x}_{(2)}))\otimes 1+\lambda\mathfrak{x}_{(1)}\varepsilon_{\mathrm{M}}(\mathfrak{x}_{(2)})\otimes 1+(\mathrm{id}\otimes P_A)(\mathfrak{x}_{(1)}\otimes\mathfrak{x}_{(2)})\Big)$$

$$=P_A(\mathfrak{x})\otimes 1+\lambda\mathfrak{x}\otimes 1+(\mathrm{id}\otimes P_A)\Delta_M(\mathfrak{x})\quad (\textit{by } \varepsilon_M \textit{ being the counit}).$$

*In other words,*

$$\Delta_M P_A = P_A\otimes 1+(\mathrm{id}\otimes P_A)\Delta_M+\lambda\mathrm{id}\otimes 1, \tag{16}$$

*which is analogue to the 1-cocycle condition in the well-known Connes-Kreimer Hopf algebra on rooted trees* [11].

**3.2. The Hopf algebraic structure.** In this last part of the paper we show that if we start with $A$ being a connected filtered bialgebra and $\lambda\in\mathbf{k}$, then the bialgebra $F_{-\lambda^2}(A)$ also has a connected filtration and hence is a Hopf algebra.

**Definition 3.6.** A bialgebra $(A,m,\mu,\Delta,\varepsilon)$ is called *filtered* if it has an increasing filtration $A_n$, $n\geq 0$, such that

$$A=\cup_{n\geq 0}A_n,\ A_pA_q\subseteq A_{p+q}\ \text{ and }\ \Delta(A_n)\subseteq\sum_{p+q=n}A_p\otimes A_q\ \text{ for } p,q,n\geq 0.$$

A filtered bialgebra $A$ is called *connected* if $A_0=\operatorname{im}\mu$ and $A=A_0\oplus\ker\varepsilon$.

The following result is well-known.

**Lemma 3.7.** [23] *A connected filtered bialgebra is a Hopf algebra.*

Our discussion in this section will be based on the following condition.

**Definition 3.8.** A $\mathbf{k}$-basis $X$ of a connected filtered bialgebra $A=\cup_{n\geq 0}A_n$ is called a *filtered basis* of $A$ if there is an increasing filtration $X=\cup_{n\geq 0}X_n$ such that

$$A_n=\mathbf{k}X_n, X\backslash\{1\}\subseteq\ker\varepsilon, X_0=\{1\}.$$

Here 1 is the identity of $A$. Elements $x\in X_n\setminus X_{n-1}$ are said to have *degree* $n$, denoted by $\deg_A(x)=n$.

Let $A$ be a connected filtered bialgebra with a filtered basis $X$. Recall that $\mathfrak{X}_\infty$ constructed in Subsection 2.1 is a $\mathbf{k}$-basis of the free modified Rota-Baxter algebra $F_{-\lambda^2}(A)$. We now define the *degree* $\deg(\mathfrak{x})$ for $\mathfrak{x}\in\mathfrak{X}_\infty$ by induction on $\operatorname{dep}(\mathfrak{x})$. For the initial step of $\operatorname{dep}(\mathfrak{x})=0$, we get $\mathfrak{x}\in X\subseteq A$ and define

$$\deg(\mathfrak{x}):=\deg_A(\mathfrak{x}). \tag{17}$$

For the inductive step of $\operatorname{dep}(\mathfrak{x})\geq 1$, if $\operatorname{bre}(\mathfrak{x})=1$, then $\mathfrak{x}=\lfloor\bar{\mathfrak{x}}\rfloor$ and we define

$$\deg(\mathfrak{x}):=\deg(\bar{\mathfrak{x}})+1; \tag{18}$$

if $\mathrm{bre}(\mathfrak{x}) \geq 2$, then write $\mathfrak{x} = \mathfrak{x}_1 \cdots \mathfrak{x}_b$ in the standard decomposition and define

$$\deg(\mathfrak{x}) := \sum_{i=1}^{b} \deg(\mathfrak{x}_i), \tag{19}$$

where each $\deg(\mathfrak{x}_i)$ is defined either in Eq. (17) or in Eq. (18) by the induction hypothesis.

**Remark 3.9.** *For later applications, we also use the notion* $\deg(c\mathfrak{x}) = \deg(\mathfrak{x})$ *for* $c \in \mathbf{k} \backslash \{0\}$.

Denote

$$\mathcal{H} := F_{-\lambda^2}(A) \text{ and } \mathcal{H}_n := \mathbf{k}\{\mathfrak{x} \in \mathfrak{X}_\infty \mid \deg(\mathfrak{x}) \leq n\} \text{ for } n \geq 0. \tag{20}$$

Then

$$\mathcal{H} = \bigcup_{n \geq 0} \mathcal{H}_n\,,\ \mathcal{H}_0 = \mathbf{k},\ \mathcal{H} = \mathbf{k}\,1 \oplus \ker \varepsilon_{\mathrm{M}} \text{ and } P_A(\mathcal{H}_n) \subseteq \mathcal{H}_{n-1}. \tag{21}$$

Now we are going to prove that $\mathcal{H}$ is a filtered bialgebra, beginning with the compatibility of the multiplication with the filtration.

**Lemma 3.10.** *For* $p, q \geq 0$, *we have*

$$\mathcal{H}_p \diamond \mathcal{H}_q \subseteq \mathcal{H}_{p+q}. \tag{22}$$

**Proof.** Let $\mathfrak{x} \in \mathcal{H}_p$ and $\mathfrak{x}' \in \mathcal{H}_q$ be two basis elements in $\mathfrak{X}_\infty$. Then

$$\deg(\mathfrak{x}) \leq p \text{ and } \deg(\mathfrak{x}') \leq q.$$

We now verify Eq. (22) by induction on the sum $s := p + q \geq 0$. When $s = 0$, then $p = q = 0$. By Eq. (21), we obtain that $\mathfrak{x} = \mathfrak{x}' = 1$ and so $\mathfrak{x} \diamond \mathfrak{x}' = 1 \in \mathcal{H}_0$. This finishes the initial step.

Given an $s \geq 0$, assume that Eq. (22) holds for $\mathfrak{x}, \mathfrak{x}'$ with $p + q = s$ and consider case $p + q = s + 1$. If $\mathfrak{x} = 1$ or $\mathfrak{x}' = 1$, without loss of generality, letting $\mathfrak{x} = 1$, then $p = 0$ and

$$\mathfrak{x} \diamond \mathfrak{x}' = \mathfrak{x}' \in \mathcal{H}_q = \mathcal{H}_{p+q}.$$

So we may suppose $\mathfrak{x}, \mathfrak{x}' \neq 1$. Write

$$\mathfrak{x} = \mathfrak{x}_1 \cdots \mathfrak{x}_b \text{ and } \mathfrak{x}' := \mathfrak{x}'_1 \cdots \mathfrak{x}'_{b'} \text{ with } b, b' \geq 1$$

in their standard decompositions. Under this condition, we proceed to prove Eq. (22) by induction on the sum $t := b + b' \geq 2$. When $t = 2$, then $b = b' = 1$. If $\mathfrak{x} \in X \subseteq A$ or $\mathfrak{x}' \in X \subseteq A$, then by Eq. (19),

$$\mathfrak{x} \diamond \mathfrak{x}' = \mathfrak{x}\mathfrak{x}' \in \mathcal{H}_{\deg(\mathfrak{x})+\deg(\mathfrak{x}')} \subseteq \mathcal{H}_{p+q}.$$

It remains to check the outstanding case of

$$\mathfrak{x} := \mathfrak{x}_1 = P_A(\bar{\mathfrak{x}}_1) \text{ and } \mathfrak{x}' := \mathfrak{x}'_1 = P_A(\bar{\mathfrak{x}}'_1),$$

where

$$\bar{\mathfrak{x}}_1, \bar{\mathfrak{x}}'_1 \in \mathfrak{X}_\infty \text{ and } \deg(\mathfrak{x}_1) + \deg(\mathfrak{x}'_1) \leq s+1.$$

Then

$$\begin{aligned}
\deg(\mathfrak{x}_1) + \deg(\bar{\mathfrak{x}}'_1) &= \deg(\mathfrak{x}_1) + \deg(\mathfrak{x}'_1) - 1 \leq s, \\
\deg(\bar{\mathfrak{x}}_1) + \deg(\mathfrak{x}'_1) &= \deg(\mathfrak{x}_1) - 1 + \deg(\mathfrak{x}'_1) \leq s, \\
\deg(\bar{\mathfrak{x}}_1) + \deg(\overline{\mathfrak{x}'_1}) &= \deg(\mathfrak{x}_1) - 1 + \deg(\mathfrak{x}'_1) - 1 \leq s-1.
\end{aligned}$$

By the induction hypothesis on $s$, we have

$$\mathfrak{x}_1 \diamond \bar{\mathfrak{x}}'_1, \bar{\mathfrak{x}}_1 \diamond \mathfrak{x}'_1 \in \mathcal{H}_s \text{ and } \overline{\mathfrak{x}_1} \diamond \bar{\mathfrak{x}}'_1 \in \mathcal{H}_{s-1},$$

which implies from Eq. (21) that

$$P_A(\mathfrak{x}_1 \diamond \bar{\mathfrak{x}}'_1), P_A(\bar{\mathfrak{x}}_1 \diamond \mathfrak{x}'_1) \in \mathcal{H}_{s+1}.$$

Hence by Eq. (7),

$$\mathfrak{x}_1 \diamond \mathfrak{x}'_1 = P_A(\mathfrak{x}_1 \diamond \bar{\mathfrak{x}}'_1) + P_A(\bar{\mathfrak{x}}_1 \diamond \mathfrak{x}'_1) - \lambda^2 \bar{\mathfrak{x}}_1 \diamond \overline{\mathfrak{x}'_1} \in \mathcal{H}_{s+1}.$$

Assume that Eq. (22) holds for $b + b' = t \geq 2$ and $p + q = s + 1$ and consider the case when $b + b' = t + 1 \geq 3$ and $p + q = s + 1$. So either $\mathfrak{x}$ or $\mathfrak{x}'$ has breadth greater than or equal to 2, giving us three cases to consider:

**Case 1.** $\mathrm{bre}(\mathfrak{x}) \geq 2$. Let $\mathfrak{x} := \mathfrak{x}_{1,1}\mathfrak{x}_{1,2}$, where $\mathfrak{x}_{1,1}, \mathfrak{x}_{1,2} \in \mathfrak{X}_\infty$ with breadths $\mathrm{bre}(\mathfrak{x}_{1,1}), \mathrm{bre}(\mathfrak{x}_{1,2}) \geq 1$ respectively. By Eq. (19), we obtain $\deg(\mathfrak{x}) = \deg(\mathfrak{x}_{1,1}) + \deg(\mathfrak{x}_{1,2})$. From Eq. (8),

$$\mathfrak{x} \diamond \mathfrak{x}' = (\mathfrak{x}_{1,1}\mathfrak{x}_{1,2}) \diamond \mathfrak{x}' = \mathfrak{x}_{1,1}(\mathfrak{x}_{1,2} \diamond \mathfrak{x}').$$

By the induction on $t$, we have

$$\mathfrak{x}_{1,2} \diamond \mathfrak{x}' \in \mathcal{H}_{\deg(\mathfrak{x}_{1,2})+\deg(\mathfrak{x}')},$$

whence by Eq. (19),

$$\mathfrak{x} \diamond \mathfrak{x}' = \mathfrak{x}_{1,1}(\mathfrak{x}_{1,2} \diamond \mathfrak{x}') \in \mathcal{H}_{\deg(\mathfrak{x}_{1,1})+\deg(\mathfrak{x}_{1,2})+\deg(\mathfrak{x}')} = \mathcal{H}_{\deg(\mathfrak{x})+\deg(\mathfrak{x}')}.$$

**Case 2**. $\mathrm{bre}(\mathfrak{x}') \geq 2$. The proof of this case is similar to Case 1.

**Case 3**. $\mathrm{bre}(\mathfrak{x}) \geq 2$ and $\mathrm{bre}(\mathfrak{x}') \geq 2$. Let $\mathfrak{x} := \mathfrak{x}_{1,1}\mathfrak{x}_{1,2}$ and $\mathfrak{x}' := \mathfrak{x}'_{1,1}\mathfrak{x}'_{1,2}$, where $\mathfrak{x}_{1,1}, \mathfrak{x}_{1,2}, \mathfrak{x}'_{1,1}, \mathfrak{x}'_{1,2} \in \mathfrak{X}_\infty$ with breadths $\mathrm{bre}(\mathfrak{x}_{1,1}), \mathrm{bre}(\mathfrak{x}_{1,2}), \mathrm{bre}(\mathfrak{x}'_{1,1}), \mathrm{bre}(\mathfrak{x}'_{1,2}) \geq 1$ respectively. By Eq. (19), we obtain

$$\deg(\mathfrak{x}) = \deg(\mathfrak{x}_{1,1}) + \deg(\mathfrak{x}_{1,2}) \text{ and } \deg(\mathfrak{x}') = \deg(\mathfrak{x}'_{1,1}) + \deg(\mathfrak{x}'_{1,2}).$$

Thus by Eq. (8),

$$\mathfrak{x} \diamond \mathfrak{x}' = (\mathfrak{x}_{1,1}\mathfrak{x}_{1,2}) \diamond (\mathfrak{x}'_{1,1}\mathfrak{x}'_{1,2}) = \mathfrak{x}_{1,1}(\mathfrak{x}_{1,2} \diamond \mathfrak{x}'_{1,1})\mathfrak{x}'_{1,2}.$$

By the induction on $t$, we have

$$\mathfrak{x}_{1,2} \diamond \mathfrak{x}'_{1,1} \in \mathcal{H}_{\deg(\mathfrak{x}_{1,2})+\deg(\mathfrak{x}'_{1,1})}.$$

With a similar argument to Case 1. we get

$$\mathfrak{x} \diamond \mathfrak{x}' = \mathfrak{x}_{1,1}(\mathfrak{x}_{1,2} \diamond \mathfrak{x}'_{1,1})\mathfrak{x}'_{1,2} \in \mathcal{H}_{\deg(\mathfrak{x})+\deg(\mathfrak{x}')}.$$

This finishes the proof. □

For the compatibility of the coproduct with the filtration, we have

**Lemma 3.11.** *For $n \geq 0$, we have*

$$\Delta_{\mathrm{M}}(\mathcal{H}_n) \subseteq \sum_{p+q=n} \mathcal{H}_p \otimes \mathcal{H}_q. \tag{23}$$

**Proof.** We verify Eq. (23) by showing

**Claim 3.12.** *For any $\mathfrak{x} \in \mathfrak{X}_\infty$, we have*

$$\Delta_{\mathrm{M}}(\mathfrak{x}) = \sum_{(\mathfrak{x})} \mathfrak{x}_{(1)} \otimes \mathfrak{x}_{(2)}, \tag{24}$$

*where $\mathfrak{x}_{(1)}$ and $\mathfrak{x}_{(2)}$ are non-zero linear multiples of elements of $\mathfrak{X}_\infty$ with $\deg(\mathfrak{x}_{(1)}) + \deg(\mathfrak{x}_{(2)}) \leq \deg(\mathfrak{x})$. Here we have adapted the notation in Remark 3.9.*

To prove this claim we proceed by induction on $\deg(\mathfrak{x}) \geq 0$. For the initial step of $\deg(\mathfrak{x}) = 0$, we get $\mathfrak{x} = 1$ and the result holds. Assume that Claim (3.12) holds for $\mathfrak{x} \in \mathcal{H}_k$ and consider $\mathfrak{x} \in \mathcal{H}_{k+1}$ for some $k \geq 0$.

In this case, we prove Claim (3.12) by induction on the breadth $b := \mathrm{bre}(\mathfrak{x}) \geq 1$. If $b = 1$, we have $\mathfrak{x} \in X \subseteq A$ or $\mathfrak{x} = P_A(\bar{\mathfrak{x}})$ for some $\bar{\mathfrak{x}} \in \mathfrak{X}_\infty$. For the former, Claim (3.12) holds since $\Delta_{\mathrm{M}}$ is given by $\Delta_A$ and $A$ is a connected filtered bialgebra by our hypothesis. For the latter, applying the induction hypothesis on $n$, we can write

$$\Delta_{\mathrm{M}}(\bar{\mathfrak{x}}) = \sum_{(\bar{\mathfrak{x}})} \bar{\mathfrak{x}}_{(1)} \otimes \bar{\mathfrak{x}}_{(2)},$$

where $\deg(\bar{\mathfrak{x}}_{(1)}) + \deg(\bar{\mathfrak{x}}_{(2)}) \leq \deg(\bar{\mathfrak{x}}) = k$, with the notion in Remark 3.9. By Eq. (16), we have

$$\begin{aligned}\Delta_{\mathrm{M}}(\mathfrak{x}) = \Delta_{\mathrm{M}}(P_A(\bar{\mathfrak{x}})) &= \mathfrak{x} \otimes 1 + (\mathrm{id} \otimes P_A)\Delta_{\mathrm{M}}(\bar{\mathfrak{x}}) + \lambda\bar{\mathfrak{x}} \otimes 1 \\ &= \mathfrak{x} \otimes 1 + \sum_{(\bar{\mathfrak{x}})} (\bar{\mathfrak{x}}_{(1)} \otimes P_A(\bar{\mathfrak{x}}_{(2)}) + \lambda\bar{\mathfrak{x}} \otimes 1.\end{aligned}$$

By Eq. (20), it is sufficient to show that the sum of degrees of tensor factors in each summand is less than or equal to $k+1$, which follows from

$$\begin{gathered}\deg(\mathfrak{x}) + \deg(1) = \deg(\mathfrak{x}) \leq k+1,\ \deg(\bar{\mathfrak{x}}) + \deg(1) = \deg(\bar{\mathfrak{x}}) \leq k, \\ \deg(\bar{\mathfrak{x}}_{(1)}) + \deg(P_A(\bar{\mathfrak{x}}_{(2)})) = \deg(\bar{\mathfrak{x}}_{(1)}) + \deg(\bar{\mathfrak{x}}_{(2)}) + 1 \leq k+1.\end{gathered}$$

Assume that Claim (3.12) holds for $\mathfrak{x} \in \mathcal{H}_{k+1}$ with $\mathrm{bre}(\mathfrak{x}) = b$ and consider the case $\mathfrak{x} \in \mathcal{H}_{k+1}$ with $\mathrm{bre}(\mathfrak{x}) = b+1 \geq 2$. Let $\mathfrak{x} = \mathfrak{x}_1\mathfrak{x}_2$, where $\mathfrak{x}_1, \mathfrak{x}_2 \in \mathfrak{X}_\infty$ with $\mathrm{bre}(\mathfrak{x}_1), \mathrm{bre}(\mathfrak{x}_2) \geq 1$. From Eq. (19), we have

$$\deg(\mathfrak{x}_1) + \deg(\mathfrak{x}_2) = \deg(\mathfrak{x}) \leq k+1. \tag{25}$$

Write

$$\Delta_{\mathrm{M}}(\mathfrak{x}_1) = \sum_{(\mathfrak{x}_1)} \mathfrak{x}_{1(1)} \otimes \mathfrak{x}_{1(2)} \text{ and } \Delta_{\mathrm{M}}(\mathfrak{x}_2) = \sum_{(\mathfrak{x}_2)} \mathfrak{x}_{2(1)} \otimes \mathfrak{x}_{2(2)}.$$

By the induction hypothesis on $b$, we have

$$\deg(\mathfrak{x}_{1(1)}) + \deg(\mathfrak{x}_{1(2)}) \leq \deg(\mathfrak{x}_1) \text{ and } \deg(\mathfrak{x}_{2(1)}) + \deg(\mathfrak{x}_{2(2)}) \leq \deg(\mathfrak{x}_2). \tag{26}$$

So we have

$$\begin{aligned}\Delta_{\mathrm{M}}(\mathfrak{x}) &= \Delta_{\mathrm{M}}(\mathfrak{x}_1\mathfrak{x}_2) = \Delta_{\mathrm{M}}(\mathfrak{x}_1 \diamond \mathfrak{x}_2) = \Delta_{\mathrm{M}}(\mathfrak{x}_1) \diamond \Delta_{\mathrm{M}}(\mathfrak{x}_2) \\ &= \left(\sum_{(\mathfrak{x}_1)} \mathfrak{x}_{1(1)} \otimes \mathfrak{x}_{1(2)}\right) \diamond \left(\sum_{(\mathfrak{x}_2)} \mathfrak{x}_{2(1)} \otimes \mathfrak{x}_{2(2)}\right) \\ &= \sum_{(\mathfrak{x}_1)} \sum_{(\mathfrak{x}_2)} (\mathfrak{x}_{1(1)} \diamond \mathfrak{x}_{2(1)}) \otimes (\mathfrak{x}_{1(2)} \diamond \mathfrak{x}_{2(2)}).\end{aligned}$$

By Eq. (22),

$$\mathfrak{x}_{1(1)} \diamond \mathfrak{x}_{2(1)} \in \mathcal{H}_{\deg(\mathfrak{x}_{1(1)})+\deg(\mathfrak{x}_{2(1)})} \text{ and } \mathfrak{x}_{1(2)} \diamond \mathfrak{x}_{2(2)} \in \mathcal{H}_{\deg(\mathfrak{x}_{1(2)})+\deg(\mathfrak{x}_{2(2)})},$$

which implies from Eqs. (20), (25) and (26) that Claim 3.12 holds. □

We now arrive at our last main result.

**Theorem 3.13.** *Let $A = \cup_{n\geq 0} A_n$ be a connected filtered bialgebra with a filtered basis. Then $\mathcal{H} = F_{-\lambda^2}(A)$ is also a connected filtered bialgebra, and hence a Hopf algebra.*

**Proof.** By Lemma 3.7, we just need to prove that $F_{-\lambda^2}(A)$ is a connected filtered bialgebra. This follows from Lemmas 3.10, 3.11 and Eq. (21). □

**Acknowledgement**: The authors thank the referees for valuable suggestions and comments.

## References

[1] M. Aguiar, *On the associative analog of Lie bialgebras*, J. Algebra, 244(2) (2001), 492-532.

[2] F. V. Atkinson, *Some aspects of Baxter's functional equation*, J. Math. Anal. Appl., 7 (1963), 1-30.

[3] C. Bai, O. Bellier, L. Guo and X. Ni, *Splitting of operations, Manin products, and Rota-Baxter operators*, Int. Math. Res. Not. IMRN, (2013) 3 (2013), 485-524.

[4] C. Bai, L. Guo and X. Ni, *Nonabelian generalized Lax pairs, the classical Yang-Baxter equation and PostLie algebras*, Comm. Math. Phys., 297(2) (2010), 553-596.

[5] C. Bai, L. Guo and X. Ni, *Generalizations of the classical Yang-Baxter equation and $\mathcal{O}$-operators*, J. Math. Phys., 52(6) (2011), 063515 (17 pp).

[6] C. Bai, L. Guo and X. Ni, *O-operators on associative algebras and associative Yang-Baxter equations*, Pacific J. Math., 256(2) (2012), 257-289.

[7] C. Bai, L. Guo and X. Ni, *O-operators on associative algebras, associative Yang-Baxter equations and dendriform algebras*, In "Quantized Algebra and Physics", Nankai Ser. Pure Appl. Math. Theoret. Phys., 8, World Sci. Publ., Hackensack, NJ, (2012), 10-51.

[8] G. Baxter, *An analytic problem whose solution follows from a simple algebraic identity,* Pacific J. Math., 10 (1960), 731-742.

[9] M. Bordemann, *Generalized Lax pairs, the modified classical Yang-Baxter equation, and affine geometry of Lie groups,* Comm. Math. Phys., 135(1) (1990), 201-216.

[10] P. Cartier, *On the structure of free Baxter algebras,* Advances in Math., 9 (1972), 253-265.

[11] A. Connes and D. Kreimer, *Hopf algebras, renormalization and noncommutative geometry,* Comm. Math. Phys., 199(1) (1998), 203-242.

[12] K. Ebrihimi-Fard, *Loday-type algebras and the Rota-Baxter relation,* Lett. Math. Phys., 61(2) (2002), 139-147.

[13] K. Ebrahimi-Fard and L. Guo, *Quasi-shuffles, mixable shuffles and Hopf algebras,* J. Algebraic Combin., 24(1) (2006), 83-101.

[14] K. Ebrahimi-Fard and L. Guo, *Rota-Baxter algebras and dendriform algebras,* J. Pure Appl. Algebra, 212(2) (2008), 320-339.

[15] X. Gao, L. Guo and T. Zhang, *Bialgebra and Hopf algebra structures on free Rota-Baxter algebra,* arXiv:1604.03238.

[16] L. Guo, An Introduction to Rota-Baxter Algebra, Surveys of Modern Mathematics, 4, International Press, Somerville, MA; Higher Education Press, Beijing, 2012.

[17] L. Guo and W. Keigher, *Baxter algebras and shuffle products,* Adv. Math., 150(1) (2000), 117-149.

[18] L. Guo and W. Y. Sit *Enumeration and generating functions of Rota-Baxter words,* Math. Comput. Sci., 4(2-3) (2010), 313-337.

[19] R. Jian and J. Zhang, *Rota-Baxter coalgebras,* arXiv:1409.3052.

[20] Y. Kosmann-Schwarzbach, *Lie bialgebras, Poisson Lie groups and dressing transformations,* in "Integrability of nonlinear systems", Lecture Notes in Phys., 495, Springer, Berlin, (1997), 104-170.

[21] T. Ma and L. Liu, *Rota-Baxter coalgebras and Rota-Baxter bialgebras,* Linear Multilinear Algebra, 64(5) (2016), 968-979.

[22] A. Makhlouf and D. Yau, *Rota-Baxter Hom-Lie-admissible algebras,* Comm. Algebra, 42(3) (2014), 1231-1257.

[23] D. Manchon, *Hoft algebras, from basics to applications to renormalization,* Comptes-rendus des Rencontres mathematiques de Glanon, 2001.

[24] M. Marcolli and X. Ni, *Rota-Baxter algebras, singular hypersurfaces, and renormalization on Kausz compactifications,* J. Singul., 15 (2016), 80-117.

[25] J. Pei, C. Bai and L. Guo, *Splitting of operads and Rota-Baxter operators on operads,* Appl. Categ. Structures, 25(4) (2017), 505-538.

[26] G.-C. Rota, *Baxter algebras and combinatorial identities I, II,* Bull. Amer. Math. Soc., 75 (1969), 325-329, 330-334.

[27] M. A. Semenov-Tian-Shansky, *What is a classical R-matrix?* Functional Anal. Appl. 17(4) (1983), 259-272.

[28] T. Zhang, X. Gao and L. Guo, *Hopf algebras of rooted forests, cocyles, and free Rota-Baxter algebras*, J. Math. Phys., 57(10) (2016), 101701 (16 pp).

[29] X. Zhang, X. Gao and L. Guo, *Commutative modified Rota-Baxter algebras, shuffle products and Hopf algebras*, Bull. Malays. Math. Sci. Soc., (2018), https://doi.org/10.1007/s40840-018-0648-3.

**Xigou Zhang**
College of Mathematics and Information Science
Jiangxi Normal University
Nanchang, Jiangxi 330022, China
e-mail: xyzhang@jxnu.edu.cn

**Xing Gao**
School of Mathematics and Statistics
Key Laboratory of Applied Mathematics and Complex Systems
Lanzhou University
Lanzhou, Gansu 730000, China
e-mail: gaoxing@lzu.edu.cn

**Li Guo** (Corresponding Author)
Department of Mathematics and Computer Science
Rutgers University - Newark
Newark, NJ 07102, USA
e-mail: liguo@rutgers.edu

# 2

# ON LEAVITT PATH ALGEBRAS OVER COMMUTATIVE RINGS

Pramod Kanwar, Meenu Khatkar and R. K. Sharma

Communicated by Surender K. Jain

ABSTRACT. In this article, basic ideals in a Leavitt path algebra over a commutative unital ring are studied. It is shown that for a finite acyclic graph $E$ and a commutative unital ring $R$, the Leavitt path algebra $L_R(E)$ is a direct sum of minimal basic ideals and that for a commutative ring $R$ and a graph $E$ satisfying Condition (L), the Leavitt path algebra $L_R(E)$ has no non-zero nilpotent basic ideals. Uniqueness theorems for Leavitt path algebras over commutative unital rings are also discussed.

**Keywords:** Leavitt path algebra, basic ideal, minimal basic ideal, uniqueness theorem

## 1. Introduction

Leavitt path algebras of row-finite graphs, introduced by Abrams and Aranda Pino in [2] and independently by Ara, Moreno, and Pardo in [9], have been of interest to algebraists as well as analysts due to their connections with algebraic structures such as matrix rings, Laurent polynomial rings and also with $C^*$-algebras (see for example [1], [4], [5], [6], [7], [8], [10], [13], [14], [15]). Several generalizations of these algebras have also been studied in the last decade. On one hand, Abrams and Aranda Pino generalized the concept to arbitrary graphs (see [3]) and on the other, Tomforde considered these algebras where the coefficients came from a commutative unital ring in place of a field (see [16]). It is not hard to see that several results about these algebras over a field do not remain valid if we replace field with a commutative unital ring. For example, Leavitt path algebra of a finite line graph, being a matrix ring over a field, is simple; the Leavitt path algebra over a field has no non-zero nilpotent ideals; and that the Jacobson radical of a Leavitt path algebra over a field is always 0. In the case of a commutative unital ring these results, however, are not true, in general. Tomforde showed that some of the well-known results about Leavitt path algebras over a field can be generalized to Leavitt path algebras over a commutative unital ring by suitably modifying the statement. To accomplish this, Tomforde introduced the concepts of basic ideals, basically simple Leavitt path algebras, among other things.

In this article, we continue the study of Leavitt path algebras over commutative unital rings and show that in the definition of basic ideals of Leavitt path algebra $L_R(E)$ given by Tomforde ([16]), the vertex $v$ can be replaced by any basis element of $L_R(E)$ (Theorem 3.8). For a finite acyclic graph $E$, we show that Leavitt path algebra $L_R(E)$, is a direct sum of minimal basic ideals (Theorem 3.9) and that a graph $E$ satisfy Condition (L) if and only if every non-zero basic ideal of $L_R(E)$ contains a vertex (Theorem 3.13 and Theorem 3.16). It is also shown that if $R$ is a commutative unital ring and $E$ is a graph satisfying Condition (L) then the Leavitt path algebra $L_R(E)$ has no non-zero nilpotent basic ideals (Theorem 3.10). We also prove that for a graph having no isolated vertices and for any ring homomorphism $\phi$ from $L_R(E)$ to a ring, the conditions "$\phi(rv) \neq 0$ for all $v \in E^0$ and $0 \neq r \in R$", "$\phi(re) \neq 0$ for all $e \in E^1$ and $0 \neq r \in R$", "$\phi(re^*) \neq 0$ for all $e^* \in (E^1)^*$ and $0 \neq r \in R$" and "$\phi(r\mu) \neq 0$ for all $\mu \in Path(E)$ and $0 \neq r \in R$" are equivalent and use these to give additional equivalent conditions in Graded Uniqueness Theorem (Theorem 4.6).

## 2. Preliminaries and notation

Throughout this article, a ring will mean a commutative unital ring and a graph will always mean a directed graph.

A graph with $E^0$ as the set of vertices, $E^1$, the set of edges, and the functions $r, s : E^1 \to E^0$ is denoted by $E = (E^0, E^1, r, s)$. For each edge $e \in E^1$, the vertices $s(e)$ and $r(e)$ are called the *source* and *range* of $e$ respectively. For $v \in E^0$, a *loop* at $v$ is an edge $e$ for which $r(e) = s(e)$. A vertex which does not receive any edge is called a *source*. A vertex which does not emit any edge is called a *sink*. A vertex $v \in E^0$ such that $|s^{-1}(v)| = \infty$ is called an *infinite emitter*. A vertex $v$ which is either a sink or an infinite emitter is called a *singular vertex*. A vertex $v$ which is not a singular vertex is called a *regular vertex*. A vertex which is both a source and a sink is called an *isolated vertex*.

A *path* $\mu$ in a graph $E$ is a finite sequence of edges $\mu = e_1e_2\cdots e_n$ such that $r(e_i) = s(e_{i+1})$ for $i = 1, 2, \dots, n-1$. In this case, $s(e_1)$ is called the source of $\mu$ (denoted by $s(\mu)$), $r(e_n)$ is called the range of $\mu$ (denoted by $r(\mu)$), and $n$ is called the *length* of $\mu$. We view the elements of $E^0$ as paths of length 0. An edge $e$ is an *exit* for a path $\mu = e_1e_2\cdots e_n$ if there exists $i$ such that $s(e) = s(e_i)$ and $e \neq e_i$. If $\mu = e_1e_2\dots e_n$ then we denote the set $\{s(e_i), r(e_i) : i = 1, 2, \dots, n\}$ by $\mu^0$. If $\mu$

is a path such that $v = s(\mu) = r(\mu)$ then $\mu$ is called a *closed path* based at $v$. If $r(\mu) = s(\mu)$ and $s(e_i) \neq s(e_j)$ for every $i \neq j$ then $\mu$ is called a *cycle.* A graph without any cycles is called *acyclic.*

We say that a graph $E$ satisfies condition $(NE)$ if no cycle in $E$ has an exit and that it satisfies condition $(L)$ if every cycle in $E$ has an exit.

For $n \geq 2$, we denote the set of paths of length $n$ by $E^n$ and the set of all paths by $E^*$. We define a relation $\leq$ on $E^0$ by setting $v \leq w$ if there is a path $\mu \in E^*$ with $s(\mu) = v$ and $r(\mu) = w$. A subset $H$ of $E^0$ is called *hereditary* if $v \leq w$ and $v \in H$ imply $w \in H$. A hereditary set is called *saturated* if $s^{-1}(v) \neq \phi$ and $r(s^{-1}(v)) \subset H$ then $v \in H$.

A graph is called *row-finite* if every vertex emits only finite number of edges. Note that, a row-finite graph is finite if $E^0$ is finite.

Given a graph $E$, the *extended graph* of $E$ is defined as the graph $\hat{E} = (E^0, E^1 \cup (E^1)^*, r', s')$ where $(E^1)^* = \{e_i^* : \ e_i \in E^1\}$ and $r'|_{E^1} = r$, $r'(e_i{}^*) = s(e_i)$, $s'|_{E^1} = s$ and $s'(e_i{}^*) = r(e_i)$. The elements of $(E^1)^*$ are called *ghost edges.*

Let $R$ be a commutative unital ring and $E$ be a graph. Following Tomforde, we define a *Leavitt $E$-family* to be the set $\{v, e, e^* : v \in E^0, \ e \in E^1\}$ in $R$ such that

(1) $vw = \delta_{vw} v$ for all $v, w \in E^0$,
(2) $s(e)e = er(e) = e$ for all $e \in E^1$,
(3) $r(e)e^* = e^* s(e) = e^*$ for all $e \in E^1$,
(4) $e^* f = \delta_{ef} r(e)$ for all $e, f \in E^1$,
(5) $v = \sum_{e \in s^{-1}(v)} ee^*$ for every regular vertex $v \in E^0$.

Note that the Condition (2) and Condition (3) can be combined to read $s'(e)e = er'(e) = e$ for all $e \in E^1 \cup (E^1)^*$. The conditions (4) and (5) are referred to as Cuntz-Kreiger relations in the literature and are denoted by (CK1) and (CK2), respectively.

The *Leavitt path algebra of $E$ with coefficients in $R$*, denoted by $L_R(E)$, is defined as the universal $R$-algebra generated by a Leavitt $E$-family.

We remark that if $E^0$ is finite then $L_R(E)$ is unital $R$-algebra with unit as sum of all the vertices. If $E^0$ is infinite then $L_R(E)$ is an algebra with local units. Also $L_R(E)$ is a $\mathbb{Z}$-graded algebra with grading induced by degree$(v_i) = 0$, degree$(e_i) = 1$, degree$(e_i{}^*) = -1$ for all $v_i \in E^0$ and $e_i \in E^1$, that is, $L_R(E) = \bigoplus_{n \in \mathbb{Z}} L_R(E)_n$, where $L_R(E)_0 = RE^0 + A_0$, $L_R(E)_n = A_n$ for $n \neq 0$ and $A_n = \sum\{re_{i_1}e_{i_2}\cdots e_{i_\sigma}e_{j_1}{}^*e_{j_2}{}^*\cdots e_{j_\tau}{}^* : \sigma+\tau \geq 0, \ e_{i_s} \in E^1, \ e_{j_t}{}^* \in (E^1)^*, \ r \in R, \ \sigma-\tau = n\}$ for all $n$.

An element $x \in L_R(E)$ is called a normal element if $xx^* = x^*x$. The *commutative core* $M_R(E)$ of Leavitt path algebra $L_R(E)$ is defined as the commutative $R$-subalgebra of $L_R(E)$, generated by the normal elements of the form $\alpha\beta^*$, where $\alpha$ and $\beta$ are paths in $E$ having the same range.

## 3. On basic ideals of Leavitt path algebras

Let $R$ be a commutative unital ring and let $E$ be a graph. An ideal $I$ of $L_R(E)$ is called a basic ideal if whenever $r \in R \setminus \{0\}$ and $v \in E^0$, we have $rv \in I$ implies $v \in I$ (see [16]). A basic ideal of $L_R(E)$ is called a minimal basic ideal if it does not contain any non-zero basic ideal other than itself. The Leavitt path algebra $L_R(E)$ is called basically simple if the only basic ideals of $L_R(E)$ are $\{0\}$ and $L_R(E)$.

**Example 3.1.** If $R = \mathbb{Z}_4$ and $E$ is the graph

then each of the ideals $\begin{pmatrix} \mathbb{Z}_4 & \mathbb{Z}_4 \\ \mathbb{Z}_4 & \mathbb{Z}_4 \end{pmatrix}$ and $\begin{pmatrix} \mathbb{Z}_4 & \mathbb{Z}_4 & \mathbb{Z}_4 \\ \mathbb{Z}_4 & \mathbb{Z}_4 & \mathbb{Z}_4 \\ \mathbb{Z}_4 & \mathbb{Z}_4 & \mathbb{Z}_4 \end{pmatrix}$ is a basic ideal of the Leavitt path algebra $L_R(E)$ of $E$ with coefficients in $R$. We further observe that each of these is a minimal basic ideal and hence $L_R(E)$ is not basically simple. We also note that the ideal $\begin{pmatrix} 2\mathbb{Z}_4 & 2\mathbb{Z}_4 \\ 2\mathbb{Z}_4 & 2\mathbb{Z}_4 \end{pmatrix}$ is not a basic ideal of $L_R(E)$.

We now give conditions that are equivalent to the condition $rv \in I$ implies $v \in I$ in the definition of a basic ideal. To do this we use $(v)$, $(e)$, and $(e^*)$ to denote the conditions:

$(v)$ : for $r \in R \setminus \{0\}$ and $v \in E^0$, $rv \in I$ implies $v \in I$,

$(e)$ : for $r \in R \setminus \{0\}$ and $e \in E^1$, $re \in I$ implies $e \in I$, and

$(e^*)$ : for $r \in R \setminus \{0\}$ and $e^* \in (E^1)^*$, $re^* \in I$ implies $e^* \in I$

for any ideal $I$ of $L_R(E)$. Note that $I$ is basic if it satisfies Condition $(v)$.

**Proposition 3.2.** *Let $R$ be a commutative unital ring and $E$ be a graph. If $I$ is an ideal in $L_R(E)$ satisfying Condition $(v)$ then $I$ satisfies Condition $(e)$ and Condition $(e^*)$.*

**Proof.** Let $I$ satisfies Condition $(v)$ and suppose $r \in R \setminus \{0\}$ and $e \in E^1$ be such that $re \in I$. Then $re^*e \in I$, that is, $r \cdot r(e) \in I$. Since $I$ satisfies Condition $(v)$, we get $r(e) \in I$. But then $e \cdot r(e) \in I$, that is, $e \in I$. Hence $I$ satisfies Condition $(e)$.

Similarly, if $re^* \in I$ then $re^*e \in I$, that is, $r \cdot r(e) \in I$. Thus $r(e) \in I$, which is equivalent to saying that $s(e^*) \in I$. But then $s(e^*) \cdot e^* \in I$, that is, $e^* \in I$. Hence $I$ satisfies Condition $(e^*)$. □

The converse of Proposition 3.2 is not true, in general. For example, if $R = \mathbb{Z}_4$ and E is the graph

then the ideal $< 2v_1 >$ satisfy Condition $(e)$ and Condition $(e^*)$, but does not satisfy Condition $(v)$. If the graph has no isolated vertices then the converse is true as shown in the following propositions (Proposition 3.3 and Proposition 3.4).

**Proposition 3.3.** *Let $R$ be a commutative unital ring and $E$ be a graph without any isolated vertices. If $I$ is an ideal in $L_R(E)$ satisfying Condition $(e)$ then $I$ satisfies Condition $(v)$.*

**Proof.** Let $I$ satisfies Condition $(e)$ and let $r \in R \setminus \{0\}$ and $v \in E^0$ be such that $rv \in I$. First assume $v$ is a sink (not necessarily a source) and let $e$ be an edge such that $r(e) = v$. Then $rv \in I$ gives $rev \in I$. Hence $re \in I$. Since $I$ satisfies Condition $(e)$, we get $e \in I$. Thus $e^*e \in I$, that is, $v \in I$. Next assume $v$ is a regular vertex and let $e$ be an edge such that $s(e) = v$. Then $rv \in I$ gives $rve \in I$. Hence $re \in I$. Since $I$ satisfies Condition $(e)$, we get $e \in I$. But then $ee^* \in I$ for every edge $e$ with $s(e) = v$. Hence $v = \sum_{\{e \mid s(e)=v\}} ee^* \in I$. Thus $I$ satisfies Condition $(v)$. □

**Proposition 3.4.** *Let $R$ be a commutative unital ring and $E$ be a graph without any isolated vertices. If $I$ is an ideal in $L_R(E)$ satisfying Condition $(e^*)$ then $I$ satisfies Condition $(v)$.*

**Proof.** The proof is similar to that of Proposition 3.3. □

By combining Proposition 3.2, Proposition 3.3, and Proposition 3.4, we get the following theorem.

**Theorem 3.5.** *Let $R$ be a commutative unital ring and $E$ be a graph without any isolated vertices. Then for any ideal $I$ of $L_R(E)$, the following conditions are equivalent.*

(1) *$I$ satisfies Condition $(v)$.*
(2) *$I$ satisfies Condition $(e)$.*
(3) *$I$ satisfies Condition $(e^*)$.*

**Corollary 3.6.** *Let $R$ be a commutative unital ring and $E$ be a graph without any isolated vertices. Then an ideal $I$ of $L_R(E)$ is basic if and only if $I$ satisfies Condition (e) if and only if $I$ satisfies Condition (e*).*

**Proposition 3.7.** *Let $R$ be a commutative unital ring, $E$ be a graph and $I$ be an ideal in $L_R(E)$. Then for any path $\mu$ in $L_R(E)$ and $r \in R \setminus \{0\}$, $r\mu \in I$ implies $\mu \in I$ if and only if $I$ satisfies Condition (v).*

**Proof.** Suppose $I$ satisfies Condition $(v)$. Let $r \in R \setminus \{0\}$ and $\mu = e_1e_2 \ldots e_n$ be a path in $L_R(E)$ such that $r\mu \in I$. Then $r \cdot {e_1}^* e_1 e_2 \ldots e_n \in I$, which is equivalent to $r \cdot r(e_1)e_2 \ldots e_n \in I$. Since $r(e_1) = s(e_2)$, we have $r \cdot e_2 \ldots e_n \in I$. Again since $I$ is an ideal, $r \cdot {e_2}^* e_2 \ldots e_n \in I$. Repeating this process, finally we get $r \cdot {e_n}^* e_n \in I$, that is, $r \cdot r(e_n) \in I$. Since $I$ satisfies Condition $(v)$, we get $r(e_n) \in I$ and hence $e_n \cdot r(e_n) \in I$. Thus $e_n \in I$. Now using the fact that $I$ is an ideal of $L_R(E)$, we get $e_1e_2 \ldots e_n \in I$, that is, $\mu \in I$. Since vertices are paths of length 0, the other direction is clear. □

A similar argument can be used to prove that the result of Proposition 3.7 remains true if we replace $\mu$ with $pq^*$, where $p$ and $q$ are paths in $L_R(E)$. Since the elements of $E^0 \cup E^1 \cup {E^1}^*$ and elements of the form $pq^*$ where $p$, $q$ are paths in $L_R(E)$ form a basis $\mathcal{B}$ of $L_R(E)$ [14, Theorem 3.7], we have the following theorem.

**Theorem 3.8.** *Let $R$ be a commutative unital ring and $E$ be a graph. Then an ideal $I$ of $L_R(E)$ satisfies Condition (v) if and only if for $r \in R \setminus \{0\}$ and for every basis element $x$ in $\mathcal{B}$, $rx \in I$ implies $x \in I$. Equivalently, $I$ is a basic ideal if and only if for $r \in R \setminus \{0\}$ and $x \in \mathcal{B}$, $rx \in I$ implies $x \in I$.*

Recall that a subset $H$ of $E^0$ is called hereditary saturated if for $v, w \in E^0$, $v \in H$ and $v \leq w$ implies $w \in H$ and if $s^{-1}(v) \neq \phi$ for $v \in E^0$ and $r(s^{-1}(v)) \subset H$ then $v \in H$. Clearly, if $v$ is an isolated vertex in $E^0$ then $\{v\}$ is a hereditary saturated subset. Also if $v$ is a sink then $\{v\}$ is a hereditary subset of $E^0$ and $\text{Span}_R\{\alpha\beta^* : r(\alpha) = r(\beta) = v\}$ is a basic ideal that contains only one vertex $v$. We denote this basic ideal by $I_v$. By [16, Lemma 7.8] there is a saturation $H$ of $\{v\}$ such that $I_v = I_H$.

We further recall that the Leavitt path algebra $L_R(E)$ of a graph $E$ with coefficients from a commutative unital ring $R$ is basically simple if it has no basic ideals other than $\{0\}$ and $L_R(E)$. For a row-finite graph $E$, Tomforde showed that $L_R(E)$ is basically simple if and only if $E^0$ has no hereditary saturated subsets other than $\phi$ and $E^0$ and $E$ satisfies Condition $(L)$, namely, every cycle in $E$ has an exit [16, Theorem 7.20]. Since a finite acyclic graph satisfies Condition $(L)$, the Leavitt path algebra $L_R(E)$ of a finite acyclic graph is basically simple if and only if $E^0$ has no hereditary saturated subsets other than $\phi$ and $E^0$. In general, we have the following for the Leavitt path algebras of finite acyclic graphs.

**Theorem 3.9.** *Let $R$ be a commutative unital ring and $E$ be a finite acyclic graph. Then the Leavitt path algebra $L_R(E)$ is a direct sum of minimal basic ideals.*

**Proof.** Let $v_1, v_2, \ldots, v_n$ be the sinks in $E$. Then $L_R(E) \cong \bigoplus_{i=1}^{n} I_{v_i}$. It is, therefore, enough to show that if $v$ is a sink then $I_v$ is a minimal basic ideal. So let $v$ be a sink. As observed above, there exists saturation $H$ (say) of $\{v\}$ such that $I_v = I_H$ is a graded basic ideal of $L_R(E)$ [16, Lemma 7.8]. To prove the minimality of the basic ideal $I_H$, let $J$ be a non-zero basic ideal of $L_R(E)$ contained in $I_H$. Since $I_H = I_v = \mathrm{Span}_R\{\alpha\beta^* : r(\alpha) = r(\beta) = v\}$ and $J \subset I_H$, it can be seen that $0 \neq rv \in J$ for some $r \in R$. Since $J$ is basic, $v \in J$. Since $J \subset I_H$, $J^0$, the set of vertices in $J$, is a subset of $H$. Since $v \in J^0$ and $J^0$ is a hereditary saturated subset of $E^0$ [16, Lemma 7.6], minimality of $H$ (being the saturation of $\{v\}$) gives $J^0 = H$. But then $J = I_H = I_v$. Hence $I_v$ is a minimal basic ideal. It follows that $L_R(E)$ is the direct sum of minimal basic ideals. □

It is known that the Leavitt path algebra $L_K(E)$ over a field $K$ has no non-zero nilpotent ideals (see [3, Proposition 6.1]). More generally, if $R$ is a commutative unital ring with no non-zero nilpotent elements (equivalently, if $R$ is a commutative semiprime unital ring) then the Leavitt path algebra $L_R(E)$ over $R$ has no non-zero nilpotent left, right, or two-sided nilpotent ideals (see [12, Proposition 4.5]). In general, this result is not true for Leavitt path algebras over arbitrary commutative unital rings. For example, if $R = \mathbb{Z}_4$ and $E$ is the graph

then $\begin{pmatrix} 2\mathbb{Z}_4 & 2\mathbb{Z}_4 \\ 2\mathbb{Z}_4 & 2\mathbb{Z}_4 \end{pmatrix}$ is a non-zero nilpotent ideal of $L_R(E)$. We, however, have the following theorem.

**Theorem 3.10.** *Let $R$ be a commutative unital ring and $E$ be a graph satisfying Condition (L). Then the Leavitt path algebra $L_R(E)$ has no non-zero nilpotent basic ideals.*

**Proof.** Let $I$ be a non-zero basic ideal of $L_R(E)$ and let $I^2 = \{0\}$. By [16, Lemma 7.6], the set $I^0$ of vertices in $I$ is a hereditary saturated subset. By [16, Lemma 7.9], the ideal $< I^0 >$ generated by $I^0$ is a graded basic ideal. Since $I^0 \subset I$, $< I^0 > \subset I$. But $I^2 = \{0\}$. Thus $< I^0 >^2 = \{0\}$. Since $< I^0 >$, being a graded basic ideal, is idempotent, $< I^0 > = \{0\}$. Thus $I^0 = \phi$. Hence $I$ does not contain any vertex, a contradiction by Proposition 3.13. Hence $I$ is not nilpotent. □

We recall that $L_R(E)$ is $\mathbb{Z}$-graded with grading induced by degree$(v) = 0$ for any vertex $v$ in $E^0$ and degree$(e) = 1$, degree$(e^*) = -1$ for every $e \in E^1$. If $\alpha$ is a non-zero homogeneous element in $L_R(E)$ then there exits a path $\beta$ such that $0 \neq \alpha\beta$ is homogeneous and is in real edges only (see Theorem 2.2.11 and Corollary 2.2.12 in [1] and Theorem 5.1 in [11]). Thus $\alpha\beta = \sum_{i=1}^{n} r_i\gamma_i$, where $0 \neq r_i \in R$ and $\gamma_i$'s are distinct paths of equal length. Since $\gamma_1^*\gamma_1 = r(\gamma_1)$ and $\gamma_1^*\gamma_i = 0$ for $2 \leq i \leq n$, it follows that $\gamma_1^*\alpha\beta = r_1 \cdot r(\gamma_1)$. We, thus, have the following proposition.

**Proposition 3.11.** *Let $R$ be a commutative unital ring and $E$ be an arbitrary graph. Let $\alpha$ be a non-zero homogeneous element of $L_R(E)$. Then there exist $\beta, \gamma \in Path(E)$, $0 \neq r \in R$, and $v \in E^0$ such that $0 \neq \gamma^*\alpha\beta = rv$.*

Since for a basic ideal $rv \in I$ implies $v \in I$, we have the following corollary.

**Corollary 3.12.** *Let $R$ be a commutative unital ring and $E$ be an arbitrary graph. Every non-zero graded basic ideal of $L_R(E)$ contains a vertex.*

In the case of graphs satisfying Condition (L), every non-zero basic ideal contains a vertex as is shown in the following theorem.

**Theorem 3.13.** *Let $R$ be a commutative unital ring and $E$ be an arbitrary graph satisfying Condition (L). Then every non-zero basic ideal of $L_R(E)$ contains a vertex.*

**Proof.** Let $I$ be a non-zero basic ideal of $L_R(E)$ and let $\alpha$ be a non-zero element in $I$. Since $E$ satisfies Condition (L), $E$ has no cycles without exits and hence, by [11, Theorem 5.1], there exists $\mu, \eta \in Path(E)$ such that either $0 \neq \mu^*\alpha\eta = rv$ for some $0 \neq r \in R$, and $v \in E^0$. Since $I$ is a basic ideal and $0 \neq rv = \mu^*\alpha\eta \in I$, we have $v \in I$. Thus $I$ contains a vertex. □

Before we prove the converse of the Proposition 3.13 we prove a lemma which is also of independent interest. The proof of lemma is similar to that of Lemma 2.7.1 in [1]. We include it here for the sake of completeness.

**Lemma 3.14.** *Let $R$ be a commutative unital ring and $E$ be a graph. Let $c$ be a cycle without exits such that $s(c) = v$ and $\wedge_v$ be a set of paths in $E$ which ends* *in $v$ but which do not contain all edges of $c$. If $I$ is a basic ideal generated by the vertices of $c$ then $I \cong M_{\wedge_v}(R[x, x^{-1}])$.*

**Proof.** Let $J$ be the basic ideal generated by $v$. Clearly $J \subseteq I$. Now for any $w \in c^0$ if $\mu$ denotes the portion of cycle $c$ from $v$ to $w$ then $\mu$ is a path from $v$ to $w$ and $w = \mu^* v \mu$. Thus $w \in J$. Hence $J$ contains all vertices of $c$. Thus $I \subseteq J$ and hence $I = J$. Now let $A$ be the set of elements of the form $\alpha c^k \beta^*$ where $\alpha, \beta \in \wedge_v$ and $k \in \mathbb{Z}$ where $c^k = v$ for $k = 0$ and $c^k = (c^*)^{-k}$ for $k < 0$. Since the elements of the form $\alpha\beta^*$ where $\alpha, \beta$ are paths form a $R$-basis of $L_R(E)$ [14, Theorem 3.7], $A$ is a $R$-linearly independent set. Also every element of $J$ is a $R$-linear combination of elements of the form $pq^*$, where $p$, $q$ are paths in $E$ such that $r(p) = r(q) \in T(v)$ (the tree of $v$). Since $c$ has no exits, $T(v)$ is precisely the set of vertices in $c$. Hence $p = \alpha c^l$ and $q = \beta c^m$ for some $\alpha, \beta \in \wedge_v$ and $l, m \geq 0$. Thus $A$ generates the basic ideal $J$. Now let $\phi : J \to M_{\wedge_v}(R[x, x^{-1}])$ is the map defined as $\phi(\alpha c^k \beta^*) = x^k e_{\alpha,\beta}$ where $e_{\alpha,\beta}$ denotes the element of $M_{\wedge_v}(R[x, x^{-1}])$ which is 1 in the $(\alpha, \beta)$ entry and zeros otherwise. Clearly $\phi$ is an $R$-algebra isomorphism. Hence $I = J \cong M_{\wedge_v}(R[x, x^{-1}])$. □

We also observe that if $I$ is a graded basic ideal of $L_R(E)$ then by [16, Theorem 7.9], there exists a hereditary saturated subset $H$ in $E^0$ such that $I = I_H$. But then by [16, Theorem 7.15], $I$ has a set of local units. By [16, Lemma 4.14], every ideal of $I$ is also an ideal of $L_R(E)$. We, thus, have the following proposition.

**Proposition 3.15.** *Let $R$ be a commutative unital ring and $E$ be a graph. Every ideal of a graded basic ideal of the Leavitt path algebra $L_R(E)$ is also an ideal of $L_R(E)$.*

**Theorem 3.16.** *Let $R$ be a commutative unital ring and $E$ be an arbitrary graph. If every non-zero basic ideal of $L_R(E)$ contains a vertex then $E$ satisfies condition (L).*

**Proof.** On the contrary, let $c$ be a cycle without exits in $E$ and let $I$ be the graded basic ideal of $L_R(E)$ generated by the vertices of $c$. By Lemma 3.14, there exists some vertex $v$ in $c$ such that $I \cong M_\wedge(R[x, x^{-1}])$ for some set $\wedge$ of paths in $E$ which ends in $v$ but which do not contain all edges of $c$. Also by Proposition 3.15, the ideals of $I$ are ideals of $L_R(E)$. Since every non-zero basic ideal of $L_R(E)$ contains a vertex, every non-zero basic ideal of $M_\wedge(R[x, x^{-1}])$ contains a non-zero idempotent, a contradiction. Hence $E$ satisfies condition (L). □

## 4. Uniqueness theorems

We recall that for a Leavitt path algebra $L_K(E)$ of a graph $E$, a graded homomorphism $\phi : L_K(E) \to A$ is injective if $\phi(v) \neq 0$ for all $v \in E^0$ (Graded Uniqueness Theorem). In the case of Leavitt path algebras over rings, a graded homomorphism $\phi : L_R(E) \to A$ is injective if $\phi(rv) \neq 0$ for all $v \in E^0$ and $0 \neq r \in R$ [16, Theorem 5.3]. We show that the condition $\phi(rv) \neq 0$ for all $v \in E^0$ and $0 \neq r \in R$ is equivalent to $\phi(re) \neq 0$ for all $e \in E^1$ and $0 \neq r \in R$ as well as $\phi(re^*) \neq 0$ for all $e^* \in E^{1^*}$ and $0 \neq r \in R$ in case the graph $E$ has no isolated vertices. First we prove the following proposition.

**Proposition 4.1.** *Let $R$ be a commutative unital ring, $S$ be any ring, and $E$ be a graph. Let $\phi : L_R(E) \to S$ be a ring homomorphism such that $\phi(rv) \neq 0$ for all $v \in E^0$ and $0 \neq r \in R$. Then $\phi(re) \neq 0$ for all $e \in E^1$ and $0 \neq r \in R$ and $\phi(re^*) \neq 0$ for all $e^* \in (E^1)^*$ and $0 \neq r \in R$.*

**Proof.** We first prove that $\phi(re) \neq 0$ for all $e \in E^1$ and $0 \neq r \in R$. If there exists $e \in E^1$ and $0 \neq r \in R$ such that $\phi(re) = 0$ then $\phi(re^*e) = \phi(e^*)\phi(re) = 0$. Since $e^*e = r(e)$, we have $\phi(r \cdot r(e)) = 0$, a contradiction, as $\phi(rv) \neq 0$ for all $v \in E^0$ and $0 \neq r \in R$. Hence $\phi(re) \neq 0$ for all $e \in E^1$ and $0 \neq r \in R$. The other part can be proved similarly. □

**Proposition 4.2.** *Let $R$ be a commutative unital ring, $S$ be any ring, and $E$ be a graph. Let $\phi : L_R(E) \to S$ be a ring homomorphism such that $\phi(rv) \neq 0$ for all $v \in E^0$ and $0 \neq r \in R$. Then $\phi(r\mu) \neq 0$ for all $\mu \in Path(E)$ and $0 \neq r \in R$.*

**Proof.** Let, if possible, there is a path $\mu = e_1e_2 \dots e_n$ in $L_R(E)$ and $0 \neq r \in R$ such that $\phi(r\mu) = 0$, that is, $\phi(r \cdot e_1e_2 \dots e_n) = 0$. Then $\phi(e_n{}^*e_{n-1}^* \cdots e_1{}^*)\phi(re_1e_2 \dots e_n) = 0$, which is equivalent to $\phi(re_n{}^*e_{n-1}^* \cdots e_1{}^*e_1e_2 \dots e_n) = 0$. Since $r(e_1) = s(e_2)$, we have $\phi(re_n{}^*e_{n-1}^* \cdots e_2{}^*r(e_1)e_2e_3 \dots e_n) = 0$. Repeating this process, finally we get $\phi(re_n{}^*e_n) = 0$, that is, $\phi(r \cdot r(e_n)) = 0$, a contradiction, as $\phi(rv) \neq 0$ for all $v \in E^0$ and $0 \neq r \in R$. Hence $\phi(r\mu) \neq 0$ for all $\mu \in Path(E)$ and $0 \neq r \in R$. □

**Proposition 4.3.** *Let $R$ be a commutative unital ring, $S$ be any ring, and $E$ be a graph without any isolated vertices. Let $\phi : L_R(E) \to S$ be a ring homomorphism such that $\phi(re) \neq 0$ for all $e \in E^1$ and $0 \neq r \in R$. Then $\phi(rv) \neq 0$ for all $v \in E^0$ and $0 \neq r \in R$.*

**Proof.** Let, if possible, there exists $v \in E^0$ and $0 \neq r \in R$ such that $\phi(rv) = 0$. If $v$ is a sink then since $v$ is not an isolated vertex it must be the range of some edge (say $f$). Thus $\phi(rf) = \phi(rfv) = \phi(f)\phi(rv) = 0$, a contradiction, as $\phi(re) \neq 0$ for all $e \in E^1$ and $0 \neq r \in R$. Hence $\phi(rv) \neq 0$ for all $v \in E^0$ and $0 \neq r \in R$. If $v$ is a regular vertex then $s^{-1}(v)$ has at least one edge. Let $e \in s^{-1}(v)$. Then $\phi(re) = \phi(rve) = \phi(rv)\phi(e) = 0$, a contradiction again. Hence $\phi(rv) \neq 0$ for all $v \in E^0$ and $0 \neq r \in R$. □

A similar argument can be used to prove the following proposition.

**Proposition 4.4.** *Let $R$ be a commutative unital ring, $S$ be any ring, and $E$ be a graph without any isolated vertices. Let $\phi : L_R(E) \to S$ be a ring homomorphism such that $\phi(re^*) \neq 0$ for all $e^* \in (E^1)^*$ and $0 \neq r \in R$. Then $\phi(rv) \neq 0$ for all $v \in E^0$ and $0 \neq r \in R$.*

Combining Proposition 4.1, Proposition 4.2, Proposition 4.3, and Proposition 4.4, we have the following theorem.

**Theorem 4.5.** *Let $R$ be a commutative unital ring, $S$ be any ring, and $E$ be a graph without any isolated vertices. Let $\phi : L_R(E) \to S$ be a ring homomorphism. Then the following conditions are equivalent.*

(1) $\phi(rv) \neq 0$ *for all* $v \in E^0$ *and* $0 \neq r \in R$.
(2) $\phi(re) \neq 0$ *for all* $e \in E^1$ *and* $0 \neq r \in R$.
(3) $\phi(re^*) \neq 0$ *for all* $e^* \in (E^1)^*$ *and* $0 \neq r \in R$.
(4) $\phi(r\mu) \neq 0$ *for all* $\mu \in Path(E)$ *and* $0 \neq r \in R$.

Recall that for a graph $E$, $M_R(E)$ denotes the commutative $R$-subalgebra of $L_R(E)$, generated by the normal elements of the form $\alpha\beta^*$ where $\alpha, \beta$ are paths in $E$ with the same range. Using Theorem 4.5 and [11, Theorem 5.4], Graded Uniqueness Theorem takes the following form.

**Theorem 4.6.** *Let $R$ be a commutative unital ring, $S$ be a graded ring, and $E$ be a graph without any isolated vertices. Let $\phi : L_R(E) \to S$ be a graded homomorphism. Then the following conditions are equivalent.*

(1) $\phi$ *is injective.*
(2) *Restriction of* $\phi$ *over the commutative core* $M_R(E)$ *is injective.*
(3) $\phi(rv) \neq 0$ *for all* $v \in E^0$ *and* $0 \neq r \in R$.
(4) $\phi(re) \neq 0$ *for all* $e \in E^1$ *and* $0 \neq r \in R$.
(5) $\phi(re^*) \neq 0$ *for all* $e^* \in (E^1)^*$ *and* $0 \neq r \in R$.
(6) $\phi(r\mu) \neq 0$ *for all* $\mu \in Path(E)$ *and* $0 \neq r \in R$.

For graphs without isolated vertices and satisfying Condition (L), we have the following theorem.

**Theorem 4.7.** *Let $R$ be a commutative unital ring, $S$ be any ring, and $E$ be a graph without any isolated vertices such that $E$ satisfies Condition $(L)$. Let $\phi : L_R(E) \to S$ be a ring homomorphism. Then the following conditions are equivalent.*

(1) *$\phi$ is injective.*

(2) *Restriction of $\phi$ over the commutative core $M_R(E)$ is injective.*

(3) *$\phi(rv) \neq 0$ for all $v \in E^0$ and $0 \neq r \in R$.*

(4) *$\phi(re) \neq 0$ for all $e \in E^1$ and $0 \neq r \in R$.*

(5) *$\phi(re^*) \neq 0$ for all $e^* \in (E^1)^*$ and $0 \neq r \in R$.*

(6) *$\phi(r\mu) \neq 0$ for all $\mu \in Path(E)$ and $0 \neq r \in R$.*

**Proof.** Follows by Theorem 4.5 and [11, Corollary 5.3]. □

## References

[1] G. Abrams, P. Ara and M. S. Molina, Leavitt Path Algebras, Lecture Notes in Mathematics, 2191, Springer, London, 2017.

[2] G. Abrams and G. Aranda Pino, *The Leavitt path algebra of a graph*, J. Algebra, 293(2) (2005), 319-334.

[3] G. Abrams and G. Aranda Pino, *The Leavitt path algebras of arbitrary graphs*, Houston J. Math., 34(2) (2008), 423-442.

[4] G. Abrams, G. Aranda Pino and M. S. Molina, *Finite dimensional Leavitt path algebras*, J. Pure Appl. Algebra, 209(3) (2007), 753-762.

[5] G. Abrams and Z. Mesyan, *Simple Lie algebras arising from Leavitt path algebra*, J. Pure Appl. Algebra, 216(10) (2012), 2302-2313.

[6] A. Alahmedi and H. Alsulami, *On the simplicity of the Lie algebra of a Leavitt path algebra*, Comm. Algebra, 44(9) (2016), 4114-4120.

[7] A. Alahmedi, H. Alsulami, S. K. Jain and E. Zelmanov, *Leavitt path algebras of finite Gelfand-Kirillov dimension*, J. Algebra Appl., 11(6) (2012), 1250225 (6 pp).

[8] A. Alahmedi, H. Alsulami, S. K. Jain and E. Zelmanov, *Structure of Leavitt path algebras of polynomial growth*, Proc. Natl. Acad. Sci. USA, 110(38) (2013), 15222-15224.

[9] P. Ara, M. A. Moreno and E. Pardo, *Nonstable K-theory for graph algebras*, Algebr. Represent. Theory, 10(2) (2007), 157-178.

[10] G. Aranda Pino and K. Crow, *The center of a Leavitt path algebra*, Rev. Mat. Iberoam., 27(2) (2011), 621-644.

[11] C. Gil Canto and A. Nasr-Isfahani, *The maximal commutative subalgebra of a Leavitt path algebra*, arXiv:1510.03992v1 [math.RA].

[12] P. Kanwar, M. Khatkar and R. K. Sharma, *Basic one sided ideals of Leavitt path algebras over commutative rings*, preprint, submitted.

[13] H. Larki, *Ideal structure of Leavitt path algebras with coefficients in a unital commutative ring*, Comm. Algebra, 43(12) (2015), 5031-5058.

[14] V. Lopatkin and T. G. Nam, *On the homological dimensions of Leavitt path algebras with coefficients in commutative rings*, J. Algebra, 481 (2017), 273-292.

[15] Z. Mesyan, *Commutator Leavitt path algebras*, Algebr. Represent. Theory, 16(5) (2013), 1207-1232.

[16] M. Tomforde, *Leavitt path algebras with coefficients in a commutative ring*, J. Pure Appl. Algebra, 215(4) (2011), 471-484.

**Pramod Kanwar** (Corresponding Author)
Department of Mathematics
Ohio University-Zanesville
Zanesville, Ohio, USA
e-mail: kanwar@ohio.edu

**Meenu Khatkar** and **R. K. Sharma**
Department of Mathematics
Indian Institute of Technology Delhi
New Delhi, 110016, India
e-mails: meenukhatkar@gmail.com (M. Khatkar)
rksharmaiitd@gmail.com (R. K. Sharma)

# 3

# SOME RESULTS ABOUT HOM-COMODULE ALGEBRAS WITH HOM-HOPF MODULE STRUCTURE

Lihong Dong and Shuan Xue

Communicated by Abdullah Harmancı

ABSTRACT. The main subject of this paper is Hom-comodule algebras with Hom-Hopf module structure. First, we give the factorization of a class of Hom-bialgebras, which is not only Hom-module coalgebras but also Hom-comodule algebras with Hom-Hopf module structure. Next, we obtain the factorization of this class of Hom-comodule algebras. Finally, we discuss the relation between this class of Hom-comodule algebas and cleft extensions.

**Keywords**: Hom-comodule algebra, Hom-Hopf module, Hom-crossed product algebra, cleft extension

## 1. Introduction

The study of Hom-associative algebras originates with work by Hartwig, Larsson and Silvestrov in the Lie case [4], where a notion of Hom-Lie algebra was introduced in the context of studying deformations of Witt and Virasoro algebras. Later, it was extended to the associative case by Makhlouf and Silvestrov in [7]. Now the associativity is replaced by Hom-associativity $\alpha(a)(bc) = (ab)\alpha(c)$. Hom-coassociativity for a Hom-coalgebra can be considered in a similar way, see [8].

The crossed product algebra was introduced in [1], which is a generalization of the smash product algebra. In [1], Blattner, Cohen and Montgomery showed the equivalence of crossed products and cleft extensions. In [2], Blattner and Montgomery gave several characterizations of crossed products. Lu and Wang [5] generalized the results in [1] to the case of Hom-Hopf algebras. Hopf modules (see [9]) are vector spaces with both a comodule and module structure which are related in a natural way. The theory of Hopf modules accounts for some of the deeper results for Hopf algebras. Hom-Hopf modules, as the generalization of Hopf modules, are also studied by many people. Motivated by this, the main subject of this paper is the Hom-comodule algebras with Hom-Hopf module structure. In this paper we not only discuss the relation between this class of Hom-algebas and Hom-crossed product algebras, but also discuss the relation between this class of Hom-algebas and cleft extensions.

This paper is organized as follows. In Section 2, we recall some basic definitions and results, such as Hom-Hopf algebra, Hom-Hopf module, Hom-(co)module (co)algebra, Hom-crossed product algebra and so on. Next, we always assume that $H$ is a Hom-Hopf algebra, $B$ is a right $H$-Hom-Hopf module and $(B, \rho)$ is a right $H$-Hom-comodule algebra, set $A = B^{coH}$. In Section 3, let $B$ be also a Hom-module coalgebra and a Hom-bialgebra, we provide the factorization of Hom-bialgebra $B$, that is $B \simeq A\square_{\rho}^{\cdot}H$ as Hom-bialgebras (see Theorem 3.3). In Section 4, we define a weak action of $H$ on $A$, thus obtain the Hom-crossed product algebra $A\sharp_{\sigma}H$ and $B \simeq A\sharp_{\sigma}H$ as Hom-algebras (see Theorem 4.2), that is, we obtain the factorization of this class of Hom-comodule algebas. Furthermore, we discuss the relation between this class of Hom-comodule algebas and cleft extensions (see Theorem 4.7).

## 2. Preliminaries

Throughout this paper, $k$ is a fixed field. Unless otherwise stated, all vector spaces, algebras, coalgebras, maps and unadorned tensor products are over $k$.

We now recall from [7,8,10] some definitions and results about Hom-Hopf algebras, Hom-(co)modules and so on.

**2.1. Hom-Hopf algebra.** A Hom-algebra is a quadruple $(A, \mu, 1_A, \alpha)$ (abbr. $(A, \alpha)$), where $A$ is a linear space, $\mu : A \otimes A \to A$ is a linear map, with notation $\mu(a \otimes a') = aa'$, $1_A \in A$ and $\alpha \in Aut_k(A)$, such that for any $a, a', a'' \in A$,

$$\alpha(aa') = \alpha(a)\alpha(a'), \ a1_A = 1_A a = \alpha(a),$$
$$\alpha(a)(a'a'') = (aa')\alpha(a''), \ \alpha(1_A) = 1_A.$$

A Hom-coalgebra is a quadruple $(C, \Delta, \varepsilon, \beta)$ (abbr. $(C, \beta)$), where $C$ is a linear space, $\Delta : C \to C \otimes C$, $\varepsilon : C \to k$ are linear maps, and $\beta \in Aut_k(C)$, such that for any $c \in C$,

$$\beta(c_1) \otimes \beta(c_2) = \beta(c)_1 \otimes \beta(c)_2, \ \varepsilon_C(c_1)c_2 = c_1\varepsilon_C(c_2) = \beta(c),$$
$$\beta(c_1) \otimes c_{21} \otimes c_{22} = c_{11} \otimes c_{12} \otimes \beta(c_2), \ \varepsilon\beta = \varepsilon.$$

A Hom-bialgebra is a sextuple $(H, \mu, 1_H, \Delta, \varepsilon, \gamma)$ (abbr. $(H, \gamma)$), where $(H, \mu, 1_H, \gamma)$ is a Hom-algebra, and $(H, \Delta, \varepsilon, \gamma)$ is a Hom-coalgebra, such that $\Delta$, $\varepsilon$ are morphisms of Hom-algebra, i.e.

$$\Delta(hh') = \Delta(h)\Delta(h'),\ \Delta(1_H) = 1_H \otimes 1_H,\ \varepsilon(hh') = \varepsilon(h)\varepsilon(h'),\ \varepsilon(1_H) = 1.$$

Furthermore, if there exists a linear map $S : H \to H$ such that

$$S(h_1)h_2 = h_1 S(h_2) = \varepsilon(h)1_H,\ S(\gamma(h)) = \gamma(S(h)),$$

then we call $(H, \mu, 1_H, \Delta, \varepsilon, S, \gamma)$ (abbr. $(H, S, \gamma)$) a Hom-Hopf algebra.

**2.2. The fundamental theorem of Hom-Hopf module.** Let $(A, \beta)$ be a Hom-algebra, a right $(A, \beta)$-Hom-module is a triple $(M, \cdot, \alpha)$, where $M$ is a linear space, $\cdot : M \otimes A \to M$ is a linear map, and $\alpha$ is an automorphism of $M$, such that for any $a, a' \in A$ and $m \in M$,

$$\alpha(m) \cdot (aa') = (m \cdot a) \cdot \beta(a'),\ m \cdot 1_A = \alpha(m),\ \alpha(m \cdot a) = \alpha(m) \cdot \beta(a).$$

Let $(C, \beta)$ be a Hom-coalgebra, a right $(C, \beta)$-Hom-comodule is a triple $(M, \rho, \alpha)$, where $M$ is a linear space, $\rho : M \to M \otimes C$ is a linear map (write $\rho(m) = m_{(0)} \otimes m_{(1)}$), and $\alpha$ is an automorphism of $M$, such that for any $m \in M$,

$$\alpha(m_{(0)}) \otimes m_{(1)1} \otimes m_{(1)2} = m_{(0)(0)} \otimes m_{(0)(1)} \otimes \beta(m_{(1)}),$$
$$m_{(0)}\varepsilon(m_{(1)}) = \alpha(m),\ \alpha(m)_{(0)} \otimes \alpha(m)_{(1)} = \alpha(m_{(0)}) \otimes \beta(m_{(0)}).$$

Let $(H, \alpha)$ be a Hom-Hopf algebra, a right $H$-Hom-Hopf module is a quadruple $(M, \cdot, \rho, \beta)$, where $M$ is a right $H$-Hom-module and a right $H$-Hom-comodule, such that for all $m \in M$, $h \in H$,

$$\rho(m \cdot h) = m_{(0)} \cdot h_1 \otimes m_{(1)} h_2.$$

**2.3. Hom-module coalgebra and Hom-comodule algebra.** Recall from [6], let $(H, \alpha)$ be a Hom-Hopf algebra and $(C, \beta)$ be a Hom-coalgebra, if $(C, \cdot, \alpha)$ is a left $(H, \beta)$-Hom-module, for all $c \in C$, $h \in H$ the following conditions hold

$$(h \cdot c)_1 \otimes (h \cdot c)_2 = h_1 \cdot c_1 \otimes h_2 \cdot c_2,\ \varepsilon(h \cdot c) = \varepsilon(h)\varepsilon(c),$$

then $(C, \cdot, \beta)$ is called an $H$-Hom-module coalgebra.

Recall from [10], let $(H, \alpha)$ be a Hom-Hopf algebra and $(A, \beta)$ be a Hom-algebra, if $(A, \rho, \beta)$ is a left $(H, \alpha)$-Hom-comodule, for all $a, b \in A$ the following conditions hold

$$\rho(ab) = a_{-1}b_{-1} \otimes a_0 b_0,\ \rho(1_A) = 1_H \otimes 1_A,$$

then $(A, \rho, \beta)$ is called an $H$-Hom-comodule algebra.

**2.4. Hom-crossed product algebra and cleft extension.** Recall from [5], let $(H,\alpha)$ be a Hom-Hopf algebra and $(A,\beta)$ be a Hom-algebra. We say that $H$ acts weakly on $A$ from the left if there is a linear map given by $\rightharpoonup: H\otimes A\to A$, such that for all $a,b\in A$ and $h\in H$,

$$\beta(h\rightharpoonup a)=\alpha(h)\rightharpoonup\beta(a),\quad h\rightharpoonup 1=\varepsilon(h)1,$$
$$\alpha^2(h)\rightharpoonup(ab)=(h_1\rightharpoonup a)(h_2\rightharpoonup b),\quad 1\rightharpoonup a=\beta(a).$$

Let $(H,\alpha)$ be a Hom-Hopf algebra and $(A,\beta)$ be a Hom-algebra. Assume that $H$ acts weakly on $A$ from the left; Let $\sigma\in Hom(H\otimes H,A)$ be a linear map. For all $a,b\in A$ and $h,g\in H$, define $A\sharp_\sigma H$ whose underlying vector space is $A\otimes H$ with the multiplication given by

$$(a\otimes h)(b\otimes g)=a((\alpha^{-4}(h_{11})\rightharpoonup\beta^{-2}(b))\sigma(\alpha^{-3}(h_{12}),\alpha^{-2}(g_1)))\otimes\alpha^{-1}(h_2g_2).$$

We say that $(A\sharp_\sigma H,\beta\otimes\alpha)$ is a Hom-crossed product algebra if and only if

(C1) $\sigma(1,h)=\sigma(h,1)=\varepsilon(h)1,\quad \sigma(\alpha\otimes\alpha)=\beta\sigma$;

(C2) $(h_1\rightharpoonup(\alpha^{-1}(g_1)\rightharpoonup a))\sigma(\alpha(h_2),\alpha(g_2))=\sigma(\alpha(h_1),\alpha(g_1))(\alpha^{-1}(h_2g_2)\rightharpoonup\beta(a))$;

(C3) $(h_1\rightharpoonup\sigma(g_1,l_1))\sigma(\alpha(h_2),g_2l_2)=\sigma(\alpha(h_1),\alpha(g_1))\sigma(h_2g_2,\alpha^2(l))$.

Let $(H,\alpha)$ be a Hom-Hopf algebra and $(B,\rho,\beta)$ be a left $H$-Hom-comodule algebra. Denote by $A=B^{coH}=\{a\in B\mid\rho(a)=\beta(a)\otimes 1\}$, then $A\subset B$ is said to be a cleft extension if there exists a left $H$-Hom-comodule map $\gamma:H\to B$ which is convolution invertible.

## 3. The structure of Hom-bialgebras with the Hom-Hopf module structure

In this section, we mainly provide the factorization of Hom-bialgebras with Hom-Hopf module structure. On the one hand, we give the algebra factorization for right $H$-Hom-comodule algebras with Hom-Hopf module structure.

**Proposition 3.1.** *Let $(H,S,\alpha)$ be a Hom-Hopf algebra and $(B,\cdot,\rho,\beta)$ be a right $H$-Hom-Hopf module. We assume that $(B,\rho,\beta)$ is a right $H$-Hom-comodule algebra (write $\rho(b)=b_{(0)}\otimes b_{(1)}$ ), and set $A=B^{coH}=\{a\in B\mid\rho(a)=\beta(a)\otimes 1\}$. Define a multiplication on $A\otimes H$ as follows*

$$(a\otimes h)(b\otimes g)=((\beta^{-2}(a)\cdot\alpha^{-4}(h_{11}))(\beta^{-2}(b)\cdot\alpha^{-4}(g_{11})))\cdot S\alpha^{-3}(h_{12}g_{12})\otimes\alpha^{-1}(h_2g_2),$$

*for all $a,b\in A,h,g\in H$, then $(A\otimes H,\beta\otimes\alpha,1_A\otimes 1_H)$ is a Hom-algebra, we write it as $A\Box_\rho H$. Moreover, $B\simeq A\Box_\rho H$ as Hom-algebras.*

**Proof.** We first check that the multiplication is well defined. For all $a, b \in A$, $h, g \in H$, we have

$$
\begin{aligned}
&\rho(((\beta^{-2}(a)\cdot\alpha^{-4}(h_1))(\beta^{-2}(b)\cdot\alpha^{-4}(g_1)))\cdot S\alpha^{-3}(h_2g_2))\\
=\ &((\beta^{-2}(a)\cdot\alpha^{-4}(h_1))(\beta^{-2}(b)\cdot\alpha^{-4}(g_1)))_{(0)}\cdot S\alpha^{-3}(h_{22}g_{22})\\
&\otimes((\beta^{-2}(a)\cdot\alpha^{-4}(h_1))(\beta^{-2}(b)\cdot\alpha^{-4}(g_1)))_{(1)}S\alpha^{-3}(h_{21}g_{21})\\
=\ &((\beta^{-2}(a)\cdot\alpha^{-4}(h_1))_{(0)}(\beta^{-2}(b)\cdot\alpha^{-4}(g_1))_{(0)})\cdot S\alpha^{-3}(h_{22}g_{22})\\
&\otimes((\beta^{-2}(a)\cdot\alpha^{-4}(h_1))_{(1)}(\beta^{-2}(b)\cdot\alpha^{-4}(g_1))_{(1)})S\alpha^{-3}(h_{21}g_{21})\\
=\ &(((\beta^{-2}(a))_{(0)}\cdot\alpha^{-4}(h_{11}))((\beta^{-2}(b))_{(0)}\cdot\alpha^{-4}(g_{11})))\cdot S\alpha^{-3}(h_{22}g_{22})\\
&\otimes(((\beta^{-2}(a))_{(1)}\cdot\alpha^{-4}(h_{12}))_{(1)}((\beta^{-2}(b))_{(1)}\cdot\alpha^{-4}(g_{12})))S\alpha^{-3}(h_{21}g_{21})\\
=\ &((\beta^{-1}(a)\cdot\alpha^{-4}(h_{11}))(\beta^{-1}(b)\cdot\alpha^{-4}(g_{11})))\cdot S\alpha^{-3}(h_{22}g_{22})\\
&\otimes\alpha^{-3}(h_{12}g_{12})S\alpha^{-3}(h_{21}g_{21})\\
=\ &(((\beta^{-1}(a)\cdot\alpha^{-3}(h_1))(\beta^{-1}(b)\cdot\alpha^{-3}(g_1))))\cdot S\alpha^{-3}(h_{22}g_{22})\\
&\otimes\alpha^{-4}(h_{211}g_{211})S\alpha^{-4}(h_{212}g_{212})\\
=\ &(((\beta^{-1}(a)\cdot\alpha^{-3}(h_1))(\beta^{-1}(b)\cdot\alpha^{-3}(g_1))))\cdot S\alpha^{-2}(h_2g_2)\otimes 1_H\\
=\ &\beta(((\beta^{-2}(a)\cdot\alpha^{-4}(h_1))(\beta^{-2}(b)\cdot\alpha^{-4}(g_1)))\cdot S\alpha^{-3}(h_2g_2))\otimes 1_H.
\end{aligned}
$$

Thus, $((\beta^{-2}(a)\cdot\alpha^{-4}(h_1))(\beta^{-2}(b)\cdot\alpha^{-4}(g_1)))\cdot S\alpha^{-3}(h_2g_2)\in A$.

Next, we check that the associativity holds. For all $a, b, c \in A, h, g, k \in H$, we can calculate

$$
\begin{aligned}
&((a\otimes h)(b\otimes g))(\beta(c)\otimes\alpha(k))\\
=\ &((\beta^{-2}(((\beta^{-2}(a)\cdot\alpha^{-4}(h_{11}))(\beta^{-2}(b)\cdot\alpha^{-4}(g_{11})))\cdot S\alpha^{-3}(h_{12}g_{12}))\cdot\alpha^{-5}(h_{211}g_{211}))\\
&(\beta^{-1}(c)\cdot\alpha^{-3}(k_{11})))\cdot S\alpha^{-3}(\alpha^{-1}(h_{212}g_{212})\alpha(k_{12}))\otimes\alpha^{-2}(h_{22}g_{22}))k_2\\
=\ &((((\beta^{-3}(a)\cdot\alpha^{-5}(h_{11}))(\beta^{-3}(b)\cdot\alpha^{-5}(g_{11})))\cdot S\alpha^{-5}(h_{12}g_{12})\alpha^{-6}(h_{211}g_{211}))\\
&(\beta^{-1}(c)\cdot\alpha^{-3}(k_{11})))\cdot S\alpha^{-4}(h_{212}g_{212})\alpha^{-2}(k_{12})\otimes\alpha^{-2}(h_{22}g_{22}))k_2\\
=\ &((((\beta^{-3}(a)\cdot\alpha^{-4}(h_1))(\beta^{-3}(b)\cdot\alpha^{-4}(g_1)))\cdot S\alpha^{-6}(h_{211}g_{211})\alpha^{-6}(h_{212}g_{212}))\\
&(\beta^{-1}(c)\cdot\alpha^{-3}(k_{11})))\cdot S\alpha^{-4}(h_{221}g_{221})\alpha^{-2}(k_{12}))\otimes\alpha^{-3}(h_{222}g_{222}))k_2\\
=\ &(((\beta^{-2}(a)\cdot\alpha^{-3}(h_1))(\beta^{-2}(b)\cdot\alpha^{-3}(g_1)))(\beta^{-1}(c)\cdot\alpha^{-3}(k_{11})))\\
&\cdot S\alpha^{-3}(h_{21}g_{21})\alpha^{-2}(k_{12})\otimes\alpha^{-2}(h_{22}g_{22}))k_2.
\end{aligned}
$$

In a similar way, we get

$$(\beta(a)\otimes\alpha(h))((b\otimes g)(c\otimes k))$$

$$\begin{aligned}
&= ((\beta^{-1}(a)\cdot\alpha^{-3}(h_{11}))((\beta^{-2}(b)\cdot\alpha^{-3}(g_1))(\beta^{-2}(c)\cdot\alpha^{-3}(k_1))))\\
&\quad \cdot S(\alpha^{-2}(h_{12})\alpha^{-3}(g_{21}k_{21}))\otimes h_2\alpha^{-2}(g_{22}k_{22})\\
&= (((\beta^{-2}(a)\cdot\alpha^{-3}(h_1))(\beta^{-2}(b)\cdot\alpha^{-3}(g_1)))(\beta^{-1}(c)\cdot\alpha^{-3}(k_{11})))\\
&\quad \cdot S(\alpha^{-3}(h_{21}g_{21})\alpha^{-2}(k_{12}))\otimes\alpha^{-2}(h_{22}g_{22}))k_2.
\end{aligned}$$

So we get $((a\otimes h)(b\otimes g))(\beta(c)\otimes\alpha(k)) = (\beta(a)\otimes\alpha(h))((b\otimes g)(c\otimes k))$. It is easy to see that $(a\otimes h)(1_A\otimes 1_H) = (1_A\otimes 1_H)(a\otimes h) = \beta(a)\otimes\alpha(h)$. So $(A\otimes H, \beta\otimes\alpha, 1_A\otimes 1_H)$ is a Hom-algebra.

Finally, we show that $B\simeq A\Box_\rho H$ as Hom-algebras. Since $(B,\beta,\cdot,\rho)$ is a right $H$-Hom-Hopf module and $A = B^{coH}$, there is an isomorphism of right $H$-Hom-Hopf module, which is given by

$$\varphi: A\otimes H\to B,\quad \varphi(a\otimes h) = a\cdot h,\ \forall a,b\in A, h,g\in H.$$

We only need to verify that $\varphi$ is a Hom-algebra morphism. For all $a,b\in A, h,g\in H$,

$$\begin{aligned}
&\varphi((a\otimes h)(b\otimes g))\\
&= \varphi(((\beta^{-2}(a)\cdot\alpha^{-4}(h_{11}))(\beta^{-2}(b)\cdot\alpha^{-4}(g_{11})))\cdot S\alpha^{-3}(h_{12}g_{12})\otimes\alpha^{-1}(h_2g_2))\\
&= (((\beta^{-2}(a)\cdot\alpha^{-4}(h_{11}))(\beta^{-2}(b)\cdot\alpha^{-4}(g_{11})))\cdot S\alpha^{-3}(h_{12}g_{12}))\cdot\alpha^{-1}(h_2g_2)\\
&= ((\beta^{-1}(a)\cdot\alpha^{-3}(h_{11}))(\beta^{-1}(b)\cdot\alpha^{-3}(g_{11})))\cdot(S\alpha^{-3}(h_{12}g_{12})\alpha^{-2}(h_2g_2))\\
&= ((\beta^{-1}(a)\cdot\alpha^{-2}(h_1))(\beta^{-1}(b)\cdot\alpha^{-2}(g_1)))\cdot\varepsilon(h_2g_2)1_H\\
&= ((\beta^{-1}(a)\cdot\alpha^{-1}(h))(\beta^{-1}(b)\cdot\alpha^{-1}(g)))\cdot 1_H\\
&= (a\cdot h)(b\cdot g) = \varphi(a\otimes h)\varphi(b\otimes g).
\end{aligned}$$

Thus, $B\simeq A\Box_\rho H$ as Hom-algebras. □

On the other hand, we have obtained the coalgebra factorization of right $H$-Hom-module coalgebras with Hom-Hopf module structure (see [3], Theorem 4.1).

**Proposition 3.2.** *Let $(H,\alpha)$ be a Hom-Hopf algebra and $(B,\cdot,\rho,\beta)$ be a right $H$-Hom-Hopf module. We assume that $(B,\cdot,\beta)$ is a right $H$-Hom-module coalgebra, and set $A = B^{coH} = \{a\in B \mid \rho(a) = \beta(a)\otimes 1\}$. The comultiplication on $A\otimes H$ is given by*

$$\begin{aligned}
\Delta(a\otimes h) &= \beta^{-3}(a_{1(0)(0)})\cdot S\alpha^{-3}(a_{1(0)(1)})\otimes\alpha^{-2}(a_{1(1)})\alpha^{-1}(h_1)\\
&\quad \otimes\beta^{-3}(a_{2(0)(0)})\cdot S\alpha^{-3}(a_{2(0)(1)})\otimes\alpha^{-2}(a_{2(1)})\alpha^{-1}(h_2),
\end{aligned}$$

*and the counit is given by*

$$\varepsilon(a\otimes h) = \varepsilon(a)\varepsilon(h),$$

*for any* $a \in A, h \in H$, *then* $(A\otimes H, \beta\otimes\alpha)$ *is a Hom-coalgebra, we write it as* $A\Box^{\cdot} H$. *Moreover,* $B \simeq A\Box^{\cdot} H$ *as Hom-coalgebras.*

By Proposition 3.1 and Proposition 3.2, we can obtain the main result of this section.

**Theorem 3.3.** *Let* $(H, \alpha)$ *be a Hom-Hopf algebra,* $(B, \beta)$ *be a Hom-bialgebra and* $(B, \cdot, \rho, \beta)$ *be a right* $H$*-Hom-Hopf module. We assume that* $(B, \rho, \beta)$ *is a right* $H$*-Hom-comodule algebra and* $(B, \cdot, \beta)$ *is a right* $H$*-Hom-module coalgebra, set* $A = B^{coH} = \{a \in B \mid \rho(a) = \beta(a)\otimes 1\}$. *Define the following operations on* $A\otimes H$,

$$\begin{aligned}(a\otimes h)(b\otimes g) &= ((\beta^{-2}(a)\cdot\alpha^{-4}(h_{11}))(\beta^{-2}(b)\cdot\alpha^{-4}(g_{11})))\cdot S\alpha^{-3}(h_{12}g_{12})\\ &\quad\otimes\alpha^{-1}(h_2g_2),\\ \Delta(a\otimes h) &= \beta^{-3}(a_{1(0)(0)})\cdot S\alpha^{-3}(a_{1(0)(1)})\otimes\alpha^{-2}(a_{1(1)})\alpha^{-1}(h_1)\\ &\quad\otimes\beta^{-3}(a_{2(0)(0)})\cdot S\alpha^{-3}(a_{2(0)(1)})\otimes\alpha^{-2}(a_{2(1)})\alpha^{-1}(h_2),\\ \varepsilon(a\otimes h) &= \varepsilon(a)\varepsilon(h),\end{aligned}$$

*for all* $a, b \in A,\ h, g \in H$, *then* $(A\otimes H, \beta\otimes\alpha)$ *is a Hom-bialgebra, we write it as* $A\Box^{\cdot}_{\rho}H$. *Moreover,* $B \simeq A\Box^{\cdot}_{\rho}H$ *as Hom-bialgebras.*

**Proof.** By Proposition 3.1, we have $B \simeq A\Box_{\rho}H$ as Hom-algebras with the isomorphism

$$\varphi : A\otimes H \to B,\quad \varphi(a\otimes h) = a\cdot h,\ \forall a, b \in A, h, g \in H.$$

By Proposition 3.2, we have $B \simeq A\Box^{\cdot} H$ as Hom-coalgebras with the same isomorphism $\varphi$. Since $B$ is a Hom-bialgebra, it follows that $A\otimes H$ is a Hom-bialgebra with the multiplication $B \simeq A\Box_{\rho}H$ on and the comultiplication on $B \simeq A\Box^{\cdot} H$. Moreover, $B \simeq A\Box^{\cdot}_{\rho}H$ as Hom-bialgebras. □

## 4. Hom-crossed product algebras and cleft extensions

In this section, let $H$ be a Hom-Hopf algebra, $B$ be a right $H$-Hom-Hopf module and $(B, \rho)$ be a right $H$-Hom-comodule algebra. Set $A = B^{coH}$, we can define a weak action of $H$ on $A$, thus obtain the Hom-crossed product algebra $A\sharp_{\sigma}H$ and $B \simeq A\sharp_{\sigma}H$ as Hom-algebras. Next, we dicuss the relation between cleft extension and Hom-comodule algebra with Hom-Hopf module structure.

**Lemma 4.1.** *Let* $(H, S, \alpha)$ *be a Hom-Hopf algebra,* $(B, \cdot, \rho, \beta)$ *be a right* $H$*-Hom-Hopf module and* $(B, \rho)$ *be a right* $H$*-Hom-comodule algebra, set* $A = B^{coH}$, *define* *the map*

$$\rightharpoonup: H \otimes A \to A,\ h \rightharpoonup a = ((1 \cdot \alpha^{-3}(h_1))\beta^{-1}(a)) \cdot S\alpha^{-1}(h_2),\ \forall h \in H, a \in A,$$

*and an element* $\sigma \in Hom(H \otimes H, A)$, *for any* $h, g \in H$,

$$\sigma(h, g) = ((1 \cdot \alpha^{-4}(h_1))(1 \cdot \alpha^{-4}(g_1))) \cdot S\alpha^{-3}(h_2 g_2).$$

*If the following condition holds,*

$$(ab) \cdot \alpha(h) = \beta(a)(b \cdot h),\ \forall a \in A, b \in B, h \in H, \tag{4.1}$$

*then*

(1) $\rightharpoonup$ *is a weak action of* $H$ *on* $A$*;*

(2) $(C1)$, $(C2)$, $(C3)$ *are satisfied.*

**Proof.** (1) We first check that $\rightharpoonup$ is well defined. Since $(B, \cdot, \rho, \beta)$ is a right $H$-Hom-Hopf module and $(B, \rho)$ is a right $H$-Hom-comodule algebra, $A = B^{coH}$. It is easy to get

$$\rho(((1 \cdot \alpha^{-3}(h_1))\beta^{-1}(a)) \cdot S\alpha^{-1}(h_2)) = ((1 \cdot \alpha^{-2}(h_1))a) \cdot S(h_2) \otimes 1.$$

From Eq.(4.1), we can get

$$(h_1 \rightharpoonup a)(1 \cdot h_2) = (1 \cdot \alpha(h))\beta(a),\ \forall h \in H, a \in A, \tag{4.2}$$

In fact

$$\begin{aligned}
& (h_1 \rightharpoonup a)(1 \cdot h_2) \\
= \ & (((1 \cdot \alpha^{-3}(h_{11}))\beta^{-1}(a)) \cdot S\alpha^{-1}(h_{12}))(1 \cdot h_2) \\
\overset{(4.1)}{=} \ & (((1 \cdot \alpha^{-3}(h_{11}))\beta^{-1}(a)) \cdot S\alpha^{-1}(h_{12})) \cdot \alpha(h_2) \\
= \ & ((1 \cdot \alpha^{-1}(h_1))a) \cdot (S\alpha^{-1}(h_{21})\alpha^{-1}(h_{22})) \\
= \ & (1 \cdot \alpha(h))\beta(a).
\end{aligned}$$

For all $a \in A, b \in B, h \in H$, we can obtain

$$\begin{aligned}
& (h_1 \rightharpoonup a)(h_2 \rightharpoonup b) \\
= \ & (h_1 \rightharpoonup a)(((1 \cdot \alpha^{-3}(h_{21}))\beta^{-1}(b)) \cdot S\alpha^{-1}(h_{22})) \\
= \ & (((\alpha^{-3}(h_{11} \rightharpoonup \beta^{-2}(a))(1 \cdot \alpha^{-3}(h_{12})))b) \cdot S\alpha(h_2) \\
\overset{(4.2)}{=} \ & (((1 \cdot \alpha^{-2}(h_1))\beta^{-1}(a))b) \cdot S\alpha(h_2) \\
= \ & ((1 \cdot \alpha^{-1}(h_1))\beta^{-1}(ab)) \cdot S\alpha(h_2) \\
= \ & \alpha^2(h) \rightharpoonup (ab),
\end{aligned}$$

and $h \rightharpoonup 1 = \varepsilon(h)1$, $1 \rightharpoonup a = \beta(a)$. Thus $\rightharpoonup$ is a weak action of $H$ on $A$.

(2) It is obvious that $(C1)$ holds. Now we verify that $(C2)$ holds. On the one hand,

$$
\begin{aligned}
& (h_1 \rightharpoonup (\alpha^{-1}(g_1) \rightharpoonup a))\sigma(\alpha(h_2), \alpha(g_2)) \\
= & (h_1 \rightharpoonup (\alpha^{-1}(g_1) \rightharpoonup a))(((1 \cdot \alpha^{-3}(h_{21}))(1 \cdot \alpha^{-3}(g_{21}))) \cdot S\alpha^{-2}(h_{22}g_{22})) \\
= & (((\alpha^{-3}(h_{11}) \rightharpoonup (\alpha^{-4}(g_{11}) \rightharpoonup \beta^{-2}(a)))(1 \cdot \alpha^{-3}(h_{12})))(1 \cdot \alpha^{-2}(g_{12}))) \cdot S(h_2g_2) \\
= & (((1 \cdot \alpha^{-2}(h_1))(\alpha^{-3}(g_{11}) \rightharpoonup \beta^{-1}(a)))(1 \cdot \alpha^{-2}(g_{12}))) \cdot S(h_2g_2) \\
= & ((1 \cdot \alpha^{-1}(h_1))((1 \cdot \alpha^{-2}(g_1))a)) \cdot S(h_2g_2).
\end{aligned}
$$

On the other hand,

$$
\begin{aligned}
& \sigma(\alpha(h_1), \alpha(g_1))(\alpha^{-1}(h_2g_2) \rightharpoonup \beta(a)) \\
= & (((1 \cdot \alpha^{-3}(h_{11}))(1 \cdot \alpha^{-3}(g_{11}))) \cdot S\alpha^{-2}(h_{12}g_{12}))(((1 \cdot \alpha^{-4}(h_{21}g_{21}))a) \cdot \\
& S\alpha^{-2}(h_{22}g_{22})) \\
= & ((((((1 \cdot \alpha^{-5}(h_{11}))(1 \cdot \alpha^{-5}(g_{11}))) \cdot S\alpha^{-4}(h_{12}g_{12}))(1 \cdot \alpha^{-4}(h_{21}g_{21})))\beta(a)) \cdot \\
& S\alpha^{-1}(h_{22}g_{22}) \\
= & ((((((1 \cdot \alpha^{-5}(h_{11}))(1 \cdot \alpha^{-5}(g_{11}))) \cdot S\alpha^{-4}(h_{12}g_{12})) \cdot \alpha^{-3}(h_{21}g_{21}))\beta(a)) \cdot \\
& S\alpha^{-1}(h_{22}g_{22}) \\
= & ((((((1 \cdot \alpha^{-4}(h_{11}))(1 \cdot \alpha^{-4}(g_{11}))) \cdot (S\alpha^{-4}(h_{12}g_{12})\alpha^{-4}(h_{21}g_{21})))\beta(a)) \cdot \\
& S\alpha^{-1}(h_{22}g_{22}) \\
= & ((1 \cdot \alpha^{-1}(h_1))((1 \cdot \alpha^{-2}(g_1))a)) \cdot S(h_2g_2).
\end{aligned}
$$

Similarly we can obtain $(C3)$ holds. The proof is completed. □

**Theorem 4.2.** *Let $(H, S, \alpha)$ be a Hom-Hopf algebra, $(B, \cdot, \rho, \beta)$ be a right $H$-Hom-Hopf module and $(B, \rho)$ be a right $H$-Hom-comodule algebra, set $A = B^{coH}$, if for all $a \in A, b \in B$ and $h \in H$, Eq.(4.1) holds. Then we get a Hom-crossed product algebra $A\sharp_\sigma H$ and $B \simeq A\sharp_\sigma H$ as Hom-algebras.*

**Proof.** By Lemma 4.1, we get a Hom-crossed product algebra $A\sharp_\sigma H$ with the multiplication given by

$$(a \otimes h)(b \otimes g) = a((\alpha^{-4}(h_{11}) \rightharpoonup \beta^{-2}(b))\sigma(\alpha^{-3}(h_{12}), \alpha^{-2}(g_1))) \otimes \alpha^{-1}(h_2g_2).$$

We note that $B \simeq A \otimes H$ as right $H$-Hom-Hopf modules, and the isomorphic map is given by

$$g : B \to A \otimes H, \ \ g(b) = \beta^{-4}(b_{(0)(0)}) \cdot S\alpha^{-4}(b_{(0)(1)}) \otimes \alpha^{-2}(b_{(1)}).$$

Then we only need to prove that $g$ is a Hom-algebra map. In fact, for all $b, b' \in B$, we have

$$\begin{aligned}
& g(b)g(b') \\
= \; & (\beta^{-4}(b_{(0)(0)}) \cdot S(\alpha^{-4}(b_{(0)(1)})) \otimes \alpha^{-2}(b_{(1)}))(\beta^{-4}(b'_{(0)(0)}) \cdot S\alpha^{-4}(b'_{(0)(1)}) \\
& \otimes \alpha^{-2}(b'_{(1)})) \\
= \; & (\beta^{-4}(b_{(0)(0)}) \cdot S(\alpha^{-4}(b_{(0)(1)})))((\alpha^{-6}(b_{(1)11}) \rightharpoonup (\beta^{-6}(b'_{(0)(0)}) \cdot S\alpha^{-6}(b'_{(0)(1)}))) \\
& \sigma(\alpha^{-5}(b_{(1)12}), \alpha^{-4}(b'_{(1)1}))) \otimes \alpha^{-3}(b_{(1)2}b'_{(1)2}) \\
\overset{(4.2)}{=} \; & (\beta^{-4}(b_{(0)(0)}) \cdot S(\alpha^{-4}(b_{(0)(1)})))((((1 \cdot \alpha^{-8}(b_{(1)11}))(\beta^{-7}(b'_{(0)(0)}) \cdot \\
& S\alpha^{-7}(b'_{(0)(1)})))(1 \cdot \alpha^{-7}(b'_{(1)11}))) \cdot S\alpha^{-6}(b_{(1)12}b'_{(1)12})) \otimes \alpha^{-3}(b_{(1)2}b'_{(1)2}) \\
= \; & (\beta^{-4}(b_{(0)(0)}) \cdot S(\alpha^{-4}(b_{(0)(1)})))(((1 \cdot \alpha^{-7}(b_{(1)11}))(\beta^{-6}(b'_{(0)(0)}) \cdot \\
& (S\alpha^{-7}(b'_{(0)(1)})\alpha^{-8}(b'_{(1)11})))) \cdot S\alpha^{-6}(b_{(1)12}b'_{(1)12})) \otimes \alpha^{-3}(b_{(1)2}b'_{(1)2}) \\
= \; & (\beta^{-3}(b_{(0)}b'_{(0)}) \cdot S\alpha^{-4}(b_{(1)1}b'_{(1)1})) \otimes \alpha^{-3}(b_{(1)2}b'_{(1)2}) \\
= \; & g(bb').
\end{aligned}$$

The proof is completed. □

Now, we discuss the relation between cleft extensions and Hom-comodule algebras with Hom-Hopf module structure.

**Proposition 4.3.** *Let $(B, \rho, \gamma, \beta)$ be a cleft extension. We assume that $H$ acts on $B$ to the right by*

$$b \cdot h = (\beta^{-2}(b_{(0)})\gamma^{-1}(\alpha^{-3}(b_{(1)1})))\gamma(\alpha^{-3}(b_{(1)2})\alpha^{-1}(h)),$$

*then*

(1) *$(B, \cdot, \beta)$ is a right $H$-Hom-module and Eq.(4.1) holds;*
(2) *$(B, \cdot, \rho, \beta)$ is a right $H$-Hom-Hopf module.*

**Proof.** Note that $\gamma$ is convolution invertible right $H$-Hom-comodule map, we have

$$\gamma^{-1}(h)_{(0)} \otimes \gamma^{-1}(h)_{(1)} = \gamma^{-1}(h_2) \otimes S(h_1).$$

(1) We first verify that $(B,\cdot,\beta)$ is a right $H$-Hom-module. For all $b\in B, h, h'\in H$, we calculate

$$
\begin{aligned}
& (b\cdot h)\cdot\alpha(h')\\
= & ((\beta^{-2}(b_{(0)})\gamma^{-1}(\alpha^{-3}(b_{(1)1})))\gamma(\alpha^{-3}(b_{(1)2})\alpha^{-1}(h)))\cdot\alpha(h')\\
= & (\beta^{-2}(((\beta^{-2}(b_{(0)})\gamma^{-1}(\alpha^{-3}(b_{(1)1})))\gamma(\alpha^{-3}(b_{(1)2})\alpha^{-1}(h)))_{(0)})\\
& \gamma^{-1}(\alpha^{-3}(((\beta^{-2}(b_{(0)})\gamma^{-1}(\alpha^{-3}(b_{(1)1})))\gamma(\alpha^{-3}(b_{(1)2})\alpha^{-1}(h)))_{(1)1})))\\
& \gamma(\alpha^{-3}(((\beta^{-2}(b_{(0)})\gamma^{-1}(\alpha^{-3}(b_{(1)1})))\gamma(\alpha^{-3}(b_{(1)2}))\alpha^{-1}(h)))_{(1)2})h')\\
= & (\beta^{-2}((\beta^{-2}(b_{(0)})_{(0)}\gamma^{-1}(\alpha^{-3}(b_{(1)1}))_{(0)})\gamma(\alpha^{-3}(b_{(1)2})\alpha^{-1}(h))_{(0)})\\
& \gamma^{-1}(\alpha^{-3}((\beta^{-2}(b_{(0)})_{(1)}\gamma^{-1}(\alpha^{-3}(b_{(1)1}))_{(1)})\gamma(\alpha^{-3}(b_{(1)2})\alpha^{-1}(h))_{(1)})_{1}))\\
& \gamma(\alpha^{-3}((\beta^{-2}(b_{(0)})_{(1)}\gamma^{-1}(\alpha^{-3}(b_{(1)1}))_{(1)})\gamma(\alpha^{-3}(b_{(1)2})\alpha^{-1}(h))_{(1)})_{2}h')\\
= & (\beta^{-2}((\beta^{-2}(b_{(0)(0)})\gamma^{-1}(\alpha^{-3}(b_{(1)12})))\gamma(\alpha^{-3}(b_{(1)21})\alpha^{-1}(h_{1})))\\
& \gamma^{-1}(\alpha^{-3}((\beta^{-2}(b_{(0)(1)})S\alpha^{-3}(b_{(1)11}))(\alpha^{-3}(b_{(1)22})\alpha^{-1}(h_{2})))_{1}))\\
& \gamma((\alpha^{-3}((\beta^{-2}(b_{(0)(1)})S\alpha^{-3}(b_{(1)11}))(\alpha^{-3}(b_{(1)22})\alpha^{-1}(h_{2})))_{2})h')\\
= & (\beta^{-2}((\beta^{-2}(b_{(0)(0)})\gamma^{-1}(\alpha^{-3}(b_{(1)21})))\gamma(\alpha^{-4}(b_{(1)221})\alpha^{-1}(h_{1})))\\
& \gamma^{-1}(\alpha^{-3}((\beta^{-2}(b_{(0)(1)})S\alpha^{-2}(b_{(1)1}))(\alpha^{-4}(b_{(1)222})\alpha^{-1}(h_{2})))_{1}))\\
& \gamma((\alpha^{-3}((\beta^{-2}(b_{(0)(1)})S\alpha^{-2}(b_{(1)1}))(\alpha^{-4}(b_{(1)222})\alpha^{-1}(h_{2})))_{2})h')\\
= & (\beta^{-2}((\beta^{-1}(b_{(0)})\gamma^{-1}(\alpha^{-4}(b_{(1)221})))\gamma(\alpha^{-5}(b_{(1)2221})\alpha^{-1}(h_{1})))\\
& \gamma^{-1}(\alpha^{-3}((\beta^{-2}(b_{(1)1})S\alpha^{-3}(b_{(1)21}))(\alpha^{-5}(b_{(1)2222})\alpha^{-1}(h_{2})))_{1}))\\
& \gamma((\alpha^{-3}((\beta^{-2}(b_{(1)1})S\alpha^{-3}(b_{(1)21}))(\alpha^{-5}(b_{(1)2222})\alpha^{-1}(h_{2})))_{2})h')\\
= & (\beta^{-2}((\beta^{-1}(b_{(0)})\gamma^{-1}(\alpha^{-3}(b_{(1)21})))\gamma(\alpha^{-4}(b_{(1)221})\alpha^{-1}(h_{1})))\\
& \gamma^{-1}(\alpha^{-3}((\beta^{-3}(b_{(1)11})S\alpha^{-3}(b_{(1)12}))(\alpha^{-4}(b_{(1)222})\alpha^{-1}(h_{2})))_{1}))\\
& \gamma((\alpha^{-3}((\beta^{-3}(b_{(1)11})S\alpha^{-3}(b_{(1)12}))(\alpha^{-4}(b_{(1)222})\alpha^{-1}(h_{2})))_{2})h')\\
= & (((\beta^{-3}(b_{(0)})\gamma^{-1}(\alpha^{-5}(b_{(1)21})))\gamma(\alpha^{-6}(b_{(1)221})\alpha^{-3}(h_{1})))\\
& \gamma^{-1}((\varepsilon(b_{(1)1})1_{H}(\alpha^{-7}(b_{(1)222})\alpha^{-4}(h_{2}))_{1}))\gamma(1_{H}(\alpha^{-7}(b_{(1)222})\alpha^{-4}(h_{2})))_{2})h')\\
= & (((\beta^{-3}(b_{(0)})\gamma^{-1}(\alpha^{-4}(b_{(1)1})))\gamma(\alpha^{-5}(b_{(1)21})\alpha^{-3}(h_{1})))\gamma^{-1}((\alpha^{-5}(b_{(1)22})\alpha^{-3}(h_{2}))_{1}))\\
& \gamma(((\alpha^{-5}(b_{(1)22})\alpha^{-3}(h_{2}))_{2})h')\\
= & ((\beta^{-2}(b_{(0)})\gamma^{-1}(\alpha^{-3}(b_{(1)1})))(\gamma(\alpha^{-5}(b_{(1)21})\alpha^{-3}(h_{1}))\gamma^{-1}(\alpha^{-6}(b_{(1)221})\alpha^{-4}(h_{21}))))\\
& \gamma((\alpha^{-5}(b_{(1)222})\alpha^{-3}(h_{22}))h')\\
= & ((\beta^{-2}(b_{(0)})\gamma^{-1}(\alpha^{-4}(b_{(1)11})))(\gamma(\alpha^{-5}(b_{(1)12})\alpha^{-4}(h_{11}))\gamma^{-1}(\alpha^{-5}(b_{(1)21})\alpha^{-4}(h_{12}))))\\
& \gamma((\alpha^{-4}(b_{(1)22})\alpha^{-2}(h_{2}))h')
\end{aligned}
$$

$$
\begin{aligned}
&= ((\beta^{-2}(b_{(0)})\gamma^{-1}(\alpha^{-3}(b_{(1)1})))(\varepsilon(b_{(1)21})\varepsilon(h_1)1_H))\gamma((\alpha^{-4}(b_{(1)22})\alpha^{-2}(h_2))h^{'}) \\
&= (\beta^{-1}(b_{(0)})\gamma^{-1}(\alpha^{-2}(b_{(1)1})))\gamma((\alpha^{-3}(b_{(1)2)})\alpha^{-1}(h))h^{'}) \\
&= \beta(b)\cdot(hh^{'}),
\end{aligned}
$$

and $b\cdot 1=(\beta^{-2}(b_{(0)})\gamma^{-1}(\alpha^{-3}(b_{(1)1})))\gamma(\alpha^{-2}(b_{(1)2}))=\beta(b)$. Thus $(B,\beta,\cdot)$ is a right $H$-Hom-module.

(2) Now we prove $(B,\cdot,\rho,\beta)$ is a right $H$-Hom-Hopf module.

$$
\begin{aligned}
&\rho(b\cdot h) \\
&= \rho((\beta^{-2}(b_{(0)})\gamma^{-1}(\alpha^{-3}(b_{(1)1})))\gamma(\alpha^{-3}(b_{(1)2}\alpha^{-1}(h))) \\
&= (\beta^{-2}(b_{(0)(0)})\gamma^{-1}(\alpha^{-3}(b_{(1)12})))\gamma(\alpha^{-3}(b_{(1)21})\alpha^{-1}(h_1)) \\
&\quad \otimes(\alpha^{-2}(b_{(0)(1)})S\alpha^{-3}(b_{(1)11}))(\alpha^{-3}(b_{(1)22})\alpha^{-1}(h_2)) \\
&= (\beta^{-1}(b_{(0)})\gamma^{-1}(\alpha^{-4}(b_{(1)212})))\gamma(\alpha^{-4}(b_{(1)121})\alpha^{-1}(h_1)) \\
&\quad \otimes(\alpha^{-2}(b_{(1)1})S\alpha^{-4}(b_{(1)211}))(\alpha^{-4}(b_{(1)222})\alpha^{-1}(h_2)) \\
&= (\beta^{-1}(b_{(0)})\gamma^{-1}(\alpha^{-3}(b_{(1)12})))\gamma(\alpha^{-3}(b_{(1)21})\alpha^{-1}(h_1)) \\
&\quad \otimes(\alpha^{-4}(b_{(1)111})S\alpha^{-4}(b_{(1)112}))(\alpha^{-3}(b_{(1)22})\alpha^{-1}(h_2)) \\
&= (\beta^{-1}(b_{(0)})\gamma^{-1}(\alpha^{-3}(b_{(1)12})))\gamma(\alpha^{-3}(b_{(1)21})\alpha^{-1}(h_1))\otimes\varepsilon(b_{(1)11})(\alpha^{-2}(b_{(1)22})h_2) \\
&= (\beta^{-1}(b_{(0)})\gamma^{-1}(\alpha^{-2}(b_{(1)1})))\gamma(\alpha^{-3}(b_{(1)21})\alpha^{-1}(h_1))\otimes\alpha^{-2}(b_{(1)22})h_2 \\
&= (\beta^{-1}(b_{(0)})\gamma^{-1}(\alpha^{-3}(b_{(1)11})))\gamma(\alpha^{-3}(b_{(1)12})\alpha^{-1}(h_1))\otimes\alpha^{-1}(b_{(1)2})h_2 \\
&= (\beta^{-2}(b_{(0)(0)})\gamma^{-1}(\alpha^{-3}(b_{(0)(1)1})))\gamma(\alpha^{-3}(b_{(0)(1)2)})\alpha^{-1}(h_1))\otimes b_1h_2 \\
&= b_{(0)}\cdot h_1\otimes b_{(1)}h_2.
\end{aligned}
$$

The proof is completed. □

By Theorem 4.2 and Proposition 4.3, we can get the following result.

**Corollary 4.4.** *Let $(B,\rho,\gamma,\beta)$ be a cleft extension. Then we get a Hom-crossed product algebra $A\sharp_\sigma H$ and $B\simeq A\sharp_\sigma H$ as Hom-algebras.*

In Theorem 4.2, we get $B$ is isomorphic to a Hom-crossed product algebra, that is $B\simeq A\sharp_\sigma H$ as Hom-algebras. Now, if we slightly change the condition (4.1), we can get $B$ is isomorphic to a Hom-smash product algebra.

**Lemma 4.5.** *Let $(H,S,\alpha)$ be a Hom-Hopf algebra, $(B,\cdot,\rho,\beta)$ be a right $H$-Hom-Hopf module and $(B,\rho)$ be a right $H$-Hom-comodule algebra satisfying the following condition*

$$(bb^{'})\cdot\alpha(h)=\beta(b)(b^{'}\cdot h),\ \forall b,b^{'}\in B,h\in H. \tag{4.3}$$

*Set $A = B^{coH}$, define the map*

$$\rightharpoonup: H \otimes A \to A, \quad h \rightharpoonup a = ((1 \cdot \alpha^{-3}(h_1))\beta^{-1}(a)) \cdot S\alpha^{-1}(h_2).$$

*Then $(A, \rightharpoonup)$ is a left $H$-Hom-module algebra.*

**Proof.** We only need prove $(A, \rightharpoonup)$ is a left $H$-Hom-module. $\forall a \in A, h, g \in H$,

$$\begin{aligned}
& \alpha(h) \rightharpoonup (g \rightharpoonup a) \\
= & \alpha(h) \rightharpoonup (((1 \cdot \alpha^{-3}(g_1))\beta^{-1}(a)) \cdot S\alpha^{-1}(g_2)) \\
= & ((1 \cdot \alpha^{-2}(h_1))(((1 \cdot \alpha^{-4}(g_1))\beta^{-2}(a)) \cdot S\alpha^{-2}(g_2))) \cdot S(h_2) \\
= & (((1 \cdot \alpha^{-4}(h_1 g_1))\beta^{-1}(a)) \cdot S\alpha^{-1}(g_2)) \cdot S(h_2) \\
= & ((1 \cdot \alpha^{-3}(h_1 g_1))a) \cdot S\alpha^{-1}(h_2 g_2) \\
= & (hg) \rightharpoonup \beta(a).
\end{aligned}$$

It is easy to see $1 \rightharpoonup a = \beta(a)$, so $(A, \rightharpoonup)$ is a left $H$-Hom-module. □

**Theorem 4.6.** *Let $(H, S, \alpha)$ be a Hom-Hopf algebra and $(B, \cdot, \rho, \beta)$ be a right $H$-Hom-Hopf module. Assume $(B, \rho)$ is a right $H$-Hom-comodule algebra satisfying Eq.(4.3). Set $A = B^{coH}$, then we get a Hom-smash product $A\sharp H$ with the multiplication given by*

$$(a \otimes h)(a' \otimes h') = a(h_1 \rightharpoonup \beta^{-1}(a')) \otimes \alpha^{-1}(h_2)h',$$

*and $B \simeq A\sharp H$ as Hom-algebras.*

**Proof.** It is easy to prove by Lemma 4.5. □

From Corollary 4.4, we know that if $(B, \rho, \gamma, \beta)$ is a cleft extension, then we get $B$ is isomorphic to the Hom-crossed product algebra $A\sharp_\sigma H$. Next, we give an equivalent characterization about cleft extension.

**Theorem 4.7.** *Let $(H, \alpha)$ be a Hom-Hopf algebra and $(B, \rho, \beta)$ be a right $H$-Hom-comodule algebra, and set $A = B^{coH} = \{a \in B \mid \rho(a) = \beta(a) \otimes 1\}$, then $A \subset B$ is a cleft extension if and only if $(B, \cdot, \rho, \beta)$ is a right $H$-Hom-Hopf module, and there exists a convolution invertible linear map $\gamma : H \to B$ satisfying $\gamma(h) = 1 \cdot \alpha^{-1}(h)$.*

**Proof.** $\Rightarrow$): Assume that $A \subset B$ is a cleft extension, then there exists a convolution invertible Hom-comodule map $\gamma : H \to B$ such that $\gamma(1) = 1$, so we have $\gamma^{-1}(1) = 1$. Define a map

$$\cdot : B \otimes H \to B,\ b \cdot h = (\beta^{-2}(b_{(0)})\gamma^{-1}(\alpha^{-3}(b_{(1)1})))\gamma(\alpha^{-3}(b_{(1)2})\alpha^{-1}(h)),\ \forall b \in B, h \in H,$$

clearly, $\gamma(h) = 1 \cdot \alpha^{-1}(h)$. By Proposition 4.3, we see that $(B, \cdot, \rho, \beta)$ is a right $H$-Hom-Hopf module.

$\Leftarrow$): It is sufficient to show $\gamma$ is a right Hom-comodule map, using the fact that $(B, \cdot, \rho, \beta)$ is a right $H$-Hom-Hopf module and $\gamma(h) = 1 \cdot \alpha^{-1}(h)$, we get

$$\rho\gamma(h) = \rho(1 \cdot \alpha^{-1}(h)) = 1 \cdot \alpha^{-1}(h_1) \otimes h_2 = \gamma(h_1) \otimes h_2, \quad \forall h \in H.$$

The proof is completed. $\square$

By the above results, we have the following conclusion.

**Theorem 4.8.** *Let $(H, \alpha)$ be a Hom-Hopf algebra and $(B, \rho, \gamma, \beta)$ be a cleft extension, set $A = B^{coH}$, then*

$$B \simeq A\Box_\rho H = A\sharp_\sigma H,$$

*as Hom-algebras.*

**Proof.** By Proposition 3.1 and Theorem 4.2, we have $B \simeq A\Box_\rho H$ and $B \simeq A\sharp_\sigma H$ as Hom-algebras. Since $(B, \rho, \gamma, \beta)$ is a cleft extension, we get $(B, \cdot, \rho, \beta)$ is a right $H$-Hom-Hopf module by Proposition 3.3, where the module structure is defined by

$$b \cdot h = (\beta^{-2}(b_{(0)})\gamma^{-1}(\alpha^{-3}(b_{(1)1})))\gamma(\alpha^{-3}(b_{(1)2})\alpha^{-1}(h)).$$

Now we prove $A\Box_\rho H$ has the same multiplication with $A\sharp_\sigma H$.

$$\begin{aligned}
& (a \otimes h)(b \otimes g) \\
= \;& ((\beta^{-2}(a) \cdot \alpha^{-4}(h_{11}))(\beta^{-2}(b) \cdot \alpha^{-4}(g_{11}))) \cdot S\alpha^{-3}(h_{12}g_{12}) \otimes \alpha^{-1}(h_2g_2) \\
= \;& (((\beta^{-2}(\beta^{-2}(a)_{(0)})\gamma^{-1}(\alpha^{-3}(\beta^{-2}(a)_{(1)1})))\gamma(\alpha^{-3}(\beta^{-2}(a)_{(1)2})\alpha^{-5}(h_{11}))) \\
& (\beta^{-2}(b) \cdot \alpha^{-4}(g_{11}))) \cdot S\alpha^{-3}(h_{12}g_{12}) \otimes \alpha^{-1}(h_2g_2) \\
= \;& ((\beta^{-2}(a)\gamma(\alpha^{-4}(h_{11})))(\beta^{-2}(b)\gamma(\alpha^{-4}(g_{11})))) \cdot S\alpha^{-3}(h_{12}g_{12}) \otimes \alpha^{-1}(h_2g_2) \\
= \;& (\beta^{-2}((\beta^{-2}(a)_{(0)}\gamma(\alpha^{-4}(h_{11}))_{(0)})(\beta^{-2}(b)_{(0)}\gamma(\alpha^{-4}(g_{11}))_{(0)})) \\
& \gamma^{-1}(\alpha^{-3}(((\beta^{-2}(a)_{(1)}\gamma(\alpha^{-4}(h_{11}))_{(1)})(\beta^{-2}(b)_{(1)}\gamma(\alpha^{-4}(g_{11}))_{(1)}))_1))) \\
& \gamma(\alpha^{-3}(((\beta^{-2}(a)_{(1)}\gamma(\alpha^{-4}(h_{11}))_{(1)})(\beta^{-2}(b)_{(1}\gamma(\alpha^{-4}(g_{11}))_{(1)}))_2)S\alpha^{-4}(h_{12}g_{12})) \\
& \otimes\alpha^{-1}(h_2g_2) \\
= \;& (((\beta^{-3}(a)\gamma(\alpha^{-6}(h_{111})))(\beta^{-3}(b)\gamma(\alpha^{-6}(g_{111}))))\gamma^{-1}(\alpha^{-6}(h_{1121}g_{1121}))) \\
& \gamma(\alpha^{-6}(h_{1122}g_{1122})S\alpha^{-4}(h_{12}g_{12})) \otimes \alpha^{-1}(h_2g_2) \\
= \;& (((\beta^{-3}(a)\gamma(\alpha^{-6}(h_{111})))(\beta^{-3}(b)\gamma(\alpha^{-6}(g_{111}))))\gamma^{-1}(\alpha^{-5}(h_{112}g_{112}))) \\
& \gamma(\alpha^{-4}(h_{12}g_{12})S\alpha^{-4}(h_{21}g_{21})) \otimes \alpha^{-2}(h_{22}g_{22})
\end{aligned}$$

$$
\begin{aligned}
&= (((\beta^{-3}(a)\gamma(\alpha^{-5}(h_{11})))(\beta^{-3}(b)\gamma(\alpha^{-5}(g_{11}))))\gamma^{-1}(\alpha^{-4}(h_{12}g_{12})))\\
&\quad \gamma(\alpha^{-5}(h_{211}g_{211})S\alpha^{-5}(h_{212}g_{212}))\otimes\alpha^{-2}(h_{22}g_{22})\\
&= ((\beta^{-2}(a)\gamma(\alpha^{-4}(h_{11})))(\beta^{-2}(b)\gamma(\alpha^{-4}(g_{11}))))\gamma^{-1}(\alpha^{-3}(h_{12}g_{12}))\otimes\alpha^{-1}(h_2g_2)\\
&= a(\gamma(\alpha^{-3}(h_{11}))((\beta^{-3}(b)\gamma(\alpha^{-5}(g_{11})))\gamma^{-1}(\alpha^{-5}(h_{12}g_{12}))))\otimes\alpha^{-1}(h_2g_2)\\
&= a((\alpha^{-4}(h_{11})\rightharpoonup\beta^{-2}(b))\sigma(\alpha^{-3}(h_{12}),\alpha^{-2}(g_1)))\sharp\alpha^{-1}(h_2g_2),
\end{aligned}
$$

for all $h,g\in H, a,b\in A$, we define

$$\rightharpoonup: H\otimes A\to A,\quad h\rightharpoonup a=(\gamma(\alpha^{-2}(h_1))\beta^{-1}(a))\gamma^{-1}(\alpha^{-1}(h_2)),$$

and

$$\sigma: H\otimes H\to A,\quad \sigma(h,g)=\gamma(\alpha^{-2}(h_1))(\gamma(\alpha^{-3}(g_1))\gamma^{-1}(\alpha^{-4}(h_2g_2))).$$

Therefore, $B\simeq A\Box_\rho H=A\sharp_\sigma H$ as Hom-algebras. □

**Acknowledgement.** The authors would like to thank the referee for the valuable suggestions and comments.

## References


[1] R. J. Blattner, M. Cohen and S. Montgomery, *Crossed products and inner actions of Hopf algebras*, Trans. Amer. Math. Soc., 298(2) (1986), 671-711.

[2] R. J. Blattner and S. Montgomery, *Crossed products and Galois extensions of Hopf algebras*, Pacific J. Math., 137(1) (1989), 37-54.

[3] L. H. Dong, S. Xue and X. Zhang, *The factorization for a class of Hom-coalgebras*, Adv. Math. Phys., (2017), 4653172 (14 pp).

[4] J. T. Hartwig, D. Larsson and S. D. Silvestrov, *Deformations of Lie algebras using σ-derivations*, J. Algebra, 295(2) (2006), 314-361.

[5] D. W. Lu and S. H. Wang, *Crossed product Hom-Hopf algebras and lazy 2-cocycle*, arXiv: 1509.01518v2.

[6] T. S. Ma H. Y. Li and T. Yang, *Cobraided smash product Hom-Hopf algebras*, Colloq. Math., 134(1) (2014), 75-92.

[7] A. Makhlouf and S. D. Silvestrov, *Hom-algebra structures*, J. Gen. Lie Theory Appl., 2(2) (2008), 51-64.

[8] A. Makhlouf and S. D. Silvestrov, *Hom-algebras and Hom-coalgebras*, J. Algebra Appl., 9(4) (2010), 553-589.

[9] S. Montgomery, Hopf Algebras and Their Actions on Rings, CBMS Regional Conference Series in Mathematics, 82, Published for the Conference Board

of the Mathematical Sciences, Washington, DC; by the Amer. Math. Soc., Providence, RI, 1993.

[10] D. Yau, *Hom-bialgebras and comodule Hom-algebras*, Int. Electron. J. Algebra, 8 (2010), 45-64.

**Lihong Dong** (Corresponding Author) and **Shuan Xue**
College of Mathematics and Information Science
Henan Normal University
453007 Xinxiang, Henan, China
e-mails: lihongdong2010@gmail.com (L. Dong)
1506225510@qq.com (S. Xue)

# 4

# A CATEGORICAL APPROACH TO ALGEBRAS AND COALGEBRAS

Robert Wisbauer

Communicated by Abdullah Harmancı

*Dedicated to the memory of Professor John Clark*

ABSTRACT. Algebraic and coalgebraic structures are often handled independently. In this survey we want to show that they both show up naturally when approaching them from a categorical point of view. Azumaya, Frobenius, separable, and Hopf algebras are obtained when both notions are combined. The starting point and guiding lines for this approach are given by *adjoint pairs of functors* and their elementary properties.

**Keywords**: (Co)algebras, (co)monads, Frobenius algebras, separable algebras, Hopf algebras

## 1. Introduction

In the last decades categorical techniques turned out to be very effective in algebra and representation theory. Hereby, it was a key observation that module theory of an algebra $A$ over a field $K$ is essentially the theory of the functor

$$A \otimes_K - : \mathbb{M}_K \to \mathbb{M}_K,$$

an endofunctor of the category of $K$-vector spaces. An algebra $A$ is defined by $K$-linear maps, multiplication $A \otimes_K A \to A$ and unit $e : K \to A$, subject to associativity and unitality conditions. Left $A$-modules are given by a $K$-vector space $V$ with $K$-linear maps $\varrho : A \otimes_K V \to V$, also subject to associativity and unitality conditions. Together with $A$-linear maps, this yields the category ${}_A\mathbb{M}$ of left $A$-modules. The tensor product with multiplication $(A \otimes_K V, m \otimes_K V)$ is a left $A$-module and this leads to the free and forgetful functors

$$\begin{aligned} \phi_A : &\quad \mathbb{M}_K \to {}_A\mathbb{M},\ V \mapsto (A \otimes_K V, m \otimes_K V), \\ U_A : &\quad {}_A\mathbb{M} \to \mathbb{M}_K,\ (M, \rho) \mapsto M, \end{aligned}$$

and the bijection

$$\mathrm{Hom}_A(A \otimes_K V, M) \to \mathrm{Hom}_K(V, U_A(M)),$$

that is, the functor $U_A$ is right adjoint to $A \otimes_K -$.

These basic structures can be defined for arbitrary categories $\mathbb{A}$, replacing $\mathbb{M}_K$, and any functor $F : \mathbb{A} \to \mathbb{A}$, replacing $A \otimes_K - : \mathbb{M}_K \to \mathbb{M}_K$. Multiplication and unit are replaced by natural transformations, $m : FF \to F$ and $\eta : 1 \to F$, satisfying the respective associativity and unitality conditions. This gives $F$ a *monad* structure.

An $F$-module is an object $V \in \mathbb{A}$ with a morphism $\varrho : F(V) \to V$ and for any $W \in \mathbb{A}$, $(F(W), m_W)$ is an example for this. Morphisms of $F$-modules are morphisms from $\mathbb{A}$ respecting the module structures and they yield the category $\mathbb{A}_F$ of $F$-modules with free and forgetful functors

$$\begin{aligned} \phi_F : &\quad \mathbb{A} \to \mathbb{A}_F,\ V \mapsto (F(V), m_V), \\ U_F : &\quad \mathbb{A}_F \to \mathbb{A},\ (M, \varrho) \mapsto M, \end{aligned}$$

and the bijection

$$\begin{aligned} \mathrm{Mor}_{\mathbb{A}_F}(F(V), M) &\to \mathrm{Mor}_{\mathbb{A}}(V, U_F(M)), \\ F(V) \xrightarrow{f} M &\mapsto V \xrightarrow{\eta_V} F(V) \xrightarrow{f} M, \end{aligned}$$

shows that the functor $U_F$ is right adjoint to $\phi_F$.

Thus structures from module theory can be formulated in great generality. As we will see, if the functor $F : \mathbb{A} \to \mathbb{A}$ has a right adjoint $G : \mathbb{A} \to \mathbb{A}$, then the monad structure on $F$ provides $G$ with the structure of a *comonad*. This approach leads naturally to comonads (coalgebras, bocses) and comodules and we will highlight this interplay. Notice that the categorical tools developed are also successfully applied in theoretical computer science and logic (e.g. [26], [35]).

## 2. Category theory

The idea that the role of elements in algebraic structures should be taken over by homomorphisms came up from the beginning of the last century. It has finally been poured into a solid frame 1945 by Samuel Eilenberg and Saunders Mac Lane in the seminal paper [5]. For convenience, and to fix notation, we recall the basic notions and refer to [12] for more details.

2.1. **Categories.** A *category* $\mathbb{A}$ consists of a class of objects, and

(i) for each pair $(A, A')$ of objects, there is a set of morphisms $\mathrm{Mor}_{\mathbb{A}}(A, A')$, usually denoted by arrows $A \to A'$,

(ii) for any triple $A, A', A''$ of objects there is a composition of morphisms.

$$\mathrm{Mor}_{\mathbb{A}}(A, A') \times \mathrm{Mor}_{\mathbb{A}}(A', A'') \to \mathrm{Mor}_{\mathbb{A}}(A, A'')$$

which is associative in an obvious sense,

(iii) for any object $A$, $\mathrm{Mor}_{\mathbb{A}}(A, A)$ contains an identity morphism $1_A$, leaving any composition with it unchanged.

**2.2. Functors.** A *covariant functor* $F : \mathbb{A} \to \mathbb{B}$ between categories sends

(i) an object $A$ from $\mathbb{A}$ to an object $F(A)$ in $\mathbb{B}$,

(ii) a morphism $f : A \to A'$ to $F(f) : F(A) \to F(A')$ in $\mathbb{B}$,

(iii) a composit $fg$ of morphisms in $\mathbb{A}$ to $F(fg) = F(f)F(g)$ in $\mathbb{B}$,

(iv) the identity $1_A$ for $A$ in $\mathbb{A}$, to the identity of $F(A)$, $F(1_A) = 1_{F(A)}$.

It follows from this definition that $F$ induces a set map

$$\Phi_F : \mathrm{Mor}_{\mathbb{A}}(A, A') \to \mathrm{Mor}_{\mathbb{B}}(F(A), F(A')),$$

and $F$ is called *faithful* if $\Phi_F$ is injective, and *full* if $\Phi_F$ is surjective.

$F$ defines an *equivalence of categories* provided there exists a functor $G : \mathbb{B} \to \mathbb{A}$ such that $FG$ and $GF$ both are isomorphic to the respective identities.

A fully faithful functor $F : \mathbb{A} \to \mathbb{B}$ induces an equivalence between $\mathbb{A}$ and a full subcategory of $\mathbb{B}$, the image of $F$.

**2.3. Natural transformations.** Given two functors $F, G : \mathbb{A} \to \mathbb{B}$ between categories, a *natural transformation* $\psi : F \to G$ is given by a family of morphisms $\psi_A : F(A) \to G(A)$, $A \in \mathbb{A}$, with commutative diagrams, for any morphism $h : A \to A'$ in $\mathbb{A}$,

$$\begin{array}{ccccc} A & \qquad & F(A) & \xrightarrow{\psi_A} & G(A) \\ \downarrow h & & {\scriptstyle F(h)}\downarrow & & \downarrow{\scriptstyle G(h)} \\ A', & & F(A') & \xrightarrow{\psi_{A'}} & G(A'). \end{array}$$

$\psi$ is called a *(natural) isomorphism* if all $\psi_A$ are isomorphisms in $\mathbb{B}$.

**2.4. Separable functors.** A functor $F : \mathbb{A} \to \mathbb{B}$ is said to be *separable* if, for any $A, A' \in \mathbb{A}$, the canonical map $\Phi_F$ (from 2.2) is a naturally split monomorphism, that is, there is a map

$$\Psi_F : \mathrm{Mor}_{\mathbb{B}}(F(A), F(A')) \to \mathrm{Mor}_{\mathbb{A}}(A, A'),$$

natural in $A, A'$, with $\Psi_F \cdot \Phi_F$ the identity on $\mathrm{Mor}_{\mathbb{A}}(A, A')$.

Clearly, for a separable functor $F$, $\Phi_F$ is always injective, and every fully faithful functor is separable. A survey on separability in algebra and category theory is given in [34].

It turned out that the following notion, formulated by D.M. Kan in [10] and further investigated by S. Eilenberg and J.C. Moore in [6] and others, is a milestone in category theory.

2.5. **Adjoint pair of functors.** A pair of (covariant) functors $F : \mathbb{A} \to \mathbb{B}$, $G : \mathbb{B} \to \mathbb{A}$ between any categories $\mathbb{A}$, $\mathbb{B}$, is said to be *adjoint*, we write $F \dashv G$, provided there is a bijection

$$\alpha_{A,B} : \mathrm{Mor}_{\mathbb{B}}(F(A), B) \to \mathrm{Mor}_{\mathbb{A}}(A, G(B)),$$

natural in $A \in \mathbb{A}$ and $B \in \mathbb{B}$. Such a bijection can be described by natural transformations, called *unit* and *counit*,

$$\eta : 1 \to GF, \quad \varepsilon : FG \to 1,$$

satisfying the *triangular identities*

$$F \xrightarrow{F\eta} FGF \xrightarrow{\varepsilon F} F = 1_F, \quad G \xrightarrow{\eta G} GFG \xrightarrow{G\varepsilon} G = 1_G.$$

They are obtained as images of the identities of $F(A)$ and $G(B)$, respectively, in the defining bijection.

Adjointness of *contravariant functors* $\mathbb{A} \to \mathbb{B}$ is defined by relating them with covariant functors between opposite categories.

2.6. **Remark.** The notion of adjointness can be weakened in various ways. For example, instead of being invertible one may require $\alpha$ to be *regular*, that is, there exists $\beta_{A,B} : \mathrm{Mor}_{\mathbb{A}}(A, G(B)) \to \mathrm{Mor}_{\mathbb{B}}(F(A), B)$, natural in $A \in \mathbb{A}$ and $B \in \mathbb{B}$, such that $\alpha\beta\alpha = \alpha$ (and $\beta\alpha\beta = \beta$). This yields a weaker form of the triangular identities (see [19], [32]).

A short argument shows that for adjoint functors $F \dashv G$,

- $F$ preserves epimorphisms and coproducts,
- $G$ preserves monomorphisms and products.

2.7. **Properties of units and counits.** Let $F \dashv G : \mathbb{B} \to \mathbb{A}$ be an adjoint pair of functors (notation from 2.5).

(1) $\varepsilon$ is an isomorphism if and only if $G$ is a fully faithful functor:
$G$ yields an equivalence between $\mathbb{B}$ and the image of $G$.

(2) $\eta$ is an isomorphism if and only if $F$ is a fully faithful functor: $F$ yields an equivalence between $\mathbb{A}$ and the image of $F$.

(3) $\varepsilon$ and $\eta$ are isomorphisms if and only if $F$ and $G$ both are fully faithful: $F$ and $G$ determine an equivalence between $\mathbb{A}$ and $\mathbb{B}$.

(4) $\varepsilon$ is a split epimorphism if and only if $G$ is a separable functor.

(5) $\eta$ is a split monomorphism if and only if $F$ is a separable functor.

2.8. **Rings and modules.** We follow the notation from [27]. For associative rings $R$ and $S$, denote by ${}_R\mathbb{M}$ and $\mathbb{M}_S$ the category of left and right modules, respectively. Then any bimodule ${}_RP_S$ induces an adjoint pair of functors

$$ {}_RP \otimes_S - : {}_S\mathbb{M} \to {}_R\mathbb{M}, \quad \mathrm{Hom}_R(P,-) : {}_R\mathbb{M} \to {}_S\mathbb{M} $$

with bijection, counit, and unit

$$ \mathrm{Hom}_R(P \otimes_S Y, X) \simeq \mathrm{Hom}_S(Y, \mathrm{Hom}_R(P,X)), $$

$$ \varepsilon_X : P \otimes_S \mathrm{Hom}_R(P,X) \to X, \quad p \otimes f \mapsto f(p) $$

$$ \eta_Y : Y \to \mathrm{Hom}_R(P, P \otimes_S Y), \quad y \mapsto [p \mapsto p \otimes y]. $$

Denote by Gen($P$) (Pres($P$)) the category of $P$-generated ($P$-presented) $R$-modules. Clearly the image of $P \otimes_S -$ is contained in Pres($P$).

Choose an (injective) cogenerator $Q$ in ${}_R\mathbb{M}$ and put $U = \mathrm{Hom}_R(P,Q)$. For the $S$-module $U$, denote by Cog($U$) (Cop($U$)) the category of $U$-cogenerated ($U$-copresented) $S$-modules. The image of $\mathrm{Hom}_R(P,-)$ is contained in Cog($U$). If $R$ is a cogenerator in ${}_R\mathbb{M}$ we can choose $U = P^*$.

(i) $\varepsilon$ isomorphism: $P$ is a generator in ${}_R\mathbb{M}$ and the categories ${}_R\mathbb{M}$ and Cog($U$) are equivalent.

(ii) $\eta$ isomorphism: by 2.7, $P \otimes_S -$ is full and faithful and the categories ${}_S\mathbb{M}$ and Pres($P$) are equivalent (Sato equivalence, [23, Theorem 2.1]).

(iii) $\eta$ and $\varepsilon$ are isomorphisms: the categories ${}_R\mathbb{M}$ and ${}_S\mathbb{M}$ are equivalent; $P$ is a finitely generated projective generator in ${}_R\mathbb{M}$ and $S = \mathrm{End}_R(P)$ (Morita equivalence, [21]).

(iv) $\varepsilon$ monomorph and $\eta$ epimorph: the categories Gen($P$) and Cog($U$) are equivalent ($*$-modules, e.g. [4], [24, Theorem 1.3], [28], [29]).

## 3. Monads and comonads

Monads and comonads on categories are modeled after the algebras and coalgebras on vector spaces.

**3.1. Monads and their modules.** A *monad* on any category $\mathbb{A}$ is an endofunctor $T : \mathbb{A} \to \mathbb{A}$ with natural transformations, *multiplication (product)* and *unit*,

$$m : TT \to T, \quad \eta : 1_{\mathbb{A}} \to T,$$

subject to associativity and unitality conditions (as for algebras).

$T$-modules are objects $A \in \mathbb{A}$ with a morphism $\varrho : T(A) \to A$ subject to associativity and unitality conditions (as for modules over rings).

*Morphisms* between $T$-modules $(A, \varrho)$ and $(A', \varrho')$ (or $T$*-morphisms*) are morphisms $f : A \to A'$ in $\mathbb{A}$ with commutative diagram

$$\begin{array}{ccc} T(A) & \xrightarrow{T(f)} & T(A') \\ \varrho \downarrow & & \downarrow \varrho' \\ A & \xrightarrow{f} & A'. \end{array}$$

The category determined by $T$-modules and their morphisms is called the **Eilenberg-Moore category** - or just the *module category* - of the monad $(T, m, \eta)$ and we denote it by $\mathbb{A}_T$.

For any $A \in \mathbb{A}$, $T(A)$ has a $T$-module structure by $m_A : TT(A) \to T(A)$ and this leads to the *free functor* $\phi_T$ which allows for a right adjoint, the *forgetful functor*,

$$\begin{aligned} \phi_T : \mathbb{A} \to \mathbb{A}_T, &\quad A \mapsto (T(A), m_A), \\ U_T : \mathbb{A}_T \to \mathbb{A}, &\quad (A, \varrho_A) \mapsto A. \end{aligned}$$

The adjunction $\phi_T \dashv U_T$ is given by the bijection, for $A \in \mathbb{A}$, $B \in \mathbb{A}_T$,

$$\mathrm{Mor}_{\mathbb{A}_T}(\phi_T(A), B) \xrightarrow{\simeq} \mathrm{Mor}_{\mathbb{A}}(A, U_T(B)), \quad f \mapsto f\,\eta_A. \tag{3.1}$$

Putting $B = \phi_T(A')$ with $A' \in \mathbb{A}$ and applying $U_T\phi_T = T$, we obtain

$$\mathrm{Mor}_{\mathbb{A}_T}(\phi_T(A), \phi_T(A')) \xrightarrow{\simeq} \mathrm{Mor}_{\mathbb{A}}(A, T(A')). \tag{3.2}$$

Define a new category $\widetilde{\mathbb{A}}_T$ with the objects of $\mathbb{A}$ but choosing

$$\mathrm{Mor}_{\widetilde{\mathbb{A}}_T}(A, A') = \mathrm{Mor}_{\mathbb{A}}(A, T(A'))$$

and defining the composition of $f : A \to T(A')$ and $g : A' \to T(A'')$ by

$$g \diamond f : A \xrightarrow{f} T(A') \xrightarrow{T(g)} TT(A'') \xrightarrow{m_{A''}} T(A'').$$

This is known as the **Kleisli category** of the monad $T$ and the bijection (3.1) shows that $\widetilde{\mathbb{A}}_T$ is isomorphic to the full subcategory of the Eilenberg-Moore category $\mathbb{A}_T$ determined by the objects $T(A)$, $A \in \mathbb{A}$ (free $T$-modules).

Reversing the arrows in the definitions around monads yields

3.2. **Comonads and their comodules.** A *comonad* on a category $\mathbb{A}$ is an endofunctor $S : \mathbb{A} \to \mathbb{A}$ with natural transformations, *comultipliction (coproduct)* and *counit*,

$$\delta : S \to SS, \quad \varepsilon : S \to 1_{\mathbb{A}},$$

subject to coassociativity and counitality conditions (dual to monad case).

*S-comodules* are objects $A \in \mathbb{A}$ with a morphism $\omega : A \to S(A)$ subject to coassociativity and counitality conditions.

*Morphisms between comodules* $(A, \omega)$ and $(A', \omega')$ (or *S-morphisms*), are morphisms $g : A \to A'$ in $\mathbb{A}$ with commutative diagrams

$$\begin{array}{ccc} A & \xrightarrow{g} & A' \\ {\scriptstyle \omega}\downarrow & & \downarrow{\scriptstyle \omega'} \\ S(A) & \xrightarrow{S(g)} & S(A'). \end{array}$$

The category formed by the $S$-comodules and their morphisms is called the **Eilenberg-Moore category** - or just the *comodule category* - of the comonad $(S, \delta, \varepsilon)$ and we denote it by $\mathbb{M}^S$.

For any $A \in \mathbb{A}$, the structure map $\delta_A : S(A) \to SS(A)$ makes $S(A)$ an $S$-comodule. This yields the *free functor* $\phi^S$ which is right adjoint to the forgetful functor $U^S$,

$$\phi^S : \mathbb{A} \to \mathbb{A}^S, \ A \mapsto (S(A), \delta_A),$$
$$U^S : \mathbb{A}^S \to \mathbb{A}, \ (A, \omega) \mapsto A.$$

The adjunction $U^S \dashv \phi^S$ is given by the bijection, $B \in \mathbb{A}^S$, $A' \in \mathbb{A}$,

$$\mathrm{Mor}^S(B, S(A')) \xrightarrow{\simeq} \mathrm{Mor}_{\mathbb{A}}(U^S(B), A'), \quad h \mapsto \varepsilon_{A'}\, h. \tag{3.3}$$

For $B = S(A)$ with $A \in \mathbb{A}$, using $U^S \phi^S = S$, we obtain

$$\mathrm{Mor}^S(S(A), S(A')) \xrightarrow{\simeq} \mathrm{Mor}_{\mathbb{A}}(S(A), A'). \tag{3.4}$$

Define a new category $\widetilde{\mathbb{A}}^S$ with the objects of $\mathbb{A}$ but choosing

$$\mathrm{Mor}_{\widetilde{\mathbb{A}}^S}(A, A') := \mathrm{Mor}_{\mathbb{A}}(S(A), A')$$

and take as composition, of $h : S(A) \to A'$ and $k : S(A') \to A''$,

$$k \diamond h : S(A) \xrightarrow{\delta_A} SS(A) \xrightarrow{S(h)} S(A') \xrightarrow{k} A''.$$

This is known as the **Kleisli category** of the comonad $S$. The bijection (3.4) shows that $\widetilde{\mathbb{A}}^S$ is isomorphic to the full subcategory of the Eilenberg-Moore category $\mathbb{A}^S$ determined by the objects $S(A)$, $A \in \mathbb{A}$ ((co)free $S$-comodules).

3.3. **Rings and modules.** For a ring $R$, any $(R,R)$-bimodule $A$ defines an endofunctor $A \otimes_R - : {}_R\mathbb{M} \to {}_R\mathbb{M}$. $A$ is called an **$R$-ring** if this functor allows for a monad structure. If $R$ is commutative and $ra = ar$ for $a \in A$, $r \in R$, an $R$-ring is called an **$R$-algebra**.

An $(R,R)$-bimodule $C$ is called an **$R$-coring** provided the functor $C \otimes_R -$ allows for a comonad structure. If $R$ is commutative and $rc = cr$ for $c \in C$, $r \in R$, an $R$-coring is called an **$R$-coalgebra**.

Notice that not every monad or comonad on ${}_R\mathbb{M}$ can be represented by a tensor functor. For an extensive treatment of corings refer to [3].

## 4. Adjoints and (co)monads

The notion of adjoints and (co)monads are intimately related. Using naturality of the transformations involved it is straightforward to show:

4.1. **From adjoints to (co)monads.** Let $F \dashv G : \mathbb{B} \to \mathbb{A}$ be an adjoint pair of functors with unit $\eta : 1_{\mathbb{A}} \to GF$ and counit $\varepsilon : FG \to 1_{\mathbb{B}}$.

(i) $T := GF : \mathbb{A} \to \mathbb{A}$ is an endofunctor and the natural transformations, *multiplication* and *unit*,

$$m : TT = GFGF \xrightarrow{G\varepsilon F} GF = T, \quad \eta : 1_{\mathbb{A}} \to GF = T,$$

make $(T, m, \eta)$ a **monad** on $\mathbb{A}$.

(ii) $S := FG : \mathbb{B} \to \mathbb{B}$ is an endofunctor and the natural transformations *comultiplication* and *counit*,

$$\delta : S = FG \xrightarrow{F\eta G} FGFG = SS, \quad \varepsilon : S = FG \to 1_{\mathbb{B}},$$

make $(S, \delta, \varepsilon)$ a **comonad** on $\mathbb{B}$.

Recall that the construction of module and comodule categories in 3.1 and 3.2 show the inverse direction. In fact, these structures were introduced to show that monads as well as comonads can be written as a composition of adjoint functors ([6], [11]).

4.2. **From (co)monads to adjoints.** *Let $\mathbb{A}$ be any category.*

(i) *For a monad $(T, m, \eta)$ on $\mathbb{A}$, the category $\mathbb{A}_T$ allows for an adjoint pair of functors $\phi_T \dashv U_T : \mathbb{A}_T \to \mathbb{A}$ with $T = U_T \phi_T$.*

(ii) *For a comonad $(S, \delta, \varepsilon)$ on $\mathbb{A}$, the category $\mathbb{A}^S$ allows for an adjoint pair of functors $U^S \dashv \phi^S : \mathbb{A} \to \mathbb{A}^S$ with $S = U^S \phi^S$.*

*The same assertions hold replacing the Eilenberg-Moore categories by the corresponding Kleisli categories.*

The structure of the (co)monads related to an adjunction are strongly influenced by the properties of the unit and counit.

4.3. **Separability.** Let $F \dashv G : \mathbb{B} \to \mathbb{A}$ be an adjunction as in 4.2 and $T = GF$, $S = FG$ the associated monad and comonad.

(i) Assume $G$ to be a separable functor (see 2.4). Then the monad $(T, m, \eta)$ allows for a comultiplication

$$\delta' : T = GF \xrightarrow{G\varepsilon' F} GFGF = TT,$$

where $\varepsilon'$ is a splitting for $\varepsilon$ (see 2.7(4)), with $m \cdot \delta' = 1_T$ and commutative diagrams (Frobenius conditions)

$$\begin{array}{ccc} TT & \xrightarrow{\delta' T} & TTT \\ {\scriptstyle m}\downarrow & & \downarrow{\scriptstyle Tm} \\ T & \xrightarrow{\delta'} & TT, \end{array} \qquad \begin{array}{ccc} TT & \xrightarrow{T\delta'} & TTT \\ {\scriptstyle m}\downarrow & & \downarrow{\scriptstyle mT} \\ T & \xrightarrow{\delta'} & TT. \end{array} \tag{4.1}$$

With these properties, $(T, m, \eta, \delta')$ is (called) a *separable monad.*

(ii) Assume $F$ to be a separable functor (see 2.4). Then the comcnad $(S, \delta, \varepsilon)$ allows for a multiplication

$$m' : SS = FGFG \xrightarrow{F\eta' G} FG = S,$$

where $\eta'$ is a splitting for $\eta$ (see 2.7(5)), with $m' \cdot \delta = 1_S$ and commutative diagrams as in (4.1).

$(S, \delta, \varepsilon, m')$ is (called) a *coseparable comonad.*

4.4. **Rings and modules.** As in 2.8, we consider an $(R, S)$-bimodule $P$. The related adjunction yields

$$\begin{aligned} \text{monad} \quad & \mathrm{Hom}_R(P, P \otimes_S -) : {}_S\mathbb{M} \to {}_S\mathbb{M}, \\ \text{comonad} \quad & P \otimes_S \mathrm{Hom}_R(P, -) : {}_R\mathbb{M} \to {}_R\mathbb{M}. \end{aligned}$$

Now $P^* := \mathrm{Hom}_R(P, R)$ is an $(S, R)$-bimodule and hence induces

$$\begin{aligned} \text{monad} \quad & \mathrm{Hom}_S(P^*, P^* \otimes_R -) : {}_R\mathbb{M} \to {}_R\mathbb{M}, \\ \text{comonad} \quad & P^* \otimes_R \mathrm{Hom}_S(P^*, -) : {}_S\mathbb{M} \to {}_S\mathbb{M}. \end{aligned}$$

If ${}_RP$ is finitely generated and projective, $\mathrm{Hom}_R(P, -) \simeq P^* \otimes_R -$ and the comonad $P \otimes_S \mathrm{Hom}_R(P, -) \simeq P \otimes_S P^* \otimes_R -$, that is, $P \otimes_S P^*$ is an *R-coring* (and $P^* \otimes_R P \simeq S$).

If $\eta$ is an isomorphism, $P \otimes_S -$ is (left) exact on the image of $\mathrm{Hom}_R(P,-)$ and hence the comonad

$$P \otimes_S \mathrm{Hom}_R(P,-) : {}_R\mathbb{M} \to {}_R\mathbb{M}$$

is a left exact functor commuting with products in ${}_R\mathbb{M}$. Therefore, by the dual of Watts' theorem (e.g. [9]), it is determined by the image of a cogenerator $Q \in {}_R\mathbb{M}$, that is,

$$\begin{aligned} P \otimes_S \mathrm{Hom}_R(P,-) &\simeq \mathrm{Hom}_R(P \otimes_S \mathrm{Hom}_R(P,Q),-) \\ &\simeq \mathrm{Hom}_S(\mathrm{Hom}_R(P,Q), \mathrm{Hom}_R(P,-)). \end{aligned}$$

In case $R$ is a cogenerator in ${}_R\mathbb{M}$, we put $Q = R$ to obtain

$$\begin{aligned} P \otimes_S \mathrm{Hom}_R(P,-) &\simeq \mathrm{Hom}_R(P \otimes_S P^*,-) \\ &\simeq \mathrm{Hom}_S(P^*, \mathrm{Hom}_R(P,-)). \end{aligned}$$

This is a comonad, thus $P \otimes_S P^* \otimes_R -$ is a monad (by 5.1) which means that $P \otimes_S P^*$ is an $R$-ring (see 3.3).

If unit $\eta$ and counit $\varepsilon$ are isomorphisms, the comonad $P^* \otimes_R \mathrm{Hom}_S(P^*,-) \simeq P^* \otimes_R P \otimes_S -$, that is, $P^* \otimes_R P$ is an $S$-*coring*.

As a special case, we look at a situation studied in representation theory.

4.5. **Nakayama functors.** Let $A$ be a finite dimensional algebra over a field $K$ and $A$-mod the category of finitely generated left $A$-modules. It is customary to write $D(-) = \mathrm{Hom}_K(-,K)$, so $A^* = D(A)$, and one gets $\mathrm{Hom}_K(A,-) \simeq D(A) \otimes_K -$. The functor (e.g. [9])

$$\nu = D\,\mathrm{Hom}_A(-, {}_AA) : A\text{-mod} \to A\text{-mod}$$

is called the *Nakayama functor*, and

$$\nu^- = \mathrm{Hom}_A(D(-), A_A) : A\text{-mod} \to A\text{-mod}$$

is said to be the *inverse Nakayama functor*. By the Eilenberg-Watts theorems one gets the natural isomorphisms

$$\nu(-) \simeq D(A) \otimes_A -, \quad \nu^-(-) \simeq \mathrm{Hom}_A(D(A),-),$$

thus there is an adjoint pair of endofunctors to be investigated.

From 2.8 and 4.4 one can see which monad or comonad structures may show up in this setting. The ($A$-module) structure of $D(A)$ is of course influenced by the structure of $A$.

## 5. Adjoint monads and comonads

In 3.2, *comonads* $(S, \delta, \varepsilon)$ were defined by reversing arrows in the definition of *monads* $(T, m, \eta)$. This does not mean that the two notions are strictly dual in a categorical way: in our definition we do not have anything like the "dual" of a functor $T$ (or $S$). In this section we will outline that *adjointness* provides a bijective correspondence between monads and comonads.

**5.1. Monads versus comonads** *Let $F \dashv G : \mathbb{A} \to \mathbb{A}$ be an adjoint pair of endofunctors with unit $\overline{\eta}$, counit $\overline{\varepsilon}$ and bijection, for $X, Y \in \mathbb{A}$,*

$$\alpha_{X,Y} : \mathrm{Mor}_{\mathbb{A}}(F(X), Y) \to \mathrm{Mor}_{\mathbb{A}}(X, G(Y)).$$

(i) *A monad structure on $F$ induces a comonad structure on $G$, and vice versa, such that the associated (Eilenberg-Moore) categories $\mathbb{A}_F$ and $\mathbb{A}^G$ are isomorphic.*

(ii) *A comonad structure on $F$ induces a monad structure on $G$, and vice versa, such that the associated Kleisli categories $\widetilde{\mathbb{A}}_F$ and $\widetilde{\mathbb{A}}^G$ are equivalent.*

**Proof.** (Sketch) (i) Let $(F, m, \eta)$ be a monad. The adjunction $F \dashv G$ induces the diagram

$$\begin{array}{ccc}
\mathrm{Mor}_{\mathbb{A}}(F(X), Y) & \xrightarrow{\quad \alpha_{X,Y} \quad} & \mathrm{Mor}_{\mathbb{A}}(X, G(Y)) \\
{\scriptstyle \mathrm{Mor}(m_X, Y)} \downarrow & & \vdots\, {\scriptstyle \mathrm{Mor}(X, ?)} \\
\mathrm{Mor}_{\mathbb{A}}(FF(X), Y) & \xrightarrow{\alpha_{X,G(Y)} \cdot \alpha_{F(X),Y}} & \mathrm{Mor}_{\mathbb{A}}(X, GG(Y))
\end{array}$$

where the dotted arrow exists by applying the Yoneda Lemma to the composition of the other maps ($\alpha$ is invertible) and determines a natural transformation $\delta : G \to GG$. A similar argument shows the existence of a counit $\varepsilon : G \to 1_{\mathbb{A}}$. Explicitly we get

$$\delta : G \xrightarrow{\overline{\eta} G} GFG \xrightarrow{G\overline{\eta} FG} GGFFG \xrightarrow{GGmG} GGFG \xrightarrow{GG\overline{\varepsilon}} GG,$$

$$\varepsilon : G \xrightarrow{\eta G} FG \xrightarrow{\overline{\varepsilon}} 1_{\mathbb{A}} .$$

The symmetric construction shows that a comonad structure on $G$ leads to a monad structure on $F$. The equivalence of categories is given by

$$\mathbb{A}_F \to \mathbb{A}^G, \quad F(A) \xrightarrow{h} A \ \mapsto\ A \xrightarrow{\overline{\eta}_A} GF(A) \xrightarrow{G(h)} G(A),$$

$$\mathbb{A}^G \to \mathbb{A}_F, \quad A \xrightarrow{\varrho} G(A) \ \mapsto\ F(A) \xrightarrow{F(\varrho)} FG(A) \xrightarrow{\overline{\varepsilon}_A} A.$$

(ii) By arguments symmetric to those in (i), a comonad structure on $F$ induces a monad structure on $G$, and vice versa. The equivalence of Kleisli categories $\widetilde{\mathbb{A}}^F$

and $\widetilde{\mathbb{A}}_G$ follows from the isomorphisms

$$\begin{aligned}\mathrm{Mor}_{\mathbb{A}^F}(\phi^F(A),\phi^F(A')) &\overset{(1)}{\simeq} \mathrm{Mor}_{\mathbb{A}}(F(A),A')\\ &\overset{(2)}{\simeq} \mathrm{Mor}_{\mathbb{A}}(A,G(A'))\\ &\overset{(3)}{\simeq} \mathrm{Mor}_{\mathbb{A}_G}(\phi_G(A),\phi_G(A')),\end{aligned}$$

where (1) is the isomorphism from (3.4), (2) is the adjunction, and (3) is the isomorphism from (3.2). □

5.2. **Frobenius monads.** Let $F : \mathbb{A} \to \mathbb{A}$ be a self-adjoint endofunctor, that is, $F \dashv F$. By subsection 5.1, there are equivalent:

(a) *F has a monad structure* $(F, m, \eta)$*;*

(b) *F has a comonad structure* $(F, \delta, \varepsilon)$.

With this (mutually) induced structures, $(F, m, \delta)$ satisfies the Frobenius conditions (4.1) and the categories $\mathbb{A}_F$ and $\mathbb{A}^F$ are isomorphic. These data define a *Frobenius monad.* ([18], [25]).

5.3. **Rings and modules.** Let $A$ be an $(R, R)$-bimodule and consider the adjoint functor pair of endofunctors (as in 2.8),

$$A \otimes_R -,\ \mathrm{Hom}_R(A, -) : {}_R\mathbb{M} \to {}_R\mathbb{M}.$$

(i) An $R$-ring $(A, m, \eta)$, that is, a monad $A \otimes_R -$ on ${}_R\mathbb{M}$, induces a comonad structure on $\mathrm{Hom}_R(A, -)$ with comultiplication

$$\begin{aligned}\mathrm{Hom}_R(A,-) \xrightarrow{\mathrm{Hom}(m,-)} &\ \mathrm{Hom}_R(A\otimes_R A,-)\\ &\simeq \mathrm{Hom}_R(A,\mathrm{Hom}_R(A,-)),\end{aligned}$$

and the module category $\mathbb{M}_A$ is isomorphic to the comodule category $\mathbb{M}^{\mathrm{Hom}(A,-)}$.

(ii) A coring $(A, \delta, \varepsilon)$, that is, a comonad $A \otimes_R -$ on ${}_R\mathbb{M}$, induces a monad structure on $\mathrm{Hom}_R(A, -)$ with multiplication

$$\begin{aligned}&\mathrm{Hom}_R(A,\mathrm{Hom}_R(A,-)) \simeq\\ &\qquad \mathrm{Hom}_R(A\otimes_R A,-) \xrightarrow{\mathrm{Hom}(\delta,-)} \mathrm{Hom}_R(A,-),\end{aligned}$$

and the *Kleisli categories* are isomorphic by assigning, for $X \in {}_R\mathbb{M}$,

$$\widetilde{\mathbb{M}}^A \to \widetilde{\mathbb{M}}_{\mathrm{Hom}(A,-)},\quad A\otimes_R X \mapsto \mathrm{Hom}_R(A, X).$$

Note that the objects of $\mathbb{M}_{\mathrm{Hom}(A,-)}$ are also called *contramodules* (e.g. [2, Section 4], [31]).

The $R$-ring $(A, m, \eta)$ is called a **Frobenius $R$-ring** provided $A \otimes_R -$ is a Frobenius monad on ${}_R\mathbb{M}$. This implies that ${}_RA$ is finitely generated and projective and hence $\mathrm{Hom}_R(A, -) \simeq A^* \otimes_R -$, thus $A \simeq A^*$ as left $A$-modules and $\mathbb{M}_A \simeq \mathbb{M}^A$. For commutative rings $R$ one obtains **Frobenius algebras** (as considered by F. Frobenius in [7] over fields).

## 6. Composition of monads and comonads

Over a commutative ring $R$, the tensor product of two $R$-algebras $(A, m, e)$, $(B, m', e')$ can be made an algebra defining multiplication (writing $\otimes := \otimes_R$)

$$\overline{m}_{AB} : A \otimes B \otimes A \otimes B \xrightarrow{A \otimes \mathrm{tw} \otimes B} A \otimes A \otimes B \otimes B \xrightarrow{m \otimes m'} A \otimes B,$$

with the twist map $\mathrm{tw} : B \otimes A \to A \otimes B$, $b \otimes a \mapsto a \otimes b$.

This map is no longer available if $R$ is not commutative. On the other hand, the multiplication is formally defined replacing tw by any morphism $\tau : B \otimes A \to A \otimes B$, however, it is not clear which properties it has. So one may ask which conditions are to be satisfied by $\tau$ to make $(A \otimes B, \overline{m}_{AB}, e \otimes e')$ a unital algebra. These lead to commutative diagrams

$$\begin{array}{ccc} B \otimes B \otimes A & \xrightarrow{\quad m' \otimes A \quad} & B \otimes A \\ {\scriptstyle B \otimes \tau} \downarrow & & \downarrow {\scriptstyle \tau} \\ B \otimes A \otimes B \xrightarrow{\tau \otimes B} & A \otimes B \otimes B & \xrightarrow{A \otimes m'} A \otimes B, \end{array} \qquad \begin{array}{ccc} A & \xrightarrow{e' \otimes A} & B \otimes A \\ & {\scriptstyle A \otimes e'} \searrow & \downarrow {\scriptstyle \tau} \\ & & A \otimes B, \end{array}$$

$$\begin{array}{ccc} B \otimes A \otimes A & \xrightarrow{\quad A \otimes m \quad} & B \otimes A \\ {\scriptstyle \tau \otimes B} \downarrow & & \downarrow {\scriptstyle \tau} \\ A \otimes B \otimes A \xrightarrow{B \otimes \tau} & A \otimes A \otimes B & \xrightarrow{m \otimes A} A \otimes B, \end{array} \qquad \begin{array}{ccc} B & \xrightarrow{B \otimes e} & B \otimes A \\ & {\scriptstyle e \otimes B} \searrow & \downarrow {\scriptstyle \tau} \\ & & A \otimes B, \end{array}$$

and should induce commutative diagrams of functors, with the forgetful functor $U_B : {}_B\mathbb{M} \to {}_R\mathbb{M}$,

$$\begin{array}{ccc} {}_B\mathbb{M} & \xrightarrow{\quad A \otimes B \otimes_B - \quad} & {}_B\mathbb{M} \\ {\scriptstyle U_B} \downarrow & & \downarrow {\scriptstyle U_B} \\ {}_R\mathbb{M} & \xrightarrow{\quad A \otimes_R - \quad} & {}_R\mathbb{M}. \end{array}$$

Then $A \otimes - \simeq A \otimes B \otimes_B -$ is called a lifting of $A \otimes -$ from ${}_R\mathbb{M}$ to ${}_B\mathbb{M}$.

The questions considered above for the functors $A \otimes -$ and $B \otimes -$ can be asked for endofunctors in any category.

6.1. **Liftings of endofunctors.** Let $\mathbb{A}$ be any category with endofunctors $F, G : \mathbb{A} \to \mathbb{A}$. If $(F, m, e)$ is a monad or $(G, \delta, \varepsilon)$ is a comonad, we get the two diagrams, respectively,

$$\begin{array}{ccc} \mathbb{A}_F & \xrightarrow{\overline{G}} & \mathbb{A}_F \\ {\scriptstyle U_F}\downarrow & & \downarrow{\scriptstyle U_F} \\ \mathbb{A} & \xrightarrow{G} & \mathbb{A}, \end{array} \qquad \begin{array}{ccc} \mathbb{A}^G & \xrightarrow{\overline{F}} & \mathbb{A}^G \\ {\scriptstyle U^G}\downarrow & & \downarrow{\scriptstyle U^G} \\ \mathbb{A} & \xrightarrow{F} & \mathbb{A}. \end{array}$$

If $\overline{G}$ ($\overline{F}$) exists it is called a lifting of $G$ ($F$) from $\mathbb{A}$ to $\mathbb{A}_F$ ($\mathbb{A}^G$). The following questions come up:

(i) when does a lifting $\overline{G}$ or $\overline{F}$ exist?

(ii) if $F$ and $G$ are monads, when is $\overline{G}$ a monad?

(iii) if $F$ and $G$ are comonads, when is $\overline{F}$ a monad?

(iv) if $F$ is a monad and $G$ is a comonad, when is $\overline{G}$ a comonad, when is $\overline{F}$ a monad?

All these problems can successfully be handled applying *distributive laws* as introduced and investigated in the 1970's by J. Beck [1] and others (see [30] for an overview and [2], [14], [15], et al. for more details).

Question (ii) above will lead to conditions which make the composition $FG$ a monad on $\mathbb{A}$ (as considered for algebras above) and (iii) describes the corresponding properties of comonads. Of particular interest are the questions in (iv) since they reveal an interesting interplay between monads and comonads.

6.2. **Mixed distributive laws.** Let $(F, m, \eta)$ be a monad and $(G, \delta, \varepsilon)$ a comonad on a category $\mathbb{A}$. A natural transformations $\lambda : FG \to GF$ is called a *mixed distributive law* or *(mixed) entwining* if it induces commutativity of the diagrams

$$\begin{array}{ccccc} FFG & & \xrightarrow{m_G} & & FG \\ {\scriptstyle F\lambda}\downarrow & & & & \downarrow{\scriptstyle \lambda} \\ FGF & \xrightarrow{\lambda_F} & GFF & \xrightarrow{Gm} & GF, \end{array} \qquad \begin{array}{ccccc} FG & \xrightarrow{F\delta} & FGG & \xrightarrow{\lambda_G} & GFG \\ {\scriptstyle \lambda}\downarrow & & & & \downarrow{\scriptstyle G\lambda} \\ GF & & \xrightarrow{\delta_F} & & GGF, \end{array}$$

$$\begin{array}{ccc} G & \xrightarrow{\eta_G} & FG \\ & {\scriptstyle G\eta}\searrow & \downarrow{\scriptstyle \lambda} \\ & & GF, \end{array} \qquad \begin{array}{ccc} FG & \xrightarrow{F\varepsilon} & F \\ {\scriptstyle \lambda}\downarrow & \nearrow{\scriptstyle \varepsilon_F} & \\ GF. & & \end{array}$$

The following are equivalent (with notation from 6.1):

(a) *$G$ can be lifted from $\mathbb{A}$ to $\overline{G} : \mathbb{A}_F \to \mathbb{A}_F$;*

(b) *$F$ can be lifted from $\mathbb{A}$ to $\overline{F} : \mathbb{A}^G \to \mathbb{A}^G$;*

(c) *there exists a mixed distributive law $\lambda : FG \to GF$.*

Given a monad and a comonad on $\mathbb{A}$, objects can have a module and a comodule structure. An entwining allows to require a compatibility of these structures.

6.3. **Mixed modules.** Let $(F, G, \lambda)$ be a mixed entwining and assume $A \in \mathbb{A}$ to be an $F$-module $\varrho : F(A) \to A$ and a $G$-comodule $\omega : A \to G(A)$. Then $(A, \varrho, \omega)$ is called a *mixed $(F, G)$-module* if we get commutativity of the diagram

$$\begin{array}{ccccc} F(A) & \xrightarrow{\varrho} & A & \xrightarrow{\omega} & G(A) \\ {\scriptstyle F(\omega)}\downarrow & & & & \uparrow{\scriptstyle G(\varrho)} \\ FG(A) & & \xrightarrow{\lambda_A} & & GF(A). \end{array}$$

These objects with morphisms that are module as well as comodule morphisms, form a category which we denote by $\mathbb{A}^G_F$.

In 6.1, the lifting to Eilenberg-Moore categories was considered. A corresponding construction for Kleisli categories is the following.

6.4. **Extending of endofunctors.** Let $\mathbb{A}$ be any category with endofunctors $F, G : \mathbb{A} \to \mathbb{A}$. If $(F, m, e)$ is a monad or $(G, \delta, \varepsilon)$ is a comonad, we get the two diagrams (with $\phi$ the free functors), respectively,

$$\begin{array}{ccc} \mathbb{A} & \xrightarrow{G} & \mathbb{A} \\ {\scriptstyle \phi_F}\downarrow & & \downarrow{\scriptstyle \phi_F} \\ \widetilde{\mathbb{A}}_F & \xrightarrow{\widetilde{G}} & \widetilde{\mathbb{A}}_F, \end{array} \qquad \begin{array}{ccc} \mathbb{A} & \xrightarrow{F} & \mathbb{A} \\ {\scriptstyle \phi^G}\downarrow & & \downarrow{\scriptstyle \phi^G} \\ \widetilde{\mathbb{A}}^G & \xrightarrow{\widetilde{F}} & \widetilde{\mathbb{A}}^G. \end{array}$$

$\widetilde{G}$ and $\widetilde{F}$ are called the *extensions* of $F$ and $G$, respectively. Here again distributive laws apply for further investigation but with the role of monad and comonad interchanged, that is, one needs natural transformations $\sigma : GF \to FG$ inducing commutativity of the corresponding diagrams. Notice that the role of mixed modules (as in 6.3) does not transfer to this situation.

## 7. Bimonad and Hopf monads

Of special interest are endofunctors $B$ on $\mathbb{A}$ which carry a monad structure $(B, m, \eta)$ and a comonad structure $(B, \delta, \varepsilon)$ at the same time. To make these data a **bimonad** we first require the existence of a mixed distributive law $\lambda : BB \to BB$.

The relevant commutative diagrams are either for the monad structure or else for the comonad structure - they do not relate multiplication with comultiplication, for example. To connect these we need further compatibility conditions, namely commutativity of the diagrams

$$\begin{array}{ccccc} BB & \xrightarrow{m} & B & \xrightarrow{\delta} & BB \\ {\scriptstyle B\delta}\downarrow & & & & \uparrow{\scriptstyle Bm} \\ BBB & & \xrightarrow{\quad\lambda_B\quad} & & BBB, \end{array} \tag{7.1}$$

$$\begin{array}{ccc} BB & \xrightarrow{B\varepsilon} & B \\ {\scriptstyle m}\downarrow & & \downarrow{\scriptstyle \varepsilon} \\ B & \xrightarrow{\varepsilon} & 1_{\mathbb{A}}, \end{array} \quad \begin{array}{ccc} 1_{\mathbb{A}} & \xrightarrow{\eta} & B \\ {\scriptstyle \eta}\downarrow & & \downarrow{\scriptstyle B\eta} \\ B & \xrightarrow{\delta} & BB, \end{array} \quad \begin{array}{ccc} 1_{\mathbb{A}} & \xrightarrow{\eta} & B \\ & \searrow^{=} & \downarrow{\scriptstyle \varepsilon} \\ & & 1_{\mathbb{A}}, \end{array} \tag{7.2}$$

where the bottom diagrams mean that $\eta$ is comonad morphism and $\varepsilon$ a monad morphism. Diagram (7.1) guarantees that for each $A \in \mathbb{A}$, $B(A)$ is a mixed $(B, B)$-module in the sense of 6.3, that is, $B(A) \in \mathbb{A}_B^B$. This determines a functor

$$\phi_B^B : \mathbb{A} \to \mathbb{A}_B^B, \quad A \mapsto BB(A) \xrightarrow{m_A} B(A) \xrightarrow{\delta_A} BB(A)$$

which is full and faithful by the isomorphisms (see 3.4, 3.1)

$$\mathrm{Mor}_B^B(B(A), B(A')) \simeq \mathrm{Mor}_B(B(A), A') \simeq \mathrm{Mor}_{\mathbb{A}}(A, A').$$

A natural transformation $S : B \to B$ is called an **antipode** if it induces commutativity of the diagram

$$\begin{array}{ccccc} B & \xrightarrow{\varepsilon} & 1_{\mathbb{A}} & \xrightarrow{\eta} & B \\ {\scriptstyle \delta}\downarrow & & & & \uparrow{\scriptstyle m} \\ BB & & \overset{S_B}{\underset{BS}{\rightrightarrows}} & & BB \end{array} . \tag{7.3}$$

A bimonad which allows for an antipode is called a **Hopf monad**. Such an antipode exists if and only if either of the composites

$$BB \xrightarrow{B\delta} BBB \xrightarrow{mB} BB, \quad BB \xrightarrow{\delta B} BBB \xrightarrow{Bm} BB$$

is an isomorphism ([16, 5.5]).

Under certain conditions on the category $\mathbb{A}$ and the functor $B$, the existence of an *antipode* $S$ is equivalent to $\phi_B^B : \mathbb{A} \to \mathbb{A}_B^B$ defining an equivalence ([16, 5.6, 6.11]).

7.1. **Adjoints of bimonads.** Let $(B, m, \delta, \lambda)$ be a bimonad on $\mathbb{A}$ (as above). Assume $B$ allows for a right adjoint functor $C : \mathbb{A} \to \mathbb{A}$. Then by 5.1, $C$ can be endowed with a comonad as well as a monad structure. Furthermore, there is a mixed distributive law $\lambda' : CC \to CC$ (derived from $\lambda$, see [16, 7.4]) which makes $C$ a bimonad.

The bimonad $B$ has an antipode if and only if the associated bimonad $C$ has an antipode. Thus, given an adjoint pair $B \dashv C$ of functors on $\mathbb{A}$, $B$ allows for a Hopf monad structure if and only if so does $C$.

7.2. **Bimonad on** $\mathrm{S{\scriptsize ET}}$. ([30, 5.19], [16, 7.9]) For any set $G$, the cartesian product defines an endofunctor

$$G \times - : \mathrm{Set} \to \mathrm{Set}, \quad A \mapsto G \times A,$$

which is a *comonad* with comultiplication $\delta : G \to G \times G$, $g \to (g, g)$, and is a *monad* provided $G$ is a monoid. Then it is a *bimonad* with entwining

$$\lambda : G \times G \to G \times G, \quad (g, h) \mapsto (gh, g).$$

Now $G \times -$ has an antipode, i.e., is a Hopf monad, if and only if the monoid $G$ is in fact a group. By 7.1, this is also equivalent to $\mathrm{Map}(G, -) : \mathrm{Set} \to \mathrm{Set}$, a right adjoint of $G \times -$, being a Hopf monad.

7.3. **Bialgebras.** Let $R$ be a commutative ring. An $R$-module $B$ with an algebra structure $(B, m, \eta)$ and a coalgebra structure $(B, \delta, \varepsilon)$ is called a *bialgebra* if $\delta$ and $\varepsilon$ are algebra morphisms, or, equivalently, $m$ and $\eta$ are coalgebra morphisms. These conditions require commutativity of the outer path in the diagram

$$\begin{array}{ccccc}
B \otimes B & \xrightarrow{\;m\;} & B & \xrightarrow{\;\delta\;} & B \otimes B \\
{\scriptstyle B\otimes\delta}\downarrow & & & & \uparrow{\scriptstyle B\otimes m} \\
B \otimes B \otimes B & & \overset{\omega\otimes B}{- - - - - - \rightarrow} & & B \otimes B \otimes B \\
{\scriptstyle \delta\otimes B\otimes B}\downarrow & & & & \uparrow{\scriptstyle m\otimes B\otimes B} \\
B \otimes B \otimes B \otimes B & & \xrightarrow{\;B\otimes \mathrm{tw}\otimes B\;} & & B \otimes B \otimes B \otimes B.
\end{array}$$

Defining an $R$-linear map

$$\omega : B \otimes B \xrightarrow{\delta\otimes B} B \otimes B \otimes B \xrightarrow{B\otimes \mathrm{tw}} B \otimes B \xrightarrow{m\otimes B} B \otimes B,$$

the condition reduces to commutativity of the upper rectangle. As readily checked, $\omega : B \otimes B \to B \otimes B$ is a mixed distributive law between the monad $B \otimes -$ and the comonad $B \otimes -$. Commutativity of the upper rectangle is just the compatibility condition which makes $B$ a bialgebra. Thus a bialgebra $B$ corresponds to a monad $B \otimes -$ with the associated monad and comonad structures and the specific mixed distributive law $\omega$. The mixed $(B, B)$-modules $\mathbb{M}_B^B$ are called **Hopf modules**.

The composite

$$\overline{\omega} : B \otimes B \xrightarrow{B\otimes\delta} B \otimes B \otimes B \xrightarrow{\mathrm{tw}\otimes B} B \otimes B \xrightarrow{B\otimes m} B \otimes B$$

yields a similar rectangle (sides interchanged). It gives a mixed distributive law between the related monad $-\otimes B$ and the comonad $-\otimes B$. It may also be considered as a mixed distributive law between the comonad $B\otimes -$ and the monad $B\otimes -$ (see 6.4).

An **antipode** for the bialgebra $B$ is an $R$-linear map $S : B \to B$ leading to commutativity of the diagram (7.3) for the monad $B \otimes -$.

Bialgebras $B$ with an antipode are called **Hopf algebras** and they are characterized by the fact that $\phi_B^B : \mathbb{M}_R \to \mathbb{M}_B^B$ is an equivalence, that is, every Hopf module is of the form $B \otimes X$, for some $X \in \mathbb{M}_R$ (**Fundamental Theorem**).

In case $B$ is a finite dimensional algebra over a field $K$, $B \otimes_K -$ is left adjoint to the endofunctor $\mathrm{Hom}_K(B, -) \simeq B^* \otimes_K -$, where $B^* = \mathrm{Hom}_K(B, K)$. Hence $B$ is a Hopf algebra if and only if $B^*$ is a Hopf algebra (see 7.1). This is known as the *duality principle* of Hopf algebras.

Hopf algebras were brought to light by Heinz Hopf in his seminal paper [8] in topology (1941). The algebraic essentials of this notion were extracted by Milnor-Moore in [20] (1965). As a result, interest was also directed to the more elementary notions of coalgebras and corings as building blocks for the theory. It took several years until their value for representation theory was unveiled (1980) by A.V. Roiter in [22], there a *coring* is called *bocs*, and the interest is focused on Kleisli categories. In the meantime more attention is paid to coalgebraic aspects of finite dimensional algebras. For this we refer to the recent paper [13] by R. Marczinzik and the references given there.

7.4. **Remark.** As pointed out in subsection 7.3, the mixed distributive law for a bialgebra was derived from the canonical twist of the tensor product of modules over a commutative ring. This setting was extended to *monoidal braided categories*, where a general version of such a twist map is required. Concentrating on the properties needed, for endofunctors $B$ on any category it is sufficient to define a *local braiding* $\tau : BB \to BB$ to derive the corresponding theory. For details we refer to [16, Section 6] and [17].

7.5. **Rings and modules.** The preceding sections show that *comultiplication* plays an important role in the structure theory of algebras. Summarising we consider an algebra $(A, m, \eta)$ over a commutative ring $R$ and assume $A$ to allow for a comultiplication $\delta : A \to A \otimes A$. Then $A$ becomes

(1) a *separable algebra* if $(A, m, \delta)$ satisfies the Frobenius condition and $m\cdot\delta = 1_A$;

(2) an *Azumaya algebra* if it is separable and $R \simeq C(A)$ (center of $A$); the category of $C(A)$-modules is equivalent to the category ${}_A\mathbb{M}_A$ of $(A, A)$-bimodules;

(3) a *Frobenius algebra* if $(A, m, \delta)$ satisfies the Frobenius condition and $(A, \delta)$ has a counit $\varepsilon : A \to R$; every $A$-module has an $A$-comodule structure (Frobenius (bi)module, see [33]);

(4) a *Hopf algebra* if $(A, m, \delta)$ induce commutativity of (7.1 and (7.2)) and allows for an antipode; the category of $R$-modules is equivalent to the category $\mathbb{M}_A^A$ of mixed $(A, A)$-bimodules (e.g. [3]).

**Acknowledgement.** The author is grateful to Bachuki Mesablishvili for proof-reading and helpful comments.

## References

[1] J. Beck, *Distributive laws*, in Seminar on Triples and Categorical Homology Theory, B. Eckmann (ed.), Springer, Berlin, 80 (1969), 119-140.

[2] G. Böhm, T. Brzeziński and R. Wisbauer, *Monads and comonads on module categories*, J. Algebra, 322(5) (2009), 1719-1747.

[3] T. Brzeziński and R. Wisbauer, Corings and Comodules, London Mathematical Society Lecture Note Series, 309, Cambridge University Press, Cambridge, 2003.

[4] J. Clark and R. Wisbauer, *Idempotent monads and $\star$-functors*, J. Pure Appl. Algebra, 215(2) (2011), 145-153.

[5] S. Eilenberg and S. Mac Lane, *General theory of natural equivalences*, Trans. Amer. Math. Soc., 58(2) (1945), 231-294.

[6] S. Eilenberg and J. C. Moore, *Adjoint functors and triples*, Illinois. J. Math., 9 (1965), 381-398.

[7] F. Frobenius, *Theorie der hypercomplexen Größen*, Sitz. Kön. Preuss. Akad. Wiss., (1903), 504-537; Gesammelte Abhandlungen, art. 70, 284-317.

[8] H. Hopf, *Über die Topologie der Gruppen-Mannigfaltigkeiten und ihre Verallgemeinerungen*, Ann. of Math., 42(2) (1941), 22-52.

[9] S. O. Ivanov, *Nakayama functors and Eilenberg-Watts theorems*, J. Math. Sci., 183(5) (2012), 675-680.

[10] D. M. Kan, *Adjoint functors*, Trans. Amer. Math. Soc., 87 (1958), 294-329.

[11] H. Kleisli, *Every standard construction is induced by a pair of adjoint functors*, Proc. Amer. Math. Soc., 16 (1965), 544-546.

[12] S. Mac Lane, Categories for the Working Mathematician, 2nd edn, Graduate Texts in Mathematics, 5, Springer-Verlag, New York, 1998.

[13] R. Marczinzik, *A bocs theoretic characterization of gendo-symmetric algebras*, J. Algebra, 470 (2017), 160-171.

[14] B. Mesablishvili, *Entwining structures in monoidal categories*, J. Algebra, 319(6) (2008), 2496-2517.

[15] B. Mesablishvili and R. Wisbauer, *Galois functors and entwining structures*, J. Algebra, 324(3) (2010), 464-506.

[16] B. Mesablishvili and R. Wisbauer, *Bimonads and Hopf monads on categories*, J. K-Theory, 7(2) (2011), 349-388.

[17] B. Mesablishvili and R. Wisbauer, *Notes on bimonads and Hopf monads*, Theory Appl. Categ., 26(10) (2012), 281-303.

[18] B. Mesablishvili and R. Wisbauer, *QF functors and (co)monads*, J. Algebra, 376 (2013), 101-122.

[19] B. Mesablishvili and R. Wisbauer, *The fundamental theorem for weak braided bimonads*, J. Algebra, 490 (2017), 55-103.

[20] J. W. Milnor and J. C. Moore, *On the structure of Hopf algebras*, Ann. of Math., 81(2) (1965), 211-264.

[21] K. Morita, *Duality for modules and its applications to the theory of rings with minimum condition*, Sci. Rep. Tokyo Kyoiku Daigaku Sect. A, 6 (1958), 83-142.

[22] A. V. Roiter, *Matrix problems and representations of BOCSs*, Representation Theory I, Lecture Notes in Math., 831, Springer, Berlin-New York, (1980), 288-324.

[23] M. Sato, *Fuller's theorem on equivalences*, J. Algebra, 52(1) (1978), 274-284.

[24] M. Sato, *On equivalences between module subcategories*, J. Algebra, 59(2) (1979), 412-420.

[25] R. Street, *Frobenius monads and pseudomonoids*, J. Math. Phys., 45(10) (2004), 3930-3948.

[26] D. Turi and G. Plotkin, *Towards a mathematical operational semantics*, Proceedings 12th Ann. IEEE Symp. on Logic in Computer Science, LICS'97, Warsaw, Poland, (1997).

[27] R. Wisbauer, Foundations of Module and Ring Theory, Algebra, Logic and Applications, 3, Gordon and Breach Science Publishers, Philadelphia, PA, 1991.

[28] R. Wisbauer, *Tilting in module categories*, Abelian groups, module theory, and topology (Padua, 1997), Lecture Notes in Pure and Appl. Math., 201, Dekker, New York, (1998), 421-444.

[29] R. Wisbauer, *Static modules and equivalences*, Interactions between ring theory and representations of algebras (Murcia), Lecture Notes in Pure and Appl. Math., 210, Dekker, New York, (2000), 423-449.

[30] R. Wisbauer, *Algebra versus coalgebras*, Appl. Categ. Structures, 16 (2008), 255-295.

[31] R. Wisbauer, *Comodules and contramodules*, Glasg. Math. J., 52(A) (2010), 151-162.

[32] R. Wisbauer, *Regular pairings of functors and weak (co)monads*, Algebra Discrete Math., 15(1) (2013), 127-154.

[33] R. Wisbauer, *Weak Frobenius monads and Frobenius bimodules*, Algebra Discrete Math., 21(2) (2016), 287-308.

[34] R. Wisbauer, *Separability in algebra and category theory*, Proc. Aligarh, (2016).

[35] J. Worthington, *A bialgebraic approach to automata and formal language theory*, Ann. Pure Appl. Logic, 163(7) (2012), 745-762.

**Robert Wisbauer**
Department of Mathematics
University of Düsseldorf
Germany
e-mail: wisbauer@math.uni-duesseldorf.de

# 5

# ON THE CLASSICAL PRIME SPECTRUM OF LATTICE MODULES

Pradip Girase, Vandeo Borkar and Narayan Phadatare

Communicated by Abdullah Harmancı

ABSTRACT. Let $M$ be a lattice module over a $C$-lattice $L$. A proper element $P$ of $M$ is said to be classical prime if for $a, b \in L$ and $X \in M, abX \leq P$ implies that $aX \leq P$ or $bX \leq P$. The set of all classical prime elements of $M$, $Spec^{cp}(M)$ is called as classical prime spectrum. In this article, we introduce and study a topology on $Spec^{cp}(M)$, called as Zariski-like topology of $M$. We investigate this topological space from the point of view of spectral spaces. We show that if $M$ has ascending chain condition on classical prime radical elements, then $Spec^{cp}(M)$ with the Zariski-like topology is a spectral space.

**Keywords**: Classical prime element, classical prime spectrum, classical prime radical element, Zariski-like topology, spectral space

## 1. Introduction

Zariski-like topology on the classical prime spectrum of a module is being introduce and studied by M. Behboodi and M. J. Noori in [7]. There are many generalizations of the Zariski topology over the set of all prime submodules of a $R$-module $M$ (see [5], [6], [14]). As a generalization of most of the results in [7], we introduce and study the Zariski-like topology on the classical prime spectrum of a lattice module $M$ over a $C$-lattice $L$.

A lattice $L$ is said to be *complete*, if for any subset $S$ of $L$, we have $\vee S, \wedge S \in L$. A complete lattice $L$ with least element $0_L$ and greatest element $1_L$ is said to be a *multiplicative lattice*, if there is defined a binary operation "." called multiplication on $L$ satisfying the following conditions:

(1) $a.b = b.a$, for all $a, b \in L$;

(2) $a.(b.c) = (a.b).c$, for all $a, b, c \in L$;

(3) $a.(\vee_\alpha b_\alpha) = \vee_\alpha (a.b_\alpha)$, for all $a, b_\alpha \in L$;

(4) $a.1_L = a$, for all $a \in L$.

Henceforth, $a.b$ will be simply denoted by $ab$.

An element $a$ in $L$ is called compact if $a \leq \vee_{\alpha \in I} b_\alpha$ ($I$ is an indexed set) implies $a \leq b_{\alpha_1} \vee b_{\alpha_2} \vee \cdots \vee b_{\alpha_n}$ for some subset $\{\alpha_1, \alpha_2, \cdots, \alpha_n\}$ of $I$. By a *C-lattice*, we mean a multiplicative lattice $L$, with least element $0_L$ and greatest element $1_L$ which is compact as well as multiplicative identity, that is generated under joins by a multiplicatively closed subset $C$ of compact elements of $L$. An element $a \in L$ is said to be *proper*, if $a < 1_L$.

A proper element $p$ of a multiplicative lattice $L$ is said to be *prime* if $ab \leq p$ implies $a \leq p$ or $b \leq p$ for $a, b \in L$. The collection of all prime elements of $L$ is denoted by $Spec(L)$.

The Zariski topology on the set $Spec(L)$ of all prime elements in multiplicative lattices is being studied in [21], by Thakare, Manjarekar and Maeda and in [20], by Thakare and Manjarekar as a generalization of the Zariski topology of a commutative ring with unity.

A proper element $m$ of a multiplicative lattice $L$ is said to be *maximal* if for every $x \in L$ with $m < x \leq 1_L$ implies $x = 1_L$.

A complete lattice $M$ with smallest element $0_M$ and greatest element $1_M$ is said to be a *lattice module* over the multiplicative lattice $L$ or $L$-module if there is a multiplication between elements of $M$ and $L$, denoted by $aN \in M$, for $a \in L$ and $N \in M$, which satisfies the following properties:

(1) $(ab)N = a(bN)$;
(2) $(\vee_\alpha a_\alpha)(\vee_\beta N_\beta) = (\vee_{\alpha\beta} a_\alpha N_\beta)$;
(3) $1_L N = N$;
(4) $0_L N = 0_M$; for all $a, b, a_\alpha \in L$, and for all $N, N_\beta \in M$.

Let $M$ be a lattice module over a $C$-lattice $L$. For $N \in M, b \in L$, denote $(N : b) = \vee\{K \in M | bK \leq N\}$. If $a, b \in L$, we write $(a : b) = \vee\{x \in L | bx \leq a\}$. If $A, B \in M$, then $(A : B) = \vee\{x \in L | Bx \leq A\}$. An element $N \in M$ is said to be compact if $N \leq \vee_{\alpha \in I} A_\alpha$ ($I$ is an indexed set) implies $N \leq A_{\alpha_1} \vee A_{\alpha_2} \vee \cdots \vee A_{\alpha_n}$ for some subset $\{\alpha_1, \alpha_2, \cdots, \alpha_n\}$ of $I$.

An element $N \in M$ is said to be *meet principal* (respectively *join principal*) if it satisfies the identity $A \wedge aN = (a \wedge (A : N))N$ (respectively $((aN \vee A) : N) = (a \vee (A : N)))$ for all $a \in L$ and for all $A \in M$. Also $N$ is said to be *principal* if it is both join as well as meet principal. If each element of $M$ is the join of principal (compact) elements of $M$, then $M$ is called the *principally generated (compactly generated)* lattice module.

An element $N \in M$ is said to be proper, if $N < 1_M$. A proper element $N$ of a lattice module $M$ is said to be *prime* if $aX \leq N$ implies $X \leq N$ or $a1_M \leq N$, i.e., $a \leq (N : 1_M)$ for every $a \in L$ and $X \in M$. The prime spectrum of a lattice module

$M$ is the set of all prime elements of $M$ and it is denoted by $Spec(M)$. In [4], S. Ballal and V. Kharat studied the Zariski topology over $Spec(M)$ as a generalization of the results carried out in [20], [21]. Also in [10], F. Callialp et al. studied the Zariski topology over $Spec(M)$ over multiplicative lattice $L$.

A non-zero element $N \in M$ is said to be *second*, if for $a \in L$, either $aN = N$ or $aN = 0_M$. The Zariski topology on the second spectrum of a lattice module is studied by N. Phadatare et al. in [17]. In [18], N. Phadatare and V. Kharat introduced and studied the concept of second radical elements of a lattice module $M$ over a $C$-lattice $L$ as a generalization of second socle of a submodule of $R$-module $M$. An element $N < 1_M$ of $M$ is said to be *maximal* if $N \leq B$ implies either $N = B$ or $B = 1_M, B \in M$. A non-zero element $K \neq 1_M$ of $M$ is said to be *minimal* if $0_M \leq N < K$ implies $N = 0_M, N \in M$.

Further all these concepts and for more information on multiplicative lattices, lattice modules and topology, the reader may refer ([1], [2], [13], [16], [19]).

## 2. Zariski-like topology on $Spec^{cp}(M)$

Let $M$ be a lattice module over a $C$-lattice $L$. A proper element $P \in M$ is said to be classical prime if for $a, b \in L$ and $X \in M, abX \leq P$ implies that $aX \leq P$ or $bX \leq P$ ([15]). The classical prime spectrum $Spec^{cp}(M)$ is defined to be the set of all classical prime elements of $M$, i.e., $Spec^{cp}(M) = \{P \in M \mid P \text{ is a classical prime element of } M\}$. Let $N$ be any element of $M$. Let $F^c(N)$ be the set of all classical prime elements of $M$ which contains N, i.e., $F^c(N) = \{P \in Spec^{cp}(M) | N \leq P\}$. Note that, $F^c(0_M) = Spec^{cp}(M)$ and $F^c(1_M)$ is an empty set.

**Proposition 2.1.** *Let $M$ be a lattice module over a $C$-lattice $L$ and $N, N_i, K \in M$ $(i \in I)$. Then the following statements holds.*

(1) $\cap_{i \in I} F^c(N_i) = F^c(\vee_{i \in I} N_i)$ *for any index set $I$.*
(2) $F^c(N) \cup F^c(K) \subseteq F^c(N \wedge K)$.

**Proof.** (1) Since for each $i \in I$, $N_i \leq \vee N_i$, therefore for $P \in F^c(\vee_{i \in I} N_i)$, we have $\vee_{i \in I} N_i \leq P$ and hence $N_i \leq P$ and $P \in F^c(N_i)$. Therefore $F^c(\vee_{i \in I} N_i) \subseteq \cap_{i \in I} F^c(N_i)$.

Conversely, suppose that $P \in \cap_{i \in I} F^c(N_i)$, then $P \in F^c(N_i)$ for each $i \in I$, therefore $N_i \leq P$ for each $i \in I$. Hence $\vee_{i \in I} N_i \leq P$ and so $P \in F^c(\vee_{i \in I} N_i)$. Consequently, $\cap_{i \in I} F^c(N_i) \subseteq F^c(\vee_{i \in I} N_i)$. Thus $\cap_{i \in I} F^c(N_i) = F^c(\vee_{i \in I} N_i)$.
(2) Since $N \wedge K \leq N, K$, we have $F^c(N), F^c(K) \subseteq F^c(N \wedge K)$ and therefore $F^c(N) \cup F^c(K) \subseteq F^c(N \wedge K)$. □

Let $\xi^c(M) = \{F^c(N) | N \in M\}$, then $\xi^c(M)$ contains empty set and $Spec^{cp}(M)$. By Proposition 2.1, $\xi^c(M)$ is closed under arbitrary intersections. In general $\xi^c(M)$ is not closed under finite union. A lattice module $M$ over a $C$-lattice $L$ is called top lattice module if $\xi^c(M)$ is closed under finite union. In this case, $\xi^c(M)$ induces a topology $\tau^c$ on $Spec^{cp}(M)$, we call it the Zariski topology.

Let $M$ be a lattice module over a $C$-lattice $L$. For each element $N$ of $M$, let $G^c(N) = Spec^{cp}(M) - F^c(N)$ and $\mathbb{G}^c(N) = \{G^c(N) | N \in M\}$. Let $\psi^c(M)$ be the collection of all unions of finite intersections of elements of $\mathbb{G}^c(N)$, then $\psi^c(M)$ is a topology on $Spec^{cp}(M)$ by the sub-basis $\mathbb{G}^c(N)$. We call the topology $\psi^c(M)$, a Zariski-like topology.

Note that, the set $\{G^c(N_1) \cap G^c(N_2) \cap \cdots \cap G^c(N_k) | N_i \in M, 1 \leq i \leq k, k \in \mathbb{N}\}$ is a basis for the Zariski-like topology of $M$.

Let $M$ be a lattice module over a $C$-lattice $L$ and let $Spec^{cp}(M)$ be equipped with the Zariski-like topology. Let $Y \subseteq Spec^{cp}(M)$. The closure of $Y$ in $Spec^{cp}(M)$ is denoted by $Cl(Y)$ and meet of all elements of $Y$ by $Z(Y)$. Note that, if $Y = \emptyset$, then $Z(Y) = 1_M$.

**Lemma 2.2.** *Let $M$ be a lattice module over a $C$-lattice $L$ and let $Y$ be a non-empty subset of $Spec^{cp}(M)$. Then $Cl(Y) = \cup_{P \in Y} F^c(P)$.*

**Proof.** Suppose that $Y$ is a non-empty subset of $Spec^{cp}(M)$. Clearly $Y \subseteq \cup_{P \in Y} F^c(P)$. Suppose $D$ is any closed subset of $Spec^{cp}(M)$ such that $Y \subseteq D$. Thus $D = \cap_{k \in J}(\cup_{l=1}^{n_k} F^c(N_{kl}))$, for some $N_{kl} \in M$, $k \in J$ and $n_k \in \mathbb{N}$. Let $P_1 \in \cup_{P \in Y} F^c(P)$, then there exists $P_0 \in Y$ such that $P_1 \in F^c(P_0)$ and so $P_0 \leq P_1$. Since $P_0 \in Y \subseteq D$, therefore for each $k \in J$ there exists $l \in \{1, 2, \cdots, n_k\}$ such that $N_{kl} \leq P_0$ and hence $N_{kl} \leq P_0 \leq P_1$. Therefore $P_1 \in F^c(N_{kl})$ for each $k \in J$ and hence $P_1 \in \cap_{k \in J}(\cup_{l=1}^{n_k} F^c(N_{kl})) = D$. It follows that $\cup_{P \in Y} F^c(P) \subseteq D$. Thus $\cup_{P \in Y} F^c(P)$ is the smallest closed set in $Spec^{cp}(M)$ which contains $Y$. Consequently, $Cl(Y) = \cup_{P \in Y} F^c(P)$. □

**Corollary 2.3.** *Let $M$ be a lattice module over a $C$-lattice $L$. Then*

(1) *$Cl(\{P\}) = F^c(P)$, for each $P \in Spec^{cp}(M)$.*

(2) *$P_0 \in Cl(\{P\})$ if and only if $P \leq P_0$ if and only if $F^c(P_0) \subseteq F^c(P)$, for $P_0 \in Spec^{cp}(M)$.*

(3) *The set $\{P\}$ is closed in $Spec^{cp}(M)$ if and only if $P$ is a maximal classical prime element of $M$.*

**Proof.** (1) By Lemma 2.2, for $Y \subseteq Spec^{cp}(M)$, $Cl(Y) = \cup_{P \in Y} F^c(P)$. Assume that $Y = \{P\}$, then $\cup_{P \in Y} F^c(P) = F^c(P)$. Hence $Cl(\{P\}) = F^c(P)$.
(2) Suppose that $P_0 \in Cl(\{P\})$. Then $P_0 \in Cl(\{P\}) = F^c(P)$, by part (1) and

therefore $P \leq P_0$ which implies that $F^c(P_0) \subseteq F^c(P)$. Conversely, suppose that $F^c(P_0) \subseteq F^c(P)$. Since $P_0 \in F^c(P_0) \subseteq F^c(P)$, $P \leq P_0$ and hence $P_0 \in F^c(P) = Cl(\{P\})$.
(3) Suppose that the set $\{P\}$ is closed in $Spec^{cp}(M)$ and $P$ is not maximal, then there exists $Q$ such that $P \leq Q$, which implies that $Q \in Cl(\{P\})$ by part (2). But $\{P\}$ is closed, therefore $Q \in \{P\}$ and so $P = Q$. Consequently, $P$ is a maximal classical prime element. Conversely, suppose that $P$ is a maximal classical prime element of $M$. Let $Q \in Cl(\{P\})$. Then by part (1), $Q \in F^c(P)$, therefore $P \leq Q$. But $P$ is maximal, hence $P = Q$ and therefore $Cl(\{P\}) = \{P\}$. Consequently, {P} is closed in $Spec^{cp}(M)$. □

**Lemma 2.4.** *Let $M$ be a lattice module over a $C$-lattice $L$ and let $Y$ be a non-empty closed subset of $Spec^{cp}(M)$, then $Y = \cup_{P \in Y} F^c(P)$.*

**Proof.** Let $Y$ be a non-empty closed subset of $Spec^{cp}(M)$. It is clear that, $Y \subseteq \cup_{P \in Y} F^c(P)$. By Corollary 2.3(1), for each $P \in Y$, $Cl(\{P\}) = F^c(P)$, therefore $F^c(P) = Cl(\{P\}) \subseteq Cl(Y) = Y$. Hence $\cup_{P \in Y} F^c(P) \subseteq Y$. Consequently, $Y = \cup_{P \in Y} F^c(P)$. □

A topological space $X$ is called irreducible if $X \neq \emptyset$ and every finite intersection of non-empty open sets of $X$ is non-empty. A non-empty subset $Y$ of a topological space $X$ is called an irreducible set if the subspace $Y$ of $X$ is irreducible, equivalently, for any two closed sets $Y_1$ and $Y_2$ of $X$, $Y \subseteq Y_1 \cup Y_2$ implies either $Y \subseteq Y_1$ or $Y \subseteq Y_2$ ([8]).

**Lemma 2.5.** *Let $M$ be a lattice module over a $C$-lattice $L$. Then for each $P \in Spec^{cp}(M)$, $F^c(P)$ is irreducible.*

**Proof.** Suppose that $X_1$ and $X_2$ are two closed subsets of $Spec^{cp}(M)$ and $F^c(P) \subseteq X_1 \cup X_2$. Since $P \in F^c(P)$, therefore $P \in X_1 \cup X_2$ which implies either $P \in X_1$ or $P \in X_2$. Suppose that $P \in X_1$. Since $X_1$ is closed in $Spec^{cp}(M)$, we have $X_1 = \cap_{k \in J}(\cup_{l=1}^{n_k} F^c(N_{kl}))$, for some $N_{kl} \in M$, $k \in J, n_k \in \mathbb{N}$. Thus $P \in \cup_{l=1}^{n_k} F^c(N_{kl})$, for each $k \in J$. It follows that $F^c(P) \subseteq \cup_{l=1}^{n_k} F^c(N_{kl})$, for each $k \in J$. Therefore $F^c(P) \subseteq \cap_{k \in J}(\cup_{l=1}^{n_k} F^c(N_{kl})) = X_1$. Consequently, $F^c(P)$ is irreducible. □

**Theorem 2.6.** *Let $M$ be a lattice module over a $C$-lattice $L$ and $Y \subseteq Spec^{cp}(M)$. Then*

(1) *If $Y$ is irreducible, then $Z(Y)$ is a classical prime element.*
(2) *If $Z(Y)$ is a classical prime element and $Z(Y) \in Cl(Y)$, then $Y$ is irreducible.*

**Proof.** (1) Suppose that $Y$ is an irreducible subset of $Spec^{cp}(M)$. Clearly, $Z(Y) = \wedge_{P\in Y}P < 1_M$ and $Y \subseteq F^c(Z(Y))$. Let $abX \leq Z(Y)$, for $a, b \in L$ and $X \in M$. Now for $P \in Y$, $P \in F^c(Z(Y))$, hence $Z(Y) \leq P$ and therefore $abX \leq Z(Y) \leq P$. Since $P$ is classical prime, either $aX \leq P$ or $bX \leq P$, which implies that $P \in F^c(aX)$ or $P \in F^c(bX)$ and hence $P \in F^c(aX) \cup F^c(bX)$. Therefore $Y \subseteq F^c(aX) \cup F^c(bX)$. Since $Y$ is irreducible, $Y \subseteq F^c(aX)$ or $Y \subseteq F^c(bX)$. If $Y \subseteq F^c(aX)$, then $aX \leq P$, for all $P \in Y$ and hence $aX \leq Z(Y)$. If $Y \subseteq F^c(bX)$, then $bX \leq P$, for all $P \in Y$, hence $bX \leq Z(Y)$. Consequently, Z(Y) is a classical prime element of $M$.
(2) Suppose that $Z(Y)$ is a classical prime element and $Z(Y) \in Cl(Y)$. We have $Z(Y) \leq P$, for each $P \in Y$, therefore $F^c(P) \subseteq F^c(Z(Y))$, for each $P \in Y$ by Corollary 2.3(2). Thus $Cl(Y) = \cup_{P\in Y}F^c(P) \subseteq F^c(Z(Y))$, by Lemma 2.2. On the other hand, since $Z(Y)$ is a classical prime element and $Z(Y) \in Cl(Y)$, $F^c(Z(Y)) \subseteq Cl(Y)$. Consequently, $Cl(Y) = F^c(Z(Y))$. Now suppose that $Y \subseteq Y_1 \cup Y_2$, where $Y_1$ and $Y_2$ are closed subsets of $Spec^{cp}(M)$, then $Cl(Y) \subseteq Y_1 \cup Y_2$ and hence $F^c(Z(Y)) \subseteq Y_1 \cup Y_2$. Since $Z(Y)$ is a classical prime element, $F^c(Z(Y))$ is irreducible by Lemma 2.5. Therefore we have $F^c(Z(Y)) \subseteq Y_1$ or $F^c(Z(Y)) \subseteq Y_2$. It follows that $Y \subseteq Y_1$ or $Y \subseteq Y_2$. Consequently, $Y$ is irreducible. □

**Definition 2.7.** Let $M$ be a lattice module over a $C$-lattice $L$. Let $N$ be an element of $M$. Then the classical prime radical $\sqrt[cp]{N}$ of $N$ is the meet of all classical prime elements of $M$ containing $N$, i.e., $\sqrt[cp]{N} = \wedge\{P \in Spec^{cp}(M) | N \leq P\}$.

$\sqrt[cp]{N} = 1_M$, if there is no classical prime element which contains $N$. An element $N$ is said to be classical prime radical element if $N = \sqrt[cp]{N}$. Note that, $N \leq \sqrt[cp]{N}$ and $F^c(N) = F^c(\sqrt[cp]{N})$.

**Corollary 2.8.** *Let $M$ be a lattice module over a $C$-lattice $L$ and let $N$ be any element of $M$. Then the following are equivalent:*

(1) *The subset $F^c(N)$ of $Spec^{cp}(M)$ is irreducible.*
(2) *$\sqrt[cp]{N}$ is a classical prime element.*

**Proof.** (1)⇒(2) Suppose that $F^c(N)$ is an irreducible subset of $Spec^{cp}(M)$, then $Z(F^c(N))$ is classical prime element of $M$, by Theorem 2.6(1). Now, $\sqrt[cp]{N} = \wedge\{P \in Spec^{cp}(M) | N \leq P\} = \wedge\{P \in F^c(N)\} = Z(F^c(N))$. Consequently, $\sqrt[cp]{N}$ is a classical prime element.
(2)⇒(1) Suppose that $\sqrt[cp]{N}$ is a classical prime element, then $F^c(\sqrt[cp]{N})$ is irreducible by Lemma 2.5. Since, for each $N \in M$, $F^c(N) = F^c(\sqrt[cp]{N})$, therefore $F^c(N)$ is an irreducible subset of $Spec^{cp}(M)$. □

Let $Y$ be a closed subset of a topological space. An element $y \in Y$ is called a generic point of $Y$ if $Y = Cl(\{y\})$ (see [3]). Note that, a generic point of the irreducible closed subset $Y$ of a topological space is unique if the topological space is a $T_0-$space.

**Theorem 2.9.** *Let $M$ be a lattice module over a $C$-lattice $L$. Then*

(1) *$Spec^{cp}(M)$ is always a $T_0-$space.*

(2) *Every $P \in Spec^{cp}(M)$ is a generic point of the irreducible closed subset $F^c(P)$.*

(3) *Every finite irreducible closed subset of $Spec^{cp}(M)$ has a generic point.*

**Proof.** (1) Suppose that $P_1, P_2 \in Spec^{cp}(M)$. Then by Corollary 2.3(1), $Cl(\{P_1\}) = F^c(P_1), Cl(\{P_2\}) = F^c(P_2)$ and therefore $Cl(\{P_1\}) = Cl(\{P_2\})$ if and only if $F^c(P_1) = F^c(P_2)$ if and only if $P_1 = P_2$ by Corollary 2.3(2). Now, by the fact that a topological space is a $T_0-$space if the closures of distinct points are distinct, we conclude that, $Spec^{cp}(M)$ is a $T_0-$space.
(2) For each $P \in Spec^{cp}(M)$, $F^c(P) = Cl(\{P\})$ by Corollary 2.3(1). Hence $P$ is a generic point of the irreducible closed subset $F^c(P)$.
(3) Suppose that $Y$ is an irreducible closed subset of $Spec^{cp}(M)$ and $Y = \{P_1, P_2, \cdots, P_k\}$, where $P_i \in Spec^{cp}(M)$, $k \in \mathbb{N}$. By Lemma 2.4, $Y = Cl(Y) = F^c(P_1) \cup F^c(P_2) \cup \cdots \cup F^c(P_k)$. Since $Y$ is irreducible, $Y = F^c(P_i)$, for some $i(1 \leq i \leq k)$. By part(2), $P_i$ is a generic point of $F^c(P_i) = Y$. □

**Theorem 2.10.** *Let $M$ be a lattice module over a $C$-lattice $L$ such that $M$ has ascending chain condition on classical prime radical elements. Then $Spec^{cp}(M)$ with the Zariski-like topology is a quasi-compact space.*

**Proof.** Let $M$ be a lattice module over a $C$-lattice $L$ and suppose that $M$ has ascending chain condition on classical prime radical elements. Let $\mathcal{B}$ be a family of open sets covering $Spec^{cp}(M)$ and suppose that no finite subfamily of $\mathcal{B}$ covers $Spec^{cp}(M)$. Since $F^c(\sqrt[cp]{0_M}) = F^c(0_M) = Spec^{cp}(M)$, we may use the ascending chain condition on classical prime radical elements to choose an element $N$ maximal with respect to the property that no finite subfamily of $\mathcal{B}$ covers $F^c(N)$(we may assume $N = \sqrt[cp]{N}$, because $F^c(N) = F^c(\sqrt[cp]{N})$).

Suppose that $N$ is not classical prime element of $M$. Then there exists $X \in M$ and $a, b \in L$, such that $abX \leq N$, $aX \nleq N$ and $bX \nleq N$. Thus $N < N \vee aX \leq \sqrt[cp]{N \vee aX}$ and $N < N \vee bX \leq \sqrt[cp]{N \vee bX}$. Hence, without loss of generality, there must exists a finite subfamily $\mathcal{B}'$ of $\mathcal{B}$ that covers both $F^c(N \vee aX)$ and $F^c(N \vee bX)$. Let $P \in F^c(N)$. Since $abX \leq N$, therefore $abX \leq P$ and since $P$ is classical prime, we have $aX \leq P$ or $bX \leq P$. Thus $P \in F^c(N \vee aX)$ or $P \in F^c(N \vee bX)$ and therefore $F^c(N) \subseteq F^c(N \vee aX) \cup F^c(N \vee bX)$. Thus $F^c(N)$ is covered with the finite subfamily $\mathcal{B}'$, which is contradiction. Therefore $N$ is a classical prime element of $M$.

Now choose $U \in \mathcal{B}$ such that $N \in U$. Thus $N$ must have a neighborhood $\cap_{i=1}^{n} G^c(K_i)$, for some $K_i \in M$ and $n \in \mathbb{N}$, such that $\cap_{i=1}^{n} G^c(K_i) \subseteq U$. Suppose that for each $i(1 \leq i \leq n)$, $P \in G^c(K_i \vee N) \cap F^c(N)$, then $K_i \vee N \nleq P$ and $N \leq P$. Thus $K_i \nleq P$ and $P \in G^c(K_i)$. Consequently, $N \in [G^c(K_i \vee N) \cap F^c(N)] \subseteq G^c(K_i)$ and hence for each $i(1 \leq i \leq n)$, $N \in \cap_{i=1}^{n}[G^c(K_i \vee N) \cap F^c(N)] \subseteq \cap_{i=1}^{n} G^c(K_i) \subseteq U$. Thus $[\cap_{i=1}^{n} G^c(K_i \vee N)] \cap F^c(N)$, where $N < K_i \vee N$ is a neighborhood of $N$, with $[\cap_{i=1}^{n} G^c(K_i \vee N)] \cap F^c(N) \subseteq U$.

Since for each $i(1 \leq i \leq n)$, $N < K_i \vee N$, $F^c(K_i \vee N)$ can be covered by some finite subfamily $\mathcal{B}_i'$ of $\mathcal{B}$. Now $F^c(N) - [\cup_{i=1}^{n} F^c(K_i \vee N)] = F^c(N) - [\cap_{i=1}^{n} G^c(K_i \vee N)]' = \cap_{i=1}^{n} G^c(K_i \vee N) \cap F^c(N) \subseteq U$(here $'$ denotes complement). Therefore $F^c(N)$ can be covered by $\mathcal{B}_1' \cup \mathcal{B}_2' \cup \cdots \cup \mathcal{B}_n' \cup \{U\}$, which is contradiction to our choice of $N$. Thus there must exists a finite subfamily of $\mathcal{B}$ which covers $Spec^{cp}(M)$. Therefore, $Spec^{cp}(M)$ is a quasi-compact space. □

A topological space $X$ is a spectral space if $X$ is homeomorphic to $Spec(S)$, with Zariski topology, for some commutative ring $S$. Spectral spaces have been characterized by Hochster ([12]) as the topological spaces $X$ which satisfy the following conditions.

(1) $X$ is a $T_0-$space.
(2) $X$ is a quasi-compact.
(3) The quasi-compact open subsets of $X$ are closed under finite intersection and form an open basis.
(4) Each irreducible closed subset of $X$ has a generic point.

**Theorem 2.11.** *Let $M$ be a lattice module over a $C$-lattice $L$ with finite spectrum. Then $Spec^{cp}(M)$ is a spectral space.*

**Proof.** Since $Spec^{cp}(M)$ is finite, by Theorem 2.9, $Spec^{cp}(M)$ is a $T_0-$space and every irreducible closed subset of $Spec^{cp}(M)$ has a generic point. Also, since $Spec^{cp}(M)$ is finite, it is quasi-compact and the quasi-compact open subsets of $Spec^{cp}(M)$ are closed under finite intersections and form an open basis ([9]). Hence, by Hochster's characterization, $Spec^{cp}(M)$ is a spectral space. □

## 3. Patch-like topology on $Spec^{cp}(M)$

Let $X$ be a topological space. By the patch topology on $X$ we mean the topology which has as a sub-basis for its closed sets the closed sets and compact open sets of the original space. By a patch we mean set closed in the patch topology ([11], [12]).

**Definition 3.1.** Let $M$ be a lattice module over a $C$-lattice $L$. Let $E(M)$ be the family of all subsets of $Spec^{cp}(M)$ of the form $F^c(N) \cap G^c(K)$, where $N, K \in M$. Clearly $E(M)$ contains both $Spec^{cp}(M) = F^c(0_M) \cap G^c(1_M)$ and empty set $\emptyset = F^c(1_M) \cap G^c(0_M)$. Let $T_p(M)$ be the collection $U$ of all unions of finite intersections of elements of $E(M)$. Then $T_p(M)$ is a topology on $Spec^{cp}(M)$ and is called the patch-like topology of $M$. In fact $E(M)$ is a sub-basis for the patch-like topology of $M$.

Note that, the patch-like topology on $Spec^{cp}(M)$ is finer than the Zariski-like topology on $Spec^{cp}(M)$.

**Theorem 3.2.** *Let $M$ be a lattice module over a $C$-lattice $L$. Then $Spec^{cp}(M)$ with the patch-like topology is a Hausdorff space.*

**Proof.** Suppose that $P_1$, $P_2 \in Spec^{cp}(M)$ and $P_1 \neq P_2$. Since $P_1 \neq P_2$, so either $P_1 \nleq P_2$ or $P_2 \nleq P_1$. Suppose that $P_1 \nleq P_2$. By Definition 3.1, $U_1 = G^c(1_M) \cap F^c(P_1)$ is a patch-like-neighborhood of $P_1$ and $U_2 = G^c(P_1) \cap F^c(P_2)$ is a patch-like-neighborhood of $P_2$. Clearly, $G^c(P_1) \cap F^c(P_1) = \emptyset$ and hence $U_1 \cap U_2 = \emptyset$. Therefore, $Spec^{cp}(M)$ is a Hausdorff space. □

**Theorem 3.3.** *Let $M$ be a lattice module over a $C$-lattice $L$ such that $M$ has ascending chain condition on classical prime radical elements. Then $Spec^{cp}(M)$ with the patch-like topology is a compact space.*

**Proof.** Let $M$ be a lattice module over a $C$-lattice $L$ and suppose that $M$ has ascending chain condition on classical prime radical elements. Let $\mathcal{A}$ be a family of open sets covering $Spec^{cp}(M)$ and suppose that no finite subfamily of $\mathcal{A}$ covers $Spec^{cp}(M)$. Since $F^c(\sqrt[cp]{0_M}) = F^c(0_M) = Spec^{cp}(M)$, we may use the ascending chain condition on classical prime radical elements to choose an element $N$ maximal with respect to the property that no finite subfamily of $\mathcal{A}$ covers $F^c(N)$ (we may assume $N = \sqrt[cp]{N}$, because $F^c(N) = F^c(\sqrt[cp]{N})$).

Suppose that $N$ is not classical prime element of $M$. Then there exists $X \in M$ and $a, b \in L$, such that $abX \leq N$, $aX \nleq N$ and $bX \nleq N$. Thus $N < N \vee aX \leq \sqrt[cp]{N \vee aX}$ and $N < N \vee bX \leq \sqrt[cp]{N \vee bX}$. Hence, without loss of generality, there must exists a finite subfamily $\mathcal{A}'$ of $\mathcal{A}$ that covers both $F^c(N \vee aX)$ and $F^c(N \vee bX)$. Let $P \in F^c(N)$. Since $abX \leq N$ and $N \leq P$, we have $abX \leq P$. Since $P$ is classical prime, we have either $aX \leq P$ or $bX \leq P$. Thus $N \vee aX \leq P$ or $N \vee bX \leq P$. Therefore, either $P \in F^c(N \vee aX)$ or $P \in F^c(N \vee bX)$ and hence $F^c(N) \subseteq F^c(N \vee aX) \cup F^c(N \vee bX)$. Thus $F^c(N)$ is covered with the finite subfamily $\mathcal{A}'$, which is contradiction. Therefore $N$ is a classical prime element of $M$.

Now choose $U \in \mathcal{A}$ such that $N \in U$. Thus $N$ must have a patch-like-neighborhood $\cap_{i=1}^n [G^c(K_i) \cap F^c(N_i)]$ for some $K_i, N_i \in M, n \in \mathbb{N}$ such that $\cap_{i=1}^n [G^c(K_i) \cap F^c(N_i)] \subseteq U$. Suppose that for each $i(1 \leq i \leq n)$, $P \in [G^c(K_i \vee N) \cap F^c(N)]$. Then $P \in G^c(K_i \vee N)$ and $P \in F^c(N)$ and so that $K_i \vee N \nleq P$ and $N \leq P$. Thus $K_i \nleq P$, i.e., $P \in G^c(K_i)$. On the other hand, $N \in F^c(N_i)$, therefore $N_i \leq N$ and $N_i \leq N, N \leq P$ implies $N_i \leq P$, hence $P \in F^c(N_i)$. Consequently, $N \in [G^c(K_i \vee N) \cap F^c(N)] \subseteq [G^c(K_i) \cap F^c(N_i)]$ and hence $N \in \cap_{i=1}^n [G^c(K_i \vee N) \cap F^c(N)] \subseteq \cap_{i=1}^n [G^c(K_i) \cap F^c(N_i)] \subseteq U$. Thus $[\cap_{i=1}^n G^c(K_i \vee N)] \cap F^c(N)$, where $N < K_i \vee N$, is a neighborhood of $N$, with $[\cap_{i=1}^n G^c(K_i \vee N)] \cap F^c(N) \subseteq U$. Since for each $i(1 \leq i \leq n)$, $N < K_i \vee N$, $F^c(K_i \vee N)$ is covered by some finite subfamily $\mathcal{A}_i'$ of $\mathcal{A}$. Now $F^c(N) - [\cup_{i=1}^n F^c(K_i \vee N)] = F^c(N) - [\cap_{i=1}^n G^c(K_i \vee N)]' = [\cap_{i=1}^n G^c(K_i \vee N)] \cap F^c(N) \subseteq U$(here $'$ denotes complement). Hence $F^c(N)$ can be covered by $\mathcal{A}_1' \cup \mathcal{A}_2' \cup \cdots \cup \mathcal{A}_n' \cup \{U\}$, which is contradiction to our choice of $N$. Thus there must exists a finite subfamily of $\mathcal{A}$ which covers $Spec^{cp}(M)$. Therefore, $Spec^{cp}(M)$ is compact in the patch-like topology of $M$. □

We require the following evident Lemma.

**Lemma 3.4.** *Let $\tau_1$ and $\tau_2$ be two topologies on $X$ such that $\tau_1 \subseteq \tau_2$. If $X$ is quasi-compact (i.e. any open cover of it has a finite subcover) in $\tau_2$, then $X$ is also quasi-compact in $\tau_1$.*

**Theorem 3.5.** *Let $M$ be a lattice module over a $C$-lattice $L$ such that $M$ has ascending chain condition on classical prime radical elements. Then for each $n \in \mathbb{N}$ and elements $N_i (1 \leq i \leq n)$ of $M$, $G^c(N_1) \cap G^c(N_2) \cap \cdots \cap G^c(N_n)$ is a quasi-compact subset of $Spec^{cp}(M)$ with the Zariski-like topology. Consequently, $Spec^{cp}(M)$ has a basis of quasi-compact open subsets.*

**Proof.** By Definition 3.1, for each element $N$ of $M$, $F^c(N) = F^c(N) \cap G^c(1_M)$ is an open subset of $Spec^{cp}(M)$ with the patch-like topology and $G^c(N)$ is a complement of $F^c(N)$. Therefore for each $N \in M$, $G^c(N)$ is a closed subset in $Spec^{cp}(M)$. Thus for each $n \in \mathbb{N}$ and elements $N_i (1 \leq i \leq n)$ of $M$, $G^c(N_1) \cap G^c(N_2) \cap \cdots \cap G^c(N_n)$ is also a closed subset in $Spec^{cp}(M)$ with the patch-like topology. Since every closed subset of a compact space is compact, therefore $G^c(N_1) \cap G^c(N_2) \cap \cdots \cap G^c(N_n)$ is compact in $Spec^{cp}(M)$ with the patch-like topology and by Lemma 3.4, it is quasi-compact in $Spec^{cp}(M)$ with the Zariski-like topology. Now, $Spec^{cp}(M) = G^c(1_M)$ and $\mathbb{B} = \{G^c(N_1) \cap G^c(N_2) \cap \cdots \cap G^c(N_n) \mid N_i \in M, 1 \leq i \leq n, n \in \mathbb{N}\}$ is a basis for the Zariski-like topology of $M$. Consequently, $Spec^{cp}(M)$ is quasi-compact and has a basis of quasi-compact open subsets. □

**Corollary 3.6.** *Let $M$ be a lattice module over a $C$-lattice $L$ such that $M$ has ascending chain condition on classical prime radical elements. Then quasi-compact open sets of $Spec^{cp}(M)$ (with the Zariski-like topology) are closed under finite intersections.*

**Proof.** Let $U_1$ and $U_2$ be two quasi-compact open sets of $Spec^{cp}(M)$ and let $U = U_1 \cap U_2$. Each of $U_1$ and $U_2$ is a finite union of members of $\mathbb{B} = \{G^c(N_1) \cap G^c(N_2) \cap \cdots \cap G^c(N_n) \mid N_i \in M, 1 \leq i \leq n, n \in \mathbb{N}\}$ and hence $U = \cup_{i=1}^{m}(\cap_{j=1}^{n_i} G^c(N_j))$. Let $\Pi$ be any open cover of $U$. Then $\Pi$ also covers each $\cap_{j=1}^{n_i} G^c(N_j)$ which is quasi-compact by Theorem 3.5. Hence each $\cap_{j=1}^{n_i} G^c(N_j)$ has a finite subcover of $\Pi$ and therefore $U$ has also a finite subcover of $\Pi$. Thus $U$ is quasi-compact, as required. □

**Lemma 3.7.** *Let $M$ be a lattice module over a $C$-lattice $L$ such that $M$ has ascending chain condition on classical prime radical elements. Then every irreducible closed subset of $Spec^{cp}(M)$ (with the Zariski-like topology) has a generic point.*

**Proof.** Suppose that $Y$ is an irreducible closed subset of $Spec^{cp}(M)$ with the Zariski-like topology. By Lemma 2.4, we have $Y = \cup_{P \in Y} F^c(P)$. By Definition 3.1, for each $P \in Y$, $F^c(P)$ is an open subset of $Spec^{cp}(M)$ with the patch-like topology. Now, since $Y \subseteq Spec^{cp}(M)$ is closed with the Zariski-like topology, the complement of $Y$ is open by this topology and therefore the complement of $Y$ is open with the patch-like topology. Hence $Y \subseteq Spec^{cp}(M)$ is closed with the patch-like topology. By Theorem 3.3, $Spec^{cp}(M)$ is compact with the patch-like topology and since $Y \subseteq Spec^{cp}(M)$ is closed, therefore $Y$ is also compact. Now, since $Y = \cup_{P \in Y} F^c(P)$ and each $F^c(P)$ is patch-like open, therefore there exists a finite subset $X$ of $Y$ such that $Y = \cup_{P \in X} F^c(P)$. Since $Y$ is irreducible, $Y = F^c(P)$ for some $P \in X$. Therefore, we have $Y = F^c(P) = Cl(\{P\})$ for some $P \in Y$. Consequently, $P$ is a generic point for $Y$. □

We conclude this section by proving the main theorem .

**Theorem 3.8.** *Let $M$ be a lattice module over a $C$-lattice $L$ such that $M$ has ascending chain condition on classical prime radical elements. Then $Spec^{cp}(M)$ with the Zariski-like topology is a spectral space.*

**Proof.** By Theorem 2.9, $Spec^{cp}(M)$ is a $T_0$−space. Since $M$ satisfies ascending chain condition on classical prime radical elements, therefore by Theorem 2.10,

$Spec^{cp}(M)$ is quasi-compact. By Theorem 3.5, $Spec^{cp}(M)$ has a basis of quasi-compact open subsets and by Corollary 3.6, the family of quasi-compact open subsets of $Spec^{cp}(M)$ are closed under finite intersections. Finally, by Lemma 3.7, each irreducible closed subset of $Spec^{cp}(M)$ has a generic point. Thus, by Hochster's characterization, $Spec^{cp}(M)$ is a spectral space. □

**Acknowledgement.** The authors would like to thank the referee for the valuable suggestions and comments.

## References

[1] F. Alarcon, D. D. Anderson and C. Jayaram, *Some results on abstract commutative ideal theory*, Period. Math. Hungar., 30(1) (1995), 1-26.

[2] E. A. Al-Khouja, *Maximal elements and prime elements in lattice modules*, Damascus Univ. J. Basic Sci., 19(2) (2003), 9-21.

[3] M. F. Atiyah and I. G. MacDonald, Introduction to Commutative Algebra, Addison-Wesley Publishing Co., Reading, Mass.-London-Don Mills, Ont. 1969.

[4] S. Ballal and V. Kharat, *Zariski topology on lattice modules*, Asian-Eur. J. Math., 8(4) (2015), 1550066 (10 pp).

[5] M. Behboodi and M. R. Haddadi, *Classical Zariski topology of modules and spectral spaces I*, Int. Electron. J. Algebra, 4 (2008), 104-130.

[6] M. Behboodi and M. R. Haddadi, *Classical Zariski topology of modules and spectral spaces II*, Int. Electron. J. Algebra, 4 (2008), 131-148.

[7] M. Behboodi and M. J. Noori, *Zariski-like topology on the classical prime spectrum of a module*, Bull. Iranian Math. Soc., 35(1) (2009), 253-269.

[8] N. Bourbaki, Algebre Commutative, Chap 1-2, Hermann, Paris, 1961.

[9] N. Bourbaki, Elements of Mathematics, General topology, Part 1, Hermann, Paris; Addison-Wesley Publishing Co., Reading, Mass.-London-Don Mills, Ont. 1966.

[10] F. Callialp, G. Ulucak and U. Tekir, *On the Zariski topology over an L-module M*, Turkish J. Math., 41(2) (2017), 326-336.

[11] K. R. Goodearl and R. B. Warfield, Jr., An Introduction to Noncommutative Noetherian Rings, Second edition, London Math. Soc. Student Texts, 16, Cambridge Univ. Press, Cambridge, 2004.

[12] M. Hochster, *Prime ideal structure in commutative rings*, Trans. Amer. Math. Soc., 142 (1969), 43-60.

[13] J. A. Johnson, *A-adic completions of Noetherian lattice modules*, Fund. Math., 66 (1970), 347-373.

[14] C. P. Lu, *The Zariski topology on the prime spectrum of a module*, Houston J. Math., 25(3) (1999), 417-432.

[15] C. S. Manjarekar and A. V. Bingi, *Absorbing elements in lattice modules*, Int. Electron. J. Algebra, 19 (2016), 58-76.

[16] J. R. Munkres, Topology: a First Course, Prentice-Hall, Inc. Eglewood Cliffs, New Jersey, 1975.

[17] N. Phadatare, S. Ballal and V. Kharat, *On the second spectrum of lattice modules*, Discuss. Math. Gen. Algebra Appl., 37(1) (2017), 59-74.

[18] N. Phadatare and V. Kharat, *On the second radical elements of lattice modules*, Tbilisi Math. J., 11(4) (2018), 165-173.

[19] N. Phadatare, V. Kharat and S. Ballal, *On the maximal spectrum of lattice modules*, Southeast Asian Bull. Math., Accepted.

[20] N. K. Thakare and C. S. Manjarekar, *Abstract spectral theory: Multiplicative lattices in which every character is contained in a unique maximal character*, Algebra and its applications (New Delhi, 1981), Lecture Notes in Pure and Appl. Math., 91, Dekker, New York, (1984), 265-276.

[21] N. K. Thakare, C. S. Manjarekar and S. Maeda, *Abstract spectral theory II: minimal characters and minimal spectrums of multiplicative lattices*, Acta Sci. Math. (Szeged), 52(1-2) (1988), 53-67.

**Pradip Girase**
Department of Mathematics
K. K. M. College, Manwath
Dist-Parbhani(M.S.)-431505, India
e-mail: pgpradipmaths22@gmail.com

**Vandeo Borkar** (Corresponding Author)
Department of Mathematics
Yeshwant Mahavidyalaya
Nanded(M.S.)-431602, India
e-mail: borkarvc@gmail.com

**Narayan Phadatare**
Department of Mathematics
Savitribai Phule Pune University
Pune(M.S.), India
e-mail: a9999phadatare@gmail.com

# 6

# ANNIHILATOR CONDITIONS OF MULTIPLICATIVE REVERSE DERIVATIONS ON PRIME RINGS

Gurninder S. Sandhu and Deepak Kumar

Communicated by A. Çiğdem Ozcan

ABSTRACT. Let $R$ be a ring, a mapping $F : R \to R$ together with a mapping $d : R \to R$ is called a multiplicative (generalized)-reverse derivation if $F(xy) = F(y)x + yd(x)$ for all $x, y \in R$. The aim of this note is to investigate the commutativity of prime rings admitting multiplicative (generalized)-reverse derivations. Precisely, it is proved that for some nonzero element $a$ in $R$ the conditions: $a(F(xy) \pm xy) = 0$, $a(F(x)F(y) \pm xy) = 0$, $a(F(xy) \pm F(y)F(x)) = 0$, $a(F(x)F(y) \pm yx) = 0$, $a(F(xy) \pm yx) = 0$ are sufficient for the commutativity of $R$. Moreover, we describe the possible forms of generalized reverse derivations of prime rings.

**Keywords**: Prime ring, multiplicative (generalized)-reverse derivation, generalized reverse derivation, annihilator conditions

## 1. Introduction

During the time (1940-1950), when the general structure theory of rings was in progress, a significant amount of work was done by Jacobson, Herstein, Bell, Kezlan, Abu-Khuzam (see [19] and references therein) on certain polynomial constraints that force a ring to be commutative. Nowadays, there has been an ongoing interest in the investigation of polynomial constrains involving various types of derivations of rings; such constrains are usually known as differential identities. Although the notion of derivation has been existing since long back, but it took more attention of the algebraists after Posner, who established two remarkable results on derivations of prime rings (see [23, Theorem 1-2]) and hence initiated the study of relationships between the behaviour of derivations and the structure of rings. Further, the concept of derivation has been generalized in many directions for last few decades. One of these directions reports to generalized derivation, which was introduced by Brešar [7]. The study of centralizing and commuting derivations goes back to 1957, when Posner [23] proved the first notable result on centralizing (or commuting) derivations, which states that the existence of a nonzero centralizing derivation of a prime ring $R$ implies that $R$ is commutative. Since then, there has been a considerable interest in the study of centralizing and commuting maps of prime and semiprime rings. For this sort of work, we refer the reader to [2], [4], [5], [6], [11], [21], [22], [24], [26] and references therein.

Let us turn to the brief background investigation of multiplicative derivations. The study of multiplicative derivations was initiated by Daif [10] inspired by the extraordinary work of Martindale [20] on the additivity of multiplicative isomorphisms. Later on, the complete description of these mappings was given by Goldmann and Šemrl [15]. In [12], Daif et al. generalized the idea of multiplicative derivation to multiplicative generalized derivation. Recently, Dhara and Ali [13] introduced the notion of multiplicative (generalized)-derivation which is a slight generalization of multiplicative generalized derivation. Moreover, Camci and Aydin [8] explored the nature of the mappings associated with multiplicative (generalized)-derivations of semiprime rings. One may easily observe that the idea of multiplicative (generalized)-derivation seems more appropriate as it covers both the concepts of multiplicative derivation and multiplicative generalized derivation. During the last few years many results involving multiplicative (generalized)-derivations in prime and semiprime rings has been obtained. For a good cross section of this subject, one may see [3], [13], [16], [27], [28] and references therein.

In 1957, Herstein [17] initiated the study of reverse derivations in prime rings, where he proved that a prime ring possessing a reverse derivation is a commutative integral domain and hence reverse derivation behaves like an ordinary derivation. Amid the most recent couple of years, many authors investigated reverse derivations and extended this notion in several ways (see [1], [14], [25], [28]). Recently, Aboubakr et. al. [1] generalized the concept of reverse derivation to generalized reverse derivation and provided a study of relationship between generalized reverse derivations and generalized derivations. Inspired by this, Tiwari et. al. [28] gave the notion of multiplicative (generalized)-reverse derivation, which is a further generalization of so called $l-$generalized reverse derivation.

Throughout, $R$ is an associative ring and $Z(R)$ stands for the center of $R$. In this paper, we study the action of multiplicative (generalized)-reverse derivations on prime rings with annihilator conditions and obtain several commutativity theorems.

## 2. Preliminaries

**Definition 2.1.** (i) If $R$ is a ring such that $aRb = (0)$(or $aRa = (0)$) implies either $a = 0$ or $b = 0$ (or $a = 0$) for all $a, b \in R$, then $R$ is said to be a *prime (or semi-prime)* ring.

(ii) For any nonempty subset $S$ of $R$, the *right annihilator* $r_R(S)$ of $S$ in $R$ is the set of all $r \in R$ such that $Sr = (0)$. Accordingly, the *left annihilator* $l_R(S)$ is the set of all $r \in R$ such that $rS = (0)$. The intersection of right and left annihilators of $S$ in $R$ is called *annihilator* of $S$ in $R$, that is

$$Ann_R(S) = \{r \in R : sr = 0 \text{ and } rs = 0 \text{ for all } s \in S\}$$

(iii) Let $H$ be a nonempty subset of $R$. Then a mapping $f : H \to R$ is said to be *centralizing* on $H$ if $[f(x), x] \in Z(R)$ for all $x \in H$. In particular, if $[f(x), x] = 0$ for all $x \in H$, then $f$ is called *commuting* on $H$.

(iv) Let $R$ be a ring. An additive mapping $d : R \to R$ is said to be a *derivation* of $R$ if $d(xy) = d(x)y + xd(y)$ for all $x, y \in R$. For some fixed $a \in R$, a derivation $d_a$ is called the *inner derivation* of $R$ induced by $a$ if $d_a(x) = [a, x]$ for all $x \in R$.

(v) For some fixed $a, b \in R$, a mapping $F_{a,b} : R \to R$ such that $x \mapsto ax + xb$ is called the *generalized inner derivation* of $R$. Immediately it follows that, if $F_{a,b}$ is generalized inner derivation, then $F_{a,b}(xy) = F_{a,b}(x)y + xd_{-b}(y)$, where $d_{-b}$ is the inner derivation of $R$ induced by $-b$. This observation of Brešar [7] gave rise to the notion of *generalized derivation*: An additive mapping $F : R \to R$, which is uniquely determined by a derivation $d$ of $R$ such that $F(xy) = F(x)y + xd(y)$ for all $x, y \in R$. Therefore, generalized derivations are derived from derivations in a very natural way.

(vi) A derivation $d$ which is not necessarily additive is said to be a *multiplicative derivation* of $R$ [10].

(vii) A mapping $F : R \to R$ (not necessarily additive) is called *multiplicative (generalized)-derivation* of $R$ if there exists a multiplicative derivation $d$ such that $F(xy) = F(x)y + xd(y)$ for all $x, y \in R$ [8] and [13].

(viii) If $d : R \to R$ is an additive mapping such that $d(xy) = d(y)x + yd(x)$ for all $x, y \in R$, then $R$ is called *reverse derivation* of $R$ [17]. Further, if we relax the assumption of additivity, then $d$ is called *multiplicative reverse derivation* of $R$.

(ix) An additive mapping $F : R \to R$ is said to be the *l−generalized reverse derivation* of $R$ if there exists a reverse derivation $d$ of $R$ such that $F(xy) = F(y)x + yd(x)$ for all $x, y \in R$ [1].

(x) A mapping $F : R \to R$ (not necessarily additive) together with any mapping $d : R \to R$ is said to be a *multiplicative (generalized)-reverse derivation* of $R$ if $F(xy) = F(y)x + yd(x)$ for all $x, y \in R$ [28].

Before going further, we construct an example to show that a reverse derivation need not to be a derivation and vice-versa.

**Example 2.2.** *Let us consider* $R = \left\{ \begin{pmatrix} 0 & a & b & c \\ 0 & 0 & 0 & -b \\ 0 & 0 & 0 & a \\ 0 & 0 & 0 & 0 \end{pmatrix} : a, b, c \in \mathbb{Z} \right\}$, *where* $\mathbb{Z}$ *is the ring of integers. We define a map* $\lambda : R \to R$ *such that* $\lambda \begin{pmatrix} 0 & a & b & c \\ 0 & 0 & 0 & -b \\ 0 & 0 & 0 & a \\ 0 & 0 & 0 & 0 \end{pmatrix} = \begin{pmatrix} 0 & 0 & -b & c \\ 0 & 0 & 0 & b \\ 0 & 0 & 0 & 0 \\ 0 & 0 & 0 & 0 \end{pmatrix}$. *It is easy to verify that* $\lambda$ *is a reverse derivation on* $R$ *but not a derivation. Now, let* $\phi : R \to R$ *be a map defined by* $\phi \begin{pmatrix} 0 & a & b & c \\ 0 & 0 & 0 & -b \\ 0 & 0 & 0 & a \\ 0 & 0 & 0 & 0 \end{pmatrix} = \begin{pmatrix} 0 & 0 & -b & -c \\ 0 & 0 & 0 & b \\ 0 & 0 & 0 & 0 \\ 0 & 0 & 0 & 0 \end{pmatrix}$. *This map* $\phi$ *is a derivation on* $R$ *but not a reverse derivation.*

For $x, y \in R$, the symbol $[x, y]$ stands for the commutator of $x, y$ defined by $xy - yx$. In this paper, we shall use the basic commutator identities very frequently:

$$[x, yz] = y[x, z] + [x, y]z, \; [xy, z] = x[y, z] + [x, z]y$$

for all $x, y, z \in R$. Also we shall make use of the following well known result:

**Lemma 2.3.** [22, Lemma 3] *If a prime ring* $R$ *contains a nonzero commutative (and hence central) right ideal, then* $R$ *is commutative.*

## 3. Main results

**Theorem 3.1.** *Let $I$ be a nonzero ideal of a prime ring $R$ and $F : R \to R$ a multiplicative (generalized)-reverse derivation associated with a map $d : R \to R$. If for some $0 \neq a \in R$, $a(F(xy) \pm xy) = 0$ for all $x, y \in I$, then $R$ is commutative.*

**Proof.** Let us begin with

$$a(F(xy) + xy) = 0 \ \forall \ x, y \in I. \tag{1}$$

Substitute $yz$ instead of $y$ in (1), we find

$$\begin{aligned} 0 =& (a(F(yz) + yz)x) + ayzd(x) + a[x, yz] \\ =& ayzd(x) + a[x, yz] \end{aligned}$$

$\forall \ x, y, z \in I$. That is

$$ayzd(x) + ay[x, z] + a[x, y]z = 0 \ \forall \ x, y, z \in I. \tag{2}$$

Substituting $y$ with $ay$, in (2), we get

$$a^2yzd(x) + a^2y[x, z] + a^2[x, y]z + a[x, a]yz = 0 \ \forall \ x, y, z \in I. \tag{3}$$

Left multiply (2) by $a$ and subtract from (3), we get $a[x, a]yz = 0$ where $x, y, z \in I$. Since $R$ is a prime ring and $I$ is a nonzero ideal of $R$, we obtain

$$a[x, a] = 0 \ \forall \ x \in I. \tag{4}$$

Replace $x$ by $xr$, where $r \in R$ in (4), we obtain $a[x, a]r + ax[r, a] = 0$ for all $x \in I$ and $r \in R$. Application of (4) reduces it to $ax[r, a] = 0$. Primeness of $R$ implies that $[r, a] = 0$ as $a \neq 0$ for all $r \in R$. That means $a \in Z(R)$. But center of a prime ring does not contain a zero-divisor, so (2) yields

$$yzd(x) + y[x, z] + [x, y]z = 0 \ \forall \ x, y, z \in I. \tag{5}$$

Replace $y$ by $ty$ in (5), we obtain

$$tyzd(x) + ty[x, z] + t[x, y]z + [x, t]yz = 0 \ \forall \ x, y, z, t \in I. \tag{6}$$

Use of (5) gives $[x, t]yz = 0$ for all $x, y, z, t \in I$. Again primeness of $R$ yields that $[I, I] = (0)$. Hence, $I$ is commutative and so by Lemma 2.3, $R$ is commutative.

By doing all as above, one can prove the same conclusion for $a(F(xy) - xy) = 0$ for all $x, y \in I$. □

**Theorem 3.2.** *Let $I$ be a nonzero ideal of a prime ring $R$ and $F : R \to R$ a multiplicative (generalized)-reverse derivation associated with a map $d : R \to R$. If for some $0 \neq a \in R$, $a(F(x)F(y) \pm xy) = 0$ for all $x, y \in I$, then $R$ is commutative.*

**Proof.** Let us begin with

$$a(F(x)F(y) + xy) = 0 \;\forall\; x, y \in I. \tag{7}$$

Substituting $y$ with $yt$ in (7), we get $a(F(x)F(t)+xt)y+aF(x)td(y)+a(xyt-xty) = 0$ where $x, y, t \in I$. By hypothesis, we infer

$$aF(x)td(y) + ax[y, t] = 0 \;\forall\; x, y, t \in I. \tag{8}$$

Replace $t$ by $rt$, where $r \in R$ in (8), we obtain

$$aF(x)rtd(y) + axr[y, t] + ax[y, r]t = 0 \;\forall\; x, y, t \in I, r \in R. \tag{9}$$

Replace $x$ by $rx$ in (8), we have

$$aF(x)rtd(y) + arx[y, t] + axd(r)td(y) = 0 \;\forall\; x, y, t \in I, r \in R. \tag{10}$$

Subtraction of (9) from (10) gives

$$a[r, x][y, t] + axd(r)td(y) - ax[y, r]t = 0. \tag{11}$$

Substituting $x$ with $ax$ in (11), we get

$$a^2[r, x][y, t] + a[r, a]x[y, t] + a^2xd(r)td(y) - a^2x[y, r]t = 0. \tag{12}$$

Relation (11) reduces it to

$$a[r, a]x[y, t] = 0 \;\forall\; x, y, t \in I, r \in R. \tag{13}$$

Since $R$ is prime, either $a[R, a] = (0)$ or $[I, I] = (0)$, that is either $a \in Z(R)$ or $R$ is commutative. Because $0 \neq a \in R$, so by (11), we find

$$[r, x][y, t] + xd(r)td(y) - x[y, r]t = 0 \;\forall\; x, y, t \in I, r \in R. \tag{14}$$

In particular, for $r = y$ in (14), we have

$$[y, x][y, t] + xd(y)td(y) = 0 \;\forall\; x, y, t \in I. \tag{15}$$

Substituting $x$ with $zx$ in (15) and then using this relation, we get $[y, z]x[y, t] = 0$ for all $x, y, t \in I$, which implies $[I, I] = (0)$, implying $R$ is commutative.

By doing all as above, one can prove the same conclusion for $a(F(x)F(y)-xy) = 0$ for all $x, y \in I$. □

**Theorem 3.3.** *Let $I$ be a nonzero ideal of a prime ring $R$ and $F : R \to R$ a multiplicative (generalized)-reverse derivation associated with a map $d : R \to R$. If for some $0 \neq a \in R$, $a(F(xy) \pm F(y)F(x)) = 0$ for all $x, y \in I$, then either $d = 0$ or $R$ is commutative.*

**Proof.** Let us begin with

$$a(F(xy) \pm F(y)F(x)) = 0 \ \forall \ x, y \in I. \tag{16}$$

Replace $x$ by $zx$ in (16), we find $a(F(xy) \pm F(y)F(x))z + a(xyd(z) \pm F(y)xd(z)) = 0$ for any $x, y, z \in I$. Our hypothesis yields

$$a(xyd(z) \pm F(y)xd(z)) = 0. \tag{17}$$

Substitute $ry$ instead of $y$ in (17), we find

$$a(xryd(z) \pm F(y)rxd(z) \pm yd(r)xd(z)) = 0 \ \forall \ x, y, z \in I, r \in R. \tag{18}$$

Replace $x$ by $rx$ in (17) and subtract it from (18), we obtain

$$a([x, r]yd(z) \pm yd(r)xd(z)) = 0. \tag{19}$$

Substituting $x$ with $xr$ in (19), we get

$$a([x, r]ryd(z) \pm yd(r)xrd(z)) = 0. \tag{20}$$

Substituting $y$ with $ry$ in (19), we obtain

$$a([x, r]ryd(z) \pm ryd(r)xd(z)) = 0. \tag{21}$$

Combining (21) and (20), we have

$$a([yd(r)x, r]d(z)) = 0. \tag{22}$$

Replace $y$ by $ay$ in (22), we find $a^2[yd(r)x, r]d(z) + a[a, r]yd(r)xd(z) = 0$ for all $x, y, z \in I$ and $r \in R$. Application of (22) reduces it to $a[a, r]yd(r)xd(z) = 0$ where $x, y, z \in I$ and $r \in R$. Primeness of $R$ implies either $a[a, R] = (0)$ or $Id(I) = (0)$.

Firstly, we assume that $Id(I) = (0)$ where $z \in I$. Since $I$ is a nonzero ideal of $R$ and $R$ is a prime ring, so it implies that $d(I) = (0)$. In this case $F(xy) = F(y)x$ for any $x, y \in I$. Therefore, our hypothesis yields

$$aF(y)(x \pm F(x)) = 0 \ \forall \ x, y \in I. \tag{23}$$

Replace $y$ by $ty$ in (23) to obtain $aF(y)t(x \pm F(x)) + ayd(t)(x \pm F(x)) = 0$ for all $x, y, t \in I$. Using (23), we obtain $aF(y)t(x \pm F(x)) = 0$. Since $R$ is a prime ring so either $aF(I) = (0)$ or $x \pm F(x) = 0$ for all $x \in I$. Let us suppose that

$$aF(y) = 0 \ \forall \ y \in I. \tag{24}$$

Substituting $y$ with $yr$ in (24) in order to get $aF(r)y = 0$. Again, primeness of $R$ implies that $aF(R) = (0)$, which is an interesting observation. Substituting $y$ with $ry$ in (24), we get $0 = aF(y)r + ayd(r) = ayd(r)$ for any $y \in I$ and $r \in R$. Since $R$ is a prime ring and $a \neq 0$, we obtain $d(R) = (0)$.

On the other hand, if

$$x \pm F(x) = 0 \ \forall \ x \in I. \tag{25}$$

On substituting $rx$ in the place of $x$ in (25), we find $rx \pm F(x)r \pm xd(r) = 0$ for all $x \in I, r \in I$. That is, $[r, x] + (x \pm F(x))r \pm xd(r) = 0$ where $x \in I$ and $r \in R$. In the view of (25), it becomes

$$[r, x] \pm xd(r) = 0 \ \forall \ x \in I, r \in R. \tag{26}$$

Replace $x$ by $yx$ in (26) and using it, we get $[r, y]x = 0$ for all $x, y \in I$ and $r \in R$. Since $I$ is a nonzero ideal of prime ring $R$, so $R$ contains a nonzero central ideal. In the view of Lemma 2.3, $R$ is commutative.

Next, we assume that $a[a, r] = 0$ for all $r \in R$. Substitute $rs$ instead of $r$, we find $ar[a, s] = 0$ for all $r, s \in R$. Which is $aR[a, s] = (0)$, again primeness of $R$ implies $a \in Z(R)$ as $a \neq 0$. But $R$ is a prime ring, so equation (19) forces

$$[x, r]yd(z) \pm yd(r)xd(z) = 0 \ \forall \ x, y, z \in I, r \in R. \tag{27}$$

Replace $y$ by $ty$ in (27), we get

$$[x, r]tyd(z) \pm tyd(r)xd(z) = 0 \ \forall \ x, y, z, t \in I, r \in R. \tag{28}$$

Left multiply (27) by $t$ and subtract from (28), we have $[[x, r], t]yd(z) = 0$. Since $R$ is a prime ring so either $[[x, r], t] = 0$ for all $x, t \in I$ and $r \in R$ or $Id(I) = (0)$. Now, let us consider

$$[[x, r], t] = 0 \ \forall \ x, t \in I, r \in R. \tag{29}$$

Replace $x$ by $xr$ in (29) and using it, we obtain

$$[x, r][t, r] = 0 \ \forall \ x, t \in I, r \in R. \tag{30}$$

Substitute $ty$ in place of $y$ in (30) and using it to get $[x, r]t[y, r] = 0$ for all $x, y, t \in I$ and $r \in R$. Again, primeness of $R$ implies that $R$ contains a nonzero central ideal and hence $R$ is commutative with the aid of Lemma 2.3.

Since $R$ is prime ring, the latter case $Id(I) = (0)$ implies that $d(I) = (0)$. Therefore, in this case $F(xy) = F(y)x$ for all $x, y \in I$. So our hypothesis yields that $F(x)(y \pm F(y)) = 0$ for any $x, y \in I$, by following the same procedure as in the first case we get $d(R) = 0$. □

In this sequel, it is natural to consider the following theorem:

**Theorem 3.4.** *Let $I$ be a nonzero ideal of a prime ring $R$ and $F : R \to R$ a multiplicative (generalized)-reverse derivation associated with a map $d : R \to R$. If* *for some* $0 \neq a \in R$, $a(F(xy) \pm F(x)F(y)) = 0$ *for all* $x, y \in I$, *then either* $d = 0$ *or* $R$ *is commutative.*

In order to prove this claim, we need to develop some more results. Hence, we shall give the proof in the end.

**Theorem 3.5.** *Let* $I$ *be a nonzero ideal of a prime ring* $R$ *and* $F : R \to R$ *a multiplicative (generalized)-reverse derivation associated with a map* $d : R \to R$*. If for some* $0 \neq a \in R$, $a(F(x)F(y) \pm yx) = 0$ *for all* $x, y \in I$, *then* $R$ *is commutative.*

**Proof.** Let us begin with

$$a(F(x)F(y) + yx) = 0 \;\forall\; x, y \in I. \tag{31}$$

Replace $y$ by $zy$ in (31), we get $a(F(x)F(y) + yx)z + a(F(x)yd(z) + [z, yx]) = 0$ for all $x, y, z \in I$. By hypothesis, we find

$$a(F(x)yd(z) + y[z, x] + [z, y]x) = 0 \;\forall\; x, y, z \in I. \tag{32}$$

Substitute $zx$ for $x$ in (32), we obtain

$$a(F(x)zyd(z) + xd(z)yd(z) + yz[z, x] + [z, y]zx) = 0. \tag{33}$$

Replace $y$ by $yz$ in (32), we find

$$a(F(x)yzd(z) + yz[z, x] + [z, y]yx) = 0. \tag{34}$$

Subtract (34) from (33) in order to find

$$a(F(x)[z, y]d(z) + xd(z)yd(z)) = 0 \;\forall\; x, y, z \in I. \tag{35}$$

Replace $x$ by $zx$ in (35), we obtain

$$a(F(x)z[z, y]d(z) + xd(z)[z, y]d(z) + zxd(z)yd(z)) = 0. \tag{36}$$

Substituting $y$ with $zy$ in (35), we have

$$a(F(x)z[z, y]d(z) + xd(z)zyd(z)) = 0. \tag{37}$$

Subtract (36) from (37), we get

$$a([xd(z), z]yd(z) + xd(z)[y, z]d(z)) = 0 \;\forall\; x, y, z \in I. \tag{38}$$

Substitute $ax$ instead of $x$ in (38), we obtain

$$a[axd(z), z]yd(z) + a^2xd(z)[y, z]d(z) = 0. \tag{39}$$

Left multiply (38) by $a$ and subtract it from (39), we find $[a, z]xd(z)yd(z) = 0$ for all $x, y, z \in I$. Therefore, primeness of $R$ forces either $[a, I] = (0)$ or $Id(I) = (0)$. That means either $a \in Z(R)$ or $d(I) = (0)$.

Firstly, if $d(I) = (0)$, then (32) becomes

$$ay[z, x] + a[z, y]x = 0 \ \forall \ x, y, z \in I. \tag{40}$$

Putting $y = ay$ in relation (40) in order to find

$$a^2y[z, x] + a^2[z, y]x + a[z, a]yx = 0 \ \forall \ x, y, z \in I. \tag{41}$$

Using (40) in (41), we obtain $a[z, a]yx = 0$. Since $R$ is prime ring and $I$ a nonzero ideal of $R$, we get $a[a, z] = 0$ for all $z \in I$. Replacing $z$ by $zr$, where $r \in R$, we get $aI[a, r] = (0)$. It implies that $a \in Z(R)$. But center of a prime ring does not contain zero-divisor, hence (40) implies

$$y[z, x] + [z, y]x = 0. \tag{42}$$

Replace $x$ by $xr$ in (42), where $r \in R$ and using it, we obtain $yx[z, r] = 0$ for all $x, y, z \in I$ and $r \in R$. Primeness of $R$ implies $I \subseteq Z(R)$. By Lemma 2.3, $R$ is commutative.

On the other hand, if $a \in Z(R)$ and $R$ is a prime ring so (32) yields

$$F(x)yd(z) + y[z, x] + [z, y]x = 0 \ \forall \ x, y, z \in I. \tag{43}$$

Replace $x$ by $zx$ in (43) and we have

$$F(x)zyd(z) + yz[z, x] + [z, y]xz + xd(z)yd(z) = 0. \tag{44}$$

Substitute $zy$ for $y$ in (43) and subtract from (44) to obtain

$$[y, z][z, x] + [[z, y], z]x + xd(z)yd(z) = 0. \tag{45}$$

Replace $x$ by $zx$ in (45) and we find

$$[y, z]z[z, x] + [[z, y], z]zx + zxd(z)zyd(z) = 0. \tag{46}$$

Left multiply (45) by $z$ and subtract it from (46) in order to obtain $[[y, z], z][z, x] + [[[z, y], z], z]x = 0$. By replacing $x$ with $xt$ in the last expression and using it, we obtain $[[y, z], z]x[z, t] = 0$. In particular, we get $[[y, z], z]I[[y, z], z] = (0)$. It implies that $[[y, z], z] = 0$ for all $y, z \in I$. From this it is easy to conclude that $I$ is commutative and so is $R$.

By doing all as above, one can prove the same conclusion for $a(F(x)F(y) - yx) = 0$ for all $x, y \in I$. □

**Theorem 3.6.** *Let $I$ be a nonzero ideal of a prime ring $R$ and $F : R \to R$ a multiplicative (generalized)-reverse derivation associated with a map $d : R \to R$. If for some $0 \neq a \in R$, $a(F(xy) \pm yx) = 0$ for all $x, y \in I$, then $R$ is commutative.*

**Proof.** Let us begin with

$$a(F(xy) + yx) = 0 \ \forall \ x, y \in I. \tag{47}$$

Substituting $x$ with $zx$ in (47), we get $a(F(xy)+yx)z + a(xyd(z)+y[z,x]) = 0$ for all $x, y, z \in I$. By hypothesis, we find

$$a(xyd(z) + y[z,x]) = 0. \tag{48}$$

Replace $x$ by $ax$ in (48), we obtain

$$a^2xyd(z) + aya[z,x] + ay[z,a]x = 0. \tag{49}$$

Left multiply (48) by $a$, we find

$$a^2xyd(z) + a^2y[z,x] = 0. \tag{50}$$

Subtract (50) from (49) and we infer

$$a([y,a][z,x] + y[z,a]x) = 0. \tag{51}$$

Replace $z$ by $zx$ in (51), we get

$$a([y,a][z,x]x + y[z,a]x^2 + yz[x,a]x) = 0. \tag{52}$$

Combining (51) and (52), we have

$$a(yz[x,a]x) = 0. \tag{53}$$

Putting $y = ry$ in (53), where $r \in R$, we obtain $aRyz[x,a]x = (0)$. Since $a \neq 0$ and $R$ is a prime ring, we must have $yz[x,a]x = 0$ for all $x, y, z \in I$. Again primeness of $R$ implying $[x,a]x = 0$ for all $x \in I$. Linearizing w.r.t.$x$, we get

$$[x,a]y + [y,a]x = 0 \ \forall \ x, y \in I. \tag{54}$$

Substituting $x$ with $xr$ in (54), we find

$$[x,a]ry + x[r,a]y + [y,a]xr = 0 \ \forall \ x, y \in I, r \in R. \tag{55}$$

Right multiply (55) by $r$ and subtract from (54), we get

$$[x,a][r,y] + x[r,a]y = 0. \tag{56}$$

Replace $y$ by $yz$ in (56), we obtain

$$[x,a][r,y]z + [x,a]y[r,z] + x[r,a]yz = 0. \tag{57}$$

Relation (56) reduces it to

$$[x,a]y[r,z] = 0 \ \forall \ x, y, z \in I, r \in R. \tag{58}$$

Since $R$ is a prime ring so either $[I, a] = (0)$ or $[R, I] = (0)$. That means, either $a \in Z(R)$ or $R$ contains a nonzero central ideal. It is clear from Lemma 2.3 that if $R$ contains a nonzero central ideal then $R$ is commutative. On the other hand, let us assume that $a \in Z(R)$ and $R$ is a prime ring. Therefore, relation (48) forces

$$xyd(z) + y[z, x] = 0 \ \forall \ x, y, z \in I. \tag{59}$$

Substitute $rx$ in place of $x$ in (59), we get

$$rxyd(z) + yr[z, x] + y[z, r]x = 0 \ \forall \ x, y, z \in I, r \in R. \tag{60}$$

Left multiply (59) by $r$ and subtract from (60), we find

$$[y, r][z, x] + y[z, r]x = 0. \tag{61}$$

Replace $x$ by $xr$ in (61) and using it to obtain $[y, r]x[z, r] = 0$ for all $x, y, z \in I$ and $r \in R$. Again, primeness of $R$ implies that $R$ contains a nonzero central ideal. Hence, $R$ is commutative.

By doing all as above, one can prove the same conclusion for $a(F(xy) - yx) = 0$ for all $x, y \in I$. □

In the direction of Herstein [17], we obtain the following results on multiplicative reverse derivations:

**Corollary 3.7.** *Let $I$ be a nonzero ideal of a prime ring $R$ and $F : R \to R$ a multiplicative (generalized)-reverse derivation together with a map $d : R \to R$. If any one of the following identities hold true*

1. $a(F(xy) \pm xy) = 0$
2. $a(F(xy) \pm yx) = 0$
3. $a(F(x)F(y) \pm xy) = 0$
4. $a(F(x)F(y) \pm yx) = 0$

*for all $x, y \in I$, then $R$ is a commutative integral domain and $F$ is a multiplicative (generalized)-derivation of $R$.*

**Remark 3.8.** *It is easy to see that the above conditions are not necessary for the commutativity of a prime ring. Let $F : Z_2[x] \to Z_2[x]$ be the ordinary differential operator on the polynomial ring $Z_2[x]$. Let us consider $F$ as a multiplicative (generalized)-reverse derivation associated with a map $d = F$ and we have $F(fg) = F(g)f + gd(f)$ for all $f, g \in Z_2[x]$. Now, note that the above conditions (1-4) are not always true but $Z_2[x]$ is the commutative integral domain.*

**Remark 3.9.** *Let $F$ be a multiplicative (generalized)-reverse derivation of ring $R$ together with a map $f$ and $\lambda : R^* \to R$ be a multiplicative left centralizer, where $R^*$ denotes the opposite ring of $R$. Then the mapping $G = F \pm \lambda$ is a multiplicative (generalized)-reverse derivation associated with map $f$.*

**Example 3.10.** $R = \left\{ \begin{pmatrix} 0 & m & n \\ 0 & 0 & p \\ 0 & 0 & 0 \end{pmatrix} : m, n, p \in Z \right\}$, *where $Z$ is the ring of integers. Define a mapping $d : R \to R$ such that* $d\begin{pmatrix} 0 & m & n \\ 0 & 0 & p \\ 0 & 0 & 0 \end{pmatrix} = \begin{pmatrix} 0 & m & np \\ 0 & 0 & -p \\ 0 & 0 & 0 \end{pmatrix}$. *Clearly, $d$ is a multiplicative reverse derivation. Let $F = d$ be the multiplicative (generalized)-reverse derivation associated with $d$. It is easy to check that $F$ satisfies all the polynomial constraints: $a(F(xy) \pm xy) = 0$, $a(F(xy) \pm yx) = 0$, $a(F(x)F(y) \pm xy) = 0$ and $a(F(x)F(y) \pm yx) = 0$ but $R$ is not commutative. Hence the imposed restriction of primeness on $R$ is not superfluous.*

Next we construct the following example, which is showing that some of our theorems can not be extended to semiprime rings.

**Example 3.11.** *Consider a semiprime ring $S$ and a domain $D$. Let $R = S \times D$ and $\phi$ be a multiplicative reverse derivation of $S$ so clearly $\phi(0) = 0$ . We define $F : R \to R$ and $d : R \to R$ respectively as: $F(x_1, x_2) = (\phi(x_1), x_2)$ and $d(x_1, x_2) = (\phi(x_1), 0)$. Then $F$ is a multiplicative (generalized)-reverse derivation associated with a map $d$ on $R$ and clearly $F$ is not a multiplicative (generalized)-derivation. The set $I = \{0\} \times D$ is an ideal of $R$. Now, it is easy to see that $a(F(xy) \pm xy) = 0$ and $a(F(x)F(y) \pm xy) = 0$ for all $x, y \in I$; but $R$ is not a commutative ring.*

If $R$ is a prime ring, then $Q_r(R)$ and $Q_s(R)$ denotes the right and symmetric Martindale ring of quotients respectively. It is a well known fact that centers of $Q_r(R)$ and $Q_s(R)$ coincide and denoted by $C$, which is called the extended centroid of $R$. The symbol $RC = R_C$ denotes the central closure of $R$, which is again a prime ring, hence we may construct $Q_r(R_C)$. In this section, we explore the structure of the mapping $d$ which is associated with multiplicative (generalized)-reverse derivation $F$ of $R$. Precisely, we show that in the definition of a multiplicative (generalized)-reverse derivation of a prime ring $R$, $d$ needs not necessarily be a reverse derivation of $R$.

**Proposition 3.12.** *Let $R$ be a prime ring and $F : R \to R$ the multiplicative (generalized)-reverse derivation of $R$ associated with a mapping $d : R \to R$. Then either $R$ is commutative or $d$ is the multiplicative left centralizer.*

**Proof.** For any $x, y, z \in R$, we find $F(xzy) = F(zy)x + zyd(x)$. That is,

$$F(xzy) = F(y)zx + yd(z)x + zyd(x) \tag{62}$$

On the other hand, we obtain

$$F(xzy) = F(y)xz + yd(xz) \tag{63}$$

Subtract (63) from (62), we obtain

$$F(y)[z,x] + yd(z)x + zyd(x) - yd(xz) = 0 \tag{64}$$

Replace $x$ by $xz$ in (64), we get

$$F(y)[z,x]z + yd(z)xz + zyd(xz) - yd(xz^2) = 0 \tag{65}$$

Right multiply (64) by $z$ and subtract it from (65), we obtain

$$zyd(xz) - yd(xz^2) - zyd(x)z + yd(xz)z = 0 \tag{66}$$

Substitute $ty$ instead of $y$ in (66), we find

$$ztyd(xz) - tyd(xz^2) - ztyd(x)z + tyd(xz)z = 0 \tag{67}$$

Left multiply (66) by $t$ and subtract it from (67), we get $[z,t]y(d(xz) - d(x)z) = 0$ for all $x, y, z, t \in R$. That is, $[z,t]R(d(xz) - d(x)z) = (0)$ for all $x, z, t \in R$. Hence, primeness of $R$ ensures our conclusion. □

Now, we are in the position to prove Theorem 3.4

**Proof of Theorem 3.4.** Let us assume that $R$ is noncommutative. By hypothesis, $a(F(xy) \pm F(x)F(y)) = 0$ for all $x, y \in I$. That is,

$$(F(y)x + yd(x) \pm F(x)F(y)) = 0 \tag{68}$$

Replace $y$ by $zy$ in (68), we find

$$a(F(y)zx + yd(z)x + zyd(x) \pm F(x)F(y)z \pm F(x)yd(z)) = 0 \tag{69}$$

Right multiply (68) by $z$ and subtract (69) from it, we obtain

$$a(F(y)[x,z] + [yd(x),z] - yd(z)x \pm F(x)yd(z)) = 0 \tag{70}$$

Substituting $z$ with $zx$ in (70) and applying Proposition 3.12, we get

$$a(F(y)[x,z]x + [yd(x),z]x + z[yd(x),x] - yd(z)x^2 \pm F(x)yd(z)x) = 0 \tag{71}$$

Using (70) in (71), we are left with $z[yd(x), x] = 0$ for all $x, y, z \in I$. For any $t \in I$, replace $y$ by $ty$, we find $zt[yd(x), x] + z[t, x]yd(x) = 0$. It yields $z[t, x]yd(x) = 0$ for all $x, y \in I$. Since $R$ is a prime ring so either $I$ is commutative or $d(I) = (0)$. In the view of our assumption, $d$ is zero on $I$ and hence on $R$. □

**Corollary 3.13.** *Let $R$ be a noncommutative prime ring and $F : R \to R$ be a mapping of $R$ satisfying $F(x + y) = F(x) + F(y)$ and $F(xy) = F(y)x + yd(x)$ for all $x, y \in R$ and a mapping $d$ of $R$. Then $d$ is a left centralizer of $R$.*

**Theorem 3.14.** *Every generalized reverse derivation $F$ of a noncommutative prime ring $R$ is of the form $F(x) = -xq$ for all $x \in R$ and fixed $q \in Q_r(R_C)$.*

**Proof.** Since $F$ is a generalized reverse derivation of $R$ associated with left centralizer $d$ of $R$. In the light of Hvala [18], these exist a $q \in Q_r(R_C)$ such that $d(x) = qx$ for all $x \in R$. From Eq. (64), we have $F(y)[z, x] + yqzx + zyqx - yqxz = 0$ for all $x, y, z \in R$. That gives, $(F(y) + yq)[z, x] + zyqx = 0$. Replacing $x$ by $xr$ in the last relation, we get $(F(y) + yq)R[r, y] = (0)$. Since $R$ is prime, we get $F(y) = -yq$ for all $y \in R$. □

**Corollary 3.15.** *Let $F : R \to R$ be a generalized reverse derivation of a noncommutative prime ring $R$. If $F$ is commuting on $R$, then there exists some $q \in C$ such that $F(x) = -qx$ for all $x \in R$.*

**Proof.** By hypothesis, $[F(x), x] = 0$ for all $x \in R$. In the view of theorem 3.14, we have $-[xq, x] = 0$ for some $q \in Q_r(R_C)$. That is, $x[q, x] = 0$ for all $x \in R$. Hence, by Theorem (I) of [9], we are done. □

**Acknowledgement.** We are thankful to Prof. Basudeb Dhara for his helpful suggestions. We would also like to express our sincere gratitude to the anonymous referee for his/her constructive comments that definitely improve the readability of the paper.

## References

[1] A. Aboubakr and S. González, *Reverse generalized derivations of semiprime rings*, Sib. Math. J., 56(2) (2015), 199-205.

[2] A. Ali, D. Kumar and P. Miyan, *On generalized derivations and commutativity of prime and semiprime rings*, Hacet. J. Math. Stat., 40(3) (2011), 367-374.

[3] S. Ali, B. Dhara, N. A. Dar and A. N. Khan, *On Lie ideals with multiplicative (generalized)-derivations in prime and semiprime rings*, Beitr. Algebra Geom., 56(1) (2015), 325-337.

[4] M. Ashraf, A. Ali and S. Ali, *Some commutativity theorems for rings with generalized derivations*, Southeast Asian Bull. Math., 31(3) (2007), 415-421.

[5] M. Ashraf and N. Rehman, *On derivation and commutativity in prime rings*, East-West J. Math., 3(1) (2001), 87-91.

[6] H. E. Bell and W. S. Martindale, III, *Centralizing mappings of semiprime rings*, Canad. Math. Bull., 30(1) (1987), 92-101.

[7] M. Brešar, *On the distance of composition of two derivations to the generalized derivation*, Glasgow Math. J., 33(1) (1991), 89-93.

[8] D. K. Camci and N. Aydin, *On multiplicative (generalized)-derivations in semiprime rings*, Commun. Fac. Sci. Univ. Ank. Sér. A1 Math. Stat., 66(1) (2017), 153-164.

[9] C. M. Chang and Y. C. Lin, *Derivations on one-sided ideals of prime rings*, Tamsui Oxf. J. Math. Sci., 17(2) (2001), 139-145.

[10] M. N. Daif, *When is a multiplicative derivation additive?*, Internat. J. Math. Math. Sci., 14(3) (1991), 615-618.

[11] M. N. Daif and H. E. Bell, *Remarks on derivations on semiprime rings*, Internat. J. Math. Math. Sci., 15(1) (1992), 205-206.

[12] M. N. Daif and M. S. Tammam-El-Sayiad, *Multiplicative generalized derivations which are additive*, East-West J. Math., 9(1) (2007), 31-37.

[13] B. Dhara and S. Ali, *On multiplicative (generalized)-derivations in prime and semiprime rings*, Aequationes Math., 86(1-2) (2013), 65-79.

[14] A. K. Faraj, *On generalized $(\theta,\phi)$-reverse derivations of prime rings*, Iraqi J. Sci., 52(2) (2011), 218-224.

[15] H. Goldmann and P. Šemrl, *Multiplicative derivations on C(X)*, Monatsh. Math., 121(3) (1996), 189-197.

[16] I. Gusic, *A note on generalized derivations of prime rings*, Glas. Mat. Ser. III, 40(60) (2005), 47-49.

[17] I. N. Herstein, *Jordan derivations of prime rings*, Proc. Amer. Math. Soc., 8 (1957), 1104-1110.

[18] B. Hvala, *Generalized derivations in rings*, Comm. Algebra, 26(4) (1998), 1147-1166.

[19] J. Pinter-Lucke, *Commutativity conditions for rings: 1950-2005*, Expo. Math., 25(2) (2007), 165-174.

[20] W. S. Martindale, III, *When are multiplicative mappings additive?*, Proc. Amer. Math. Soc., 21 (1969), 695-698.

[21] H. Marubayashi, M. Ashraf, N. Rehman and S. Ali, *On generalized $(\alpha, \beta)$-derivations in prime rings*, Algebra Colloq., 17(Spec. 1) (2010), 865-874.

[22] J. H. Mayne, *Centralizing mappings of prime rings*, Canad. Math. Bull., 27(1) (1984), 122-126.

[23] E. C. Posner, *Derivations in prime rings*, Proc. Amer. Math. Soc., 8 (1957), 1093-1100.

[24] N. Rehman, *On commutativity of rings with generalized derivations*, Math. J. Okayama Univ., 44 (2002), 43-49.

[25] M. Samman and N. Alyamani, *Derivations and reverse derivations in semiprime rings*, Int. Math. Forum, 2(39) (2007), 1895-1902.

[26] G. S. Sandhu and D. Kumar, *A note on derivations and Jordan ideals of prime rings*, AIMS Math., 2(4) (2017), 580-585.

[27] G. S. Sandhu and D. Kumar, *Derivable mappings and commutativity of associative rings*, Italian J. Pure Appl. Math., 40 (2018), 376-393.

[28] S. K. Tiwari, R. K. Sharma and B. Dhara, *Some theorems of commutativity on semiprime rings with mappings*, Southeast Asian Bull. Math., 42(2) (2018), 279-292.

**Gurninder S. Sandhu** (Corresponding Author)
Department of Mathematics
Punjabi University
147002 Patiala, India
and
Multani Mal Modi College
147001 Patiala, India.
e-mail: gurninder_rs@pbi.ac.in

**Deepak Kumar**
Department of Mathematics
Punjabi University
147002 Patiala, India
e-mail: deep_math1@yahoo.com

# 7

# $n$-ABSORBING MONOMIAL IDEALS IN POLYNOMIAL RINGS

Hyun Seung Choi and Andrew Walker

Communicated by Abdullah Harmancı

ABSTRACT. In a commutative ring $R$ with unity, given an ideal $I$ of $R$, Anderson and Badawi in 2011 introduced the invariant $\omega(I)$, which is the minimal integer $n$ for which $I$ is an $n$-absorbing ideal of $R$. In the specific case that $R = k[x_1, \ldots, x_n]$ is a polynomial ring over a field $k$ in $n$ variables $x_1, \ldots, x_n$, we calculate $\omega(I)$ for certain monomial ideals $I$ of $R$.

**Keywords:** $n$-Absorbing ideal, monomial ideal, Noether exponent

## 1. Introduction

Throughout this paper, we set $\mathbb{N} := \{1, 2, \ldots\}$, $\mathbb{N}_0 := \{0, 1, 2, \ldots\}$, and $R$ will denote a commutative ring with unity. Given a nonzero ideal $I$ of $R$, $\text{Ass}(R/I)$ will denote the set of associated primes of $I$ in $R$. The primary notion we are interested in this paper is the following:

**Definition 1.** Let $n \in \mathbb{N}$, $R$ a commutative ring with unity, and $I$ an ideal of $R$. $I$ is said to be an *$n$-absorbing ideal* of a ring $R$ if for any $x_1, \ldots, x_{n+1} \in R$ such that $x_1 \cdots x_{n+1} \in I$, there are $n$ of the $x_i$'s whose product is in $I$. $I$ is said to be a *strongly $n$-absorbing ideal* of a ring $R$ if for any ideals $I_1, \ldots, I_{n+1}$ of $R$ such that $I_1 \cdots I_{n+1} \subseteq I$, there are $n$ of the $I_i$'s whose product is in $I$.

(Strongly) 2-absorbing ideals were initially defined and investigated by Badawi in [3] as a generalization of prime ideals, which are precisely the proper 1-absorbing ideals. In 2011, Anderson and Badawi together generalized this further to the notion of a (strongly) $n$-absorbing ideal for any $n \in \mathbb{N}$ defined above in [1]. For an ideal $I$ in a ring $R$, we let $\omega(I)$ denote the minimal integer $n \in \mathbb{N}$ such that $I$ is $n$-absorbing. In a general ring, $I$ may not be $n$-absorbing for any $n \in \mathbb{N}$, in which case we set $\omega(I) = \infty$. Similarly, we can define the invariant $\omega^{\bullet}(I)$ to be the smallest integer $n \in \mathbb{N}$ for which an ideal $I$ is strongly $n$-absorbing, and set $\omega^{\bullet}(I) = \infty$ if no such integer exists. We set $\omega(R) = \omega^{\bullet}(R) = 0$. It is easy to see that $\omega(I) \leq \omega^{\bullet}(I)$ holds for each ideal $I$ of $R$. In fact, Anderson and Badawi in

Conjecture 1 of [1, page 1669] postulate that $\omega(I) = \omega^\bullet(I)$ holds for any ideal $I$ in an arbitrary ring $R$; that is, they conjecture that the notion of an $n$-absorbing ideal and strongly $n$-absorbing ideal coincide. As of this writing, this problem remains open. However, it is known that the conjecture holds true for any $n \in \mathbb{N}$ if $R$ is a Prüfer domain, i.e., an integral domain such that the set of ideals of $R_M$ is totally ordered under set inclusion for each maximal ideal $M$ of $R$ ([1, Corollary 6.9]), or if $R$ is a commutative algebra over an infinite field ([7]), and for any ring $R$ if $n = 2$ ([3, Theorem 2.13]). The interested reader may refer to the survey article [4, Section 5] for further information on strongly $n$-absorbing ideals. Anderson and Badawi made two more conjectures in [1] which were investigated by several researchers, and affirmative answers were given, either partial or complete. For example, the second Anderson-Badawi conjecture states that given an ideal $I$ of a ring $R$ and an indeterminate $X$, $\omega_{R[X]}(I[X]) = \omega_R(I)$ ([1, page 1661]). That is, for each $n \in \mathbb{N}$, $I$ is an $n$-absorbing ideal of $R$ if and only if $I[X]$ is an $n$-absorbing ideal of $R[X]$. This conjecture originates from the well-known result that $I$ is a prime ideal (i.e., 1-absorbing ideal) of $R$ if and only if $I[X]$ is a prime ideal of $R[X]$. Anderson and Badawi themselves proved this conjecture for an arbitrary commutative ring when $n = 2$ ([1, Theorem 4.15]), and Nasehpour proves that the second conjecture holds for every $n \in \mathbb{N}$ when $R$ belongs to certain classes of rings, including the class of Prüfer domains ([15]). In [11] Laradji independently proved that the second conjecture holds when $R$ is an arithmetical ring, i.e., when the set of ideals of $R_M$ is totally ordered under set inclusion for each maximal ideal $M$ of $R$.

Recall that for an ideal $I$ in a ring $R$, the *Noether exponent* of $I$, denoted by $e(I)$, is the minimal integer $\mu \in \mathbb{N}$ such that $(\sqrt{I})^\mu \subseteq I$. If such an integer does not exist, we set $e(I) = \infty$. We also set $e(R) = 0$. In a Noetherian ring, since $\sqrt{I}$ is finitely generated for any ideal $I$, $e(I) < \infty$. Anderson and Badawi in [1] establish a connection between $\omega^\bullet(I)$ and Noether exponents:

**Theorem 1.1.** [1, Remark 2.2, Theorem 5.3, Section 6, Paragraph 2 on page 1669] *Let $I_1, \ldots, I_r$ be ideals of a ring $R$. Then $\omega(I_1 \cap \cdots \cap I_r) \leq \omega(I_1) + \cdots + \omega(I_r)$ and $\omega^\bullet(I_1 \cap \cdots \cap I_r) \leq \omega^\bullet(I_1) + \cdots + \omega^\bullet(I_r)$. In particular, let $I$ be an ideal in a Noetherian ring $R$. If $I = Q_1 \cap \cdots \cap Q_n$, where the $Q_i$ are primary ideals, then $\omega(I) \leq \omega^\bullet(I) \leq \sum_{i=1}^{n} e(Q_i)$. Thus every ideal in a Noetherian ring is $n$-absorbing for some $n \in \mathbb{N}$.*

On the other hand, the third Anderson-Badawi conjecture claims that for each $n \in \mathbb{N}$ and an $n$-absorbing ideal $I$ of a ring $R$, $(\sqrt{I})^n \subseteq I$ ([1, Conjecture 2, page

1669]). This conjecture was proved for $n = 2$ by Badawi ([3]), for $n = 3$ by Laradji ([11]), for $n = 3, 4, 5$ by Sihem and Sana ([17]), and for arbitrary $n$ and $R$ by the authors ([6]) and Donadze ([8]), independently. We summarize this as the following theorem in terms of $\omega(I)$ and $e(I)$, along with the result concerning primary ideals ([1, Theorem 6.3(c), Theorem 6.6]).

**Theorem 1.2.** *Given an ideal $I$ of a ring $R$, $e(I) \leq \omega(I)$. If $Q$ is a primary ideal of $R$, then $\omega(Q) = \omega^\bullet(Q) = e(Q)$.*

This raises the question then if for an arbitrary ideal $I$ whether $\omega(I)$ can be described purely in terms of Noether exponents or possibly other well-known ring-theoretic invariants. This has been investigated to some extent by others in at least one case. Namely, Moghimi and Naghani [13, Theorem 2.21(1)] show that in a discrete valuation ring $R$, $\omega(I)$ is precisely the length of the $R$-module $R/I$.

In this spirit, we attempt to give in this paper a description of $\omega(I)$ in terms of other ring-theoretic invariants in the special case that $I$ is a monomial ideal of a polynomial ring over a field. In some cases, our arguments are general enough to also give the same results for $\omega^\bullet(I)$, and thus as a side-effect we can show that in some cases the notions of a $n$-absorbing ideal and a strongly $n$-absorbing ideal coincide as Anderson and Badawi conjecture.

The present paper is divided into two parts. In Section 2, we review some definitions and facts concerning $n$-absorbing ideals and monomial ideals. Using these, we calculate $\omega(I)$ for primary monomial ideals by computing Noether exponents and the standard primary decomposition of monomial ideals. These results lead to the study of how $\omega(I)$ can be explicitly computed from the generating set of $I$ when $I$ is a monomial ideal of $R = k[x_1, \ldots, x_n]$ with $n \leq 3$ in the following section.

The second part is Section 4, where we define and investigate $\omega$-linear monomial ideals, i.e., monomial ideals $I$ such that $\omega(I^m) = m\omega(I)$ for each $m \in \mathbb{N}$. We give a characterization theorem for primary $\omega$-linear monomial ideals, and in particular show that integrally closed monomial ideals in $R = k[x, y]$ are $w$-linear, as well as the edge ideal of a cycle.

## 2. Some background

As a prerequisite of the main section of this paper, we briefly review some of the basic material excerpted from [10] regarding monomial ideals, and show that $\omega(I)$ can be directly calculated from the generators of $I$ when $I$ is a primary monomial ideal.

Let $k$ be a field and $R = k[x_1, \ldots, x_n]$ be the polynomial ring with $n$ variables over $k$. An element of $R$ of the form $x_1^{a_1} \cdots x_n^{a_n}$ with $a_i \in \mathbb{N}_0$ is called a *monomial*, and an ideal of $R$ generated by monomials is called a *monomial ideal*. The degree of $f = x_1^{a_1} \cdots x_n^{a_n}$, denoted by $\deg(f)$, is defined to be $a_1 + \cdots + a_n$. $G(I)$ will denote the set of monomials in $I$ which are minimal with respect to divisibility. Any element of $R$ can be written uniquely as a $k$-linear combination of monomials; that is, given $f \in R$, we may write $f = \sum a_u u$ where the sum is taken over the monomial ideals of $R$ and $a_u \in k$ for each monomial $u$. Then the *support* of $f$, denoted by $\mathrm{supp}(f)$, is the set of monomials $u$ such that $a_u \neq 0$. An ideal $I$ of a ring $R$ is *irreducible* if there are no ideals $I_1, I_2$ of $R$ such that $I = I_1 \cap I_2$ and $I \subsetneq I_1$, $I \subsetneq I_2$. We denote by $\mathfrak{m}$ the unique maximal homogeneous ideal of $R$.

**Lemma 2.1.** [10, Chapter 1] *Let $R = k[x_1, \ldots, x_n]$ and $I$ a monomial ideal of $R$ generated by monomials $u_1, \ldots, u_r$ of $R$. Then the following hold:*

(i) *Given a monomial $f \in I$, there exists $i \in \{1, \ldots, r\}$ so $u_i | f$.*

(ii) *$G(I)$ is the unique minimal set of monomial generators of $I$.*

(iii) *$I$ can be written as a finite intersection of ideals of the form $(x_{i_1}^{a_1}, \ldots, x_{i_m}^{d_m})$. An irredundant presentation of this form is unique ($I = Q_1 \cap \cdots \cap Q_r$ is irredundant if none of the ideals $Q_i$ can be omitted).*

(iv) *$I$ is irreducible if and only if $I$ is of the form $(x_{i_1}^{d_1}, \ldots, x_{i_m}^{d_m})$. Moreover, every irreducible monomial ideal of the form $(x_{i_1}^{d_1}, \ldots, x_{i_m}^{d_m})$ is $(x_{i_1}, \ldots, x_{i_m})$-primary.*

(v) *If $J$ is another monomial ideal of $R$, then*

$$I \cap J = (\{lcm(u, v) \mid u \in G(I), v \in G(J)\}).$$

*In particular, if $a$ and $b$ are coprime monomials of $R$ and $I$ is a monomial ideal of $R$, then $(ab, I) = (a, I) \cap (b, I)$.*

(vi) *An ideal $I'$ of $R$ is monomial if and only if for each $f \in I'$, $\mathrm{supp}(f) \subseteq I'$.*

By Lemma 2.1(iv), the irredundant unique decomposition of Lemma 2.1(iii) is also a primary decomposition of $I$, which is known as the *standard decomposition* of $I$ (see [10, P. 12]). We will also need the following characterization of primary monomial ideals:

**Lemma 2.2.** [9, Exercise 3.6] *Let $R = k[x_1, \ldots, x_n]$ and $P = (x_{i_1}, \ldots, x_{i_r})$ a monomial prime ideal of $R$. Then given a $P$-primary monomial ideal $Q$, $G(Q)$ consists of monomials of the ring $k[x_{i_1}, \ldots, x_{i_r}]$ and there exists $a_1, \ldots, a_r \in \mathbb{N}$*

*so* $\{x_{i_1}^{a_1}, \ldots, x_{i_r}^{a_r}\} \subseteq G(Q)$. *Conversely, every monomial ideal of this form is a* $P$*-primary ideal.*

**Proof.** Let $f \in G(Q)$. If $f \notin k[x_{i_1}, \ldots, x_{i_r}]$, then $x_j | f$ for some $x_j \notin P$ and $g = \dfrac{f}{x_j} \in Q$ since $Q$ is a $P$-primary ideal, but this contradicts the minimality of $G(Q)$. Hence $f \in k[x_{i_1}, \ldots, x_{i_r}]$. On the other hand, given $j \in \{1, \ldots, r\}$ there exists $a_j \in \mathbb{N}$ so $x_{i_j}^{a_j} \in G(Q)$, since $\sqrt{Q} = P$.
To prove the converse, let $Q$ be a monomial ideal such that $G(Q)$ consists of monomials of the ring $k[x_{i_1}, \ldots, x_{i_r}]$ and there exists $a_1, \ldots, a_r \in \mathbb{N}$ so $\{x_{i_1}^{a_1}, \ldots, x_{i_r}^{a_r}\} \subseteq G(Q)$. Then $\sqrt{Q} = P$ by [10, Proposition 1.2.4]. On the other hand, if $P_1 \in$Ass$(R/Q) \setminus \{P\}$, then $P_1 = Q : f$ for some monomial $f$ of $R$ ([10, Corollary 1.3.10]). Now choose $d \in \{1, \ldots, n\}$ so $x_d \in P_1 \setminus P$. Then $x_d f \in Q$, and $f \in Q$ by Lemma 2.1(i). But then $P_1 = R$, a contradiction. Hence Ass$(R/Q) = \{P\}$ and $Q$ is a $P$-primary monomial ideal. □

**Corollary 2.3.** *Let* $P$ *be a prime monomial ideal and* $I, J$ *be* $P$*-primary monomial ideals of* $R$*. Then both* $I \cap J$ *and* $IJ$ *are* $P$*-primary monomial ideals. Moreover,* $I : J$ *is a* $P$*-primary monomial ideal provided* $J \not\subset I$*.*

**Proof.** This is an immediate consequence of Lemma 2.1(v) and Lemma 2.2. □

We can now calculate $\omega(I)$, where $I$ is an irreducible monomial ideal.

**Lemma 2.4.** *Let* $R = k[x_1, \ldots, x_n]$ *denote a polynomial ring over a field* $k$*. Let* $I = (x_{i_1}^{d_1}, \ldots, x_{i_m}^{d_m})$*, where* $d_1, \ldots, d_n \in \mathbb{N}$ *and* $1 \leq i_1 < i_2 < \cdots < i_m \leq n$*. Then* $\omega(I) = \omega^{\bullet}(I) = e(I) = d_1 + \cdots + d_m - m + 1$*.*

**Proof.** Since $I$ is an $(x_{i_1}, x_{i_2}, \ldots, x_{i_m})$-primary ideal by Lemma 2.2, the first two equalities follow from Theorem 1.2. Thus it suffices to show that $e(I) = r$, where $r = d_1 + \cdots + d_m - m + 1$. We have $\sqrt{I} = (x_{i_1}, \ldots, x_{i_m})$. For $N \in \mathbb{N}$, $(\sqrt{I})^N \subseteq I$ if and only if for every $c_1, \ldots, c_m \in \mathbb{N}_0$ with $c_1 + \cdots + c_m = N$, we have $x_{i_1}^{c_1} \cdots x_{i_m}^{c_m} \in I$. By Lemma 2.1(i), the latter happens precisely when $c_i \geq d_i$ for some $1 \leq i \leq m$. Thus $e(I) = r$. □

Next, we produce a way to calculate $\omega(I)$ when $I$ is a monomial primary ideal not necessarily generated by pure powers.

**Lemma 2.5.** *Let* $I$ *be an ideal of a ring* $R$*. Suppose there is* $P \in$ Spec$(R)$ *such that* $I = J_1 \cap \cdots \cap J_r$*, where* $J_i$ *are ideals of* $R$ *with* $\sqrt{J_i} = P$ *for each* $i \in \{1, \ldots, r\}$*. Then* $e(I) = \max_{1 \leq i \leq r}\{e(J_i)\}$*.*

**Proof.** Note that $\sqrt{I} = \sqrt{J_1} \cap \cdots \cap \sqrt{J_r} = P$. Thus given $\mu \in \mathbb{N}$, $(\sqrt{I})^u \subseteq I$ if and only if $(\sqrt{J_i})^\mu \subseteq J_i$ for each $i \in \{1, \ldots, r\}$, from which the conclusion of the lemma follows. □

**Corollary 2.6.** *Let $R = k[x_1, \ldots, x_n]$ denote a polynomial ring over a field $k$. If $Q$ is a monomial primary ideal of $R$ and $Q = \bigcap_{i=1}^r Q_i$ is its standard decomposition, then*

$$\omega(Q) = \omega^\bullet(Q) = \max_{1 \leq i \leq r} \{e(Q_i)\}.$$

**Example 2.7.** Let $R = k[x, y, z]$ with a field $k$ and $I = (x^4, y^3, z^2, xy, y^2 z)$. Then repeatedly applying Lemma 2.1(v), we obtain the standard decomposition $I = (x, y^2, z^2) \cap (x^4, y, z^2) \cap (x, y^3, z)$. Thus by Lemma 2.4 and Corollary 2.6,

$$\omega(I) = \omega^\bullet(I) = \max\{1 + 2 + 2 - 3 + 1, 4 + 1 + 2 - 3 + 1, 1 + 3 + 1 - 3 + 1\} = 5.$$

## 3. When $I$ is a monomial ideal of $R = k[x_1, \ldots, x_n]$ with $n \leq 3$

In this section we show that when $I$ is a monomial ideal of $R = k[x_1, \ldots, x_n]$ with $n \leq 3$, then $\omega(I)$ can be explicitly calculated from $G(I)$. We first prove a theorem analogous to [2, Theorem 2.5]. Note that by $a_1 \cdots \widehat{a_i} \cdots a_n$ we mean $\prod_{1 \leq j \leq n\ j \neq i} a_j$.

**Lemma 3.1.** *Let $R$ be a UFD and $p$ an irreducible element of $R$. Then given $n \in \mathbb{N}$, $I$ is an $n$-absorbing ideal of $R$ if and only if $pI$ is an $(n+1)$-absorbing ideal of $R$. In particular, $\omega(pI) = \omega(I) + 1$.*

**Proof.** Suppose that $I$ is $n$-absorbing. Let $f_1, \ldots, f_{n+2} \in R$ and $f_1 \cdots f_{n+2} \in pI$. Then since $p$ is irreducible, $p \mid f_i$ for some $i$. Without loss of generality, suppose that $p \mid f_1$. Then $f_1/p \in R$, and so $(f_1/p) f_2 \cdots f_{n+2} \in I$. Since $I$ is $n$-absorbing, and hence $(n+1)$-absorbing as well, we have that either $(f_1/p) f_2 \cdots \widehat{f_i} \cdots f_{n+2} \in I$ for some $i \in \{2, \ldots, n+2\}$, in which case $f_1 f_2 \cdots \widehat{f_i} \cdots f_{n+2} \in pI$ and we are done, or $f_2 \cdots f_{n+2} \in I$. This is a product of length $n+1$, so that since $I$ is $n$-absorbing, for some $j$ with $2 \leq j \leq n+2$, we have $f_2 \cdots \widehat{f_j} \cdots f_{n+2} \in I$. Thus $pf_2 \cdots \widehat{f_j} \cdots f_{n+2} \in pI$, and so $f_1 f_2 \cdots \widehat{f_j} \cdots f_{n+2} \in pI$. This shows that $pI$ is then $(n+1)$-absorbing, and $\omega(pI) \leq \omega(I) + 1$.

To show the converse, suppose that $pI$ is an $(n+1)$-absorbing ideal. If $I$ is not an $n$-absorbing ideal, then there exists $f_1, \ldots, f_{n+1} \in R$ such that $f = f_1 \cdots f_{n+1} \in I$ but $f_1 \cdots \widehat{f_i} \cdots f_{n+1} \notin I$ for each $i$. Since $pI$ is $(n+1)$-absorbing and $pf \in pI$, it follows that either $pf_1 \cdots \widehat{f_i} \cdots f_{n+1} \in pI$ for some $i$ or $f \in pI$. But the former is impossible by our choice of $f_i$'s, and without loss of generality we may assume that

$p|f_1$. Now $(f_1/p)f_2\cdots f_n \in I$, and neither $\widehat{(f_1/p)}f_2\cdots f_{n+1}$ nor $(f_1/p)f_2\cdots \widehat{f_i}\cdots f_{n+1}$ is an element of $I$ for each $i \geq 2$. Therefore, since $R$ is a UFD, we may assume that none of $f_i$ are divisible by $p$. Now $pf_1\cdots f_{n+1} \in pI$, but $pf_1\cdots \widehat{f_i}\cdots f_{n+1} \notin pI$ and $f_1\cdots f_{n+1} \notin pI$, which contradicts the assumption that $pI$ is an $(n+1)$-absorbing ideal. Hence $I$ is an $n$-absorbing ideal and $\omega(pI) \geq \omega(I) + 1$. □

The following corollary is now immediate.

**Corollary 3.2.** *Given a monomial $f$ and an ideal $I$ of $R = k[x_1, \ldots, x_n]$, $\omega(fI) = deg(f) + \omega(I)$. In particular, $\omega(fR) = deg(f)$.*

Given a monomial ideal $I$ with the standard decomposition $I = \bigcap_{\ell=1}^{t} T_\ell$, we can define an equivalence relation on $\{1, \ldots, t\}$ by defining $i \sim j$ iff $\sqrt{T_i} = \sqrt{T_j}$, and set $\{S_i\}_{i=1}^{r}$ to be the corresponding equivalence classes. Then $Q_i = \bigcap_{\ell \in S_i} T_\ell$ is a monomial primary ideal for each $i \in \{1, \ldots, r\}$, and $I = \bigcap_{i=1}^{r} Q_i$ is an irredundant primary decomposition of $I$. We will call this decomposition the *canonical primary decomposition* of $I$.

**Theorem 3.3.** *Let $R = k[x_1, \ldots, x_n]$. Let $I$ be a monomial ideal with canonical primary decomposition $I = \bigcap_{i=1}^{r} Q_i$. If there exists $k \in \{1, \ldots, r\}$ such that $\sqrt{Q_i} \subseteq \sqrt{Q_k}$ for all $i \in \{1, \ldots, r\}$, then $\omega(I) = \max\{e(Q_k), \omega(\bigcap_{1\leq i\leq r, i\neq k} Q_i)\}$ and $\omega^\bullet(I) = \max\{e(Q_k), \omega^\bullet(\bigcap_{1\leq i\leq r, i\neq k} Q_i)\}$.*

**Proof.** Let $t = \max\{e(Q_k), \omega(\bigcap_{1\leq i\leq r, i\neq k} Q_i)\}$. We will first show that $I$ is $t$-absorbing. If not, then there are $f_1, \ldots, f_{t+1} \in R$ such that $f = \prod_{j=1}^{t+1} f_j \in I$ but $g_j := f/f_j \notin I$ for each $j \in \{1, \ldots, t+1\}$. Hence given any $i \in \{1, \ldots, t+1\}$, there exists $\ell \in \{1, \ldots, r\}$ such that $g_i \notin Q_\ell$, and since $f_i g_i = f \in I \subseteq Q_\ell$, we must have $f_i \in \sqrt{Q_\ell} \subseteq \sqrt{Q_k}$. Therefore, $g_j \in (\sqrt{Q_k})^t \subseteq (\sqrt{Q_k})^{e(Q_k)} \subseteq Q_k$ for all $j \in \{1, \ldots, t+1\}$. On the other hand, $\bigcap_{1\leq i\leq r, i\neq k} Q_i$ is $t$-absorbing and $f \in \bigcap_{1\leq i\leq r, i\neq k} Q_i$, so that we conclude $g_j \in \bigcap_{1\leq i\leq r, i\neq k} Q_i$ for some $j \in \{1, \ldots, t+1\}$ and thereby $g_j \in I$, a contradiction. Thus $\omega(I) \leq t$. Next, we show that $\omega(I) \geq t$; that is, $I$ is not $(t-1)$-absorbing. We now consider two cases.

*Case* 1: $t = \omega(\bigcap_{1\leq i\leq r, i\neq k} Q_i)$. Since $\bigcap_{1\leq i\leq r, i\neq k} Q_i$ is not $(t-1)$-absorbing, there are $h_1, \ldots, h_t \in R$ such that $h = \prod_{i=1}^{t} h_i \in \bigcap_{1\leq i\leq r, i\neq k} Q_i$ and $\ell_j := h/h_j \notin \bigcap_{1\leq i\leq r, i\neq k} Q_i$ for each $j \in \{1, \ldots, t\}$. By an argument similar to the first paragraph of this proof, $h_i \in \sqrt{Q_k}$ for each $i \in \{1, \ldots, t\}$, and so $h \in Q_k$. Hence $h \in I$ and $\ell_j \notin I$ for each $j \in \{1, \ldots, t\}$, so that $I$ is not $(t-1)$-absorbing.

*Case* 2: $t = e(Q_k)$. Consider the standard decomposition of $I$, and choose an irreducible component $T$ of $I$ such that $e(T) = e(Q_k)$ and $\sqrt{T} = \sqrt{Q_k}$. Since we obtained the canonical primary decomposition $I = \bigcap_{i=1}^{r} Q_i$ from the standard decomposition, we can choose a monomial $g \in (\bigcap_{1\leq i\leq r, i\neq k} Q_i) \setminus T$ by Lemma 2.1(vi). Now $T = (x_{i_1}^{a_1}, \ldots, x_{i_l}^{a_l})$ for some $a_j \in \mathbb{N}$ and $1 \leq i_1 < \cdots < i_l \leq n$. Note that we may assume that $g = \prod_{j=1}^{l} x_{i_j}^{c_j}$ for some $c_j \in \mathbb{N}_0$ such that $c_j < a_j$ for each $j \in \{1, \ldots, l\}$. Set

$$f := x_{i_1}^{a_1-1} \cdots x_{i_l}^{a_l-1}(x_{i_1} + \cdots + x_{i_l}).$$

Then $f$ is a product of $e(T)$ elements of $\sqrt{T}$ by Lemma 2.4, and so $f \in (\sqrt{T})^{e(T)} = (\sqrt{Q_k})^{e(Q_k)} \subseteq Q_k$. Since $g \mid f$ it also follows that $f \in \bigcap_{1\leq i\leq r, i\neq k} Q_i$. Hence $f \in I$. However, given $j \in \{1, \ldots, l\}$, $\dfrac{f}{x_{i_j}} \notin T$. Indeed, $x_{i_1}^{a_1-1} \cdots x_{i_l}^{a_l-1} \in \operatorname{supp}\left(\dfrac{f}{x_{i_j}}\right) \setminus T$ by Lemma 2.1(i), and $\dfrac{f}{x_{i_j}} \notin T$ by Lemma 2.1(vi). Similarly $x_{i_1}^{a_1-1} \cdots x_{i_l}^{a_l-1} = \dfrac{f}{x_{i_1} + \cdots + x_{i_l}} \notin T$. Therefore $I$ is not $(e(Q_k)-1)$-absorbing, and $\omega(I) \geq e(Q_k) = t$. Hence we have shown that $\omega(I) = \max\{e(Q_k), \omega(\bigcap_{1\leq i\leq r, i\neq k} Q_i)\}$. The proof of $\omega^{\bullet}(I) = \max\{e(Q_k), \omega^{\bullet}(\bigcap_{1\leq i\leq r, i\neq k} Q_i)\}$ can be obtained in a similar manner, and is omitted. □

The following corollary is immediate.

**Corollary 3.4.** *Let $R = k[x_1, \ldots, x_n]$ and $I$ a monomial ideal of $R$ with standard decomposition $I = \bigcap_{i=1}^{r} T_i$. Then $\omega(I) = \omega^{\bullet}(I) = \max_{1\leq i\leq r}\{e(T_i)\}$ if $Ass(R/I)$ is totally ordered under set inclusion.*

In the next proposition, we give a characterization of when the upper bound of $\omega(I)$ from Theorem 1.1 is sharp.

**Proposition 3.5.** *Let $I$ be a monomial ideal of $R = k[x_1, \ldots, x_n]$ with an irredundant primary decomposition $I = Q_1 \cap \cdots \cap Q_r$. Then $\omega(I) = \omega^{\bullet}(I) = \sum_{i=1}^{r} e(Q_i)$ if and only if $I$ has no embedded associated primes.*

**Proof.** Set $P_i = \sqrt{Q_i}$ for each $i = 1, \ldots, r$.
$\Rightarrow$: We prove the contrapositive; assume that $P_1, \ldots, P_r$ are not incomparable prime ideals. Then without loss of generality we may assume that $P_1 \subsetneq P_2$, and we have $\omega(Q_1 \cap Q_2) = \max\{e(Q_1), e(Q_2)\}$ by Corollary 3.4. Therefore by Theorem

1.1 we have

$$\begin{aligned}\omega(I) &= \omega\Big(Q_1 \cap Q_2 \cap \big(\bigcap_{i\neq 1,2} Q_i\big)\Big)\\ &\leq \omega\Big(Q_1 \cap Q_2\Big) + \omega\Big(\bigcap_{i\neq 1,2} Q_i\Big)\\ &= \max\{e(Q_1), e(Q_2)\} + \omega\Big(\bigcap_{i\neq 1,2} Q_i\Big)\\ &\leq \max\{e(Q_1), e(Q_2)\} + \sum_{i\neq 1,2} e(Q_i)\\ &< \sum_{i=1}^{r} e(Q_i).\end{aligned}$$

$\Leftarrow$: Assume that $P_1, \ldots, P_r$ are incomparable prime ideals. The case when $r = 1$ follows from Theorem 1.2, so we may assume that $r \geq 2$. Since $\omega(I) \leq \omega^{\bullet}(I) \leq \sum_{i=1}^{r} e(Q_i)$ by Theorem 1.1, it suffices to show that $I$ is not $(\sum_{i=1}^{r} e(Q_i) - 1)$-absorbing. Now given $i \in \{1, \ldots, r\}$, choose $T_i$ to be an irreducible component of $I$ with $\sqrt{T_i} = P_i$ and $e(T_i) = e(Q_i)$. Write $T_i = (x_{i_1}^{a_1}, \ldots, x_{i_{s_i}}^{a_{s_i}})$ with $1 \leq i_1 < \cdots < i_{s_i} \leq n$ and $a_1, \ldots, a_{s_i} \in \mathbb{N}$. For $i \in \{1, \ldots, r\}$ and $j \in \{1, \ldots, s_i\}$, set

$$f_{i,j} = x_{i_j} + \sum_{t\neq j} x_{i_t}^2 \text{ and } f_i = \Big(\sum_{l=1}^{s_i} x_{i_l}\Big)\Big(\prod_{j=1}^{s_i} f_{i,j}^{a_j - 1}\Big).$$

It follows that $f_i \in P_i^{e(T_i)} = (\sqrt{Q_i})^{e(Q_i)} \subseteq Q_i$. Thus $f := \prod_{i=1}^{r} f_i \in I$, and $f$ is a product of $\sum_{i=1}^{r} e(Q_i)$ elements of $R$. We wish to show that $\dfrac{f}{\sum_{l=1}^{s_i} x_{i_l}} \notin I$ and $\dfrac{f}{f_{i,j}} \notin I$ for each $i \in \{1, \ldots, r\}$ and $j \in \{1, \ldots, s_i\}$. Without loss of generality, we let $i = 1$. Note that $\dfrac{f_1}{f_{1,j}} \notin T_1$, since $\prod_{t=1}^{s_1} x_{1_t}^{a_t - 1} \in \operatorname{supp}\Big(\dfrac{f_1}{f_{1,j}}\Big) \setminus T_1$. On the other hand, $\sum_{l=1}^{s_i} x_{i_l} \notin P_1$ and $f_{i,l} \notin P_1$ for each $i \neq 1$ and $l \in \{1, \ldots, s_i\}$. Therefore $f_i \notin P_1$ for each $i \neq 1$, and $\dfrac{f}{f_1} = \prod_{i=2}^{r} f_i \notin P_1$. Hence $\dfrac{f}{f_{1,j}} = \Big(\dfrac{f}{f_1}\Big)\Big(\dfrac{f_1}{f_{1,j}}\Big) \notin Q_1$. The proof that $\dfrac{f}{\sum_{l=1}^{s_1} x_{1_l}} \notin Q_1$ follows similarly. Hence we have $\omega(I) = \sum_{i=1}^{r} e(Q_i)$. $\square$

Theorem 3.3 and Proposition 3.5 yield the following corollary.

**Corollary 3.6.** *Let $I$ be a monomial ideal of $R = k[x_1, \dots, x_n]$ with $dim(R/I) = 1$. Let $I = \bigcap_{i=1}^{r} Q_i$ be the canonical primary decomposition of $I$. Then*

$$\omega(I) = \omega^\bullet(I) = \begin{cases} \max\{e(Q_k), \sum_{i \neq k} e(Q_i)\} & \text{if } \sqrt{Q_k} = \mathfrak{m} \text{ for some } k \in \{1, \dots, r\}. \\ \sum_{i=1}^{r} e(Q_i) & \text{otherwise.} \end{cases}$$

**Corollary 3.7.** *Let $f$ be a monomial of $R$. Then $\omega^\bullet(fR) = deg(f)$. In particular, $\omega(fR) = \omega^\bullet(fR)$.*

**Proof.** Let $f = \prod_{k=1}^{r} x_{i_k}^{a_k}$ for some $a_1, \dots, a_r \in \mathbb{N}$ and $1 \leq i_1 < i_2 < \cdots < i_r \leq n$. Then $fR = x_{i_1}^{a_1} R \cap \cdots \cap x_{i_r}^{a_r} R$, and by Lemma 2.4 and Proposition 3.5 we have $\omega(fR) = \omega^\bullet(fR) = \sum_{i=1}^{r} e(x_{i_k}^{a_k} R) = \sum_{i=1}^{r} a_k = \text{deg}(f)$. □

Given a monomial ideal $I$ of $R = k[x, y, z]$, we can produce an algorithm that can compute $\omega(I)$. If $I$ is principal, then Corollary 3.7 says that $\omega(I)$ is equal to the degree of a generator for $I$. Otherwise, $I = hJ$ for some monomial $h$ and a monomial ideal $J$ with $\dim(R/J) \leq 1$. Now, $\omega(J)$ can be calculated explicitly using Corollary 2.6 or Corollary 3.6 after obtaining a canonical primary decomposition of $J$, and we have $\omega(I) = deg(h) + \omega(J)$ by Corollary 3.2.

**Example 3.8.** Let $R = k[x, y, z]$ and $I = (x^3y^4, x^2y^5, x^4y^3z^2, x^5y^3z, x^2y^4z^2)$. Then $I = x^2y^3J$ with canonical primary decomposition $J = (x^2, y) \cap (y, z) \cap (x^3, y^2, z^2, xy)$. By Lemma 2.4 and Corollary 2.6, the standard decomposition $(x^3, y^2, z^2, xy) = (x, y^2, z^2) \cap (x^3, y, z^2)$ yields that $e((x^3, y^2, z^2, xy)) = 4$. Thus by Corollary 3.6,

$$\begin{aligned} \omega(I) &= deg(x^2y^3) + \omega(J) \\ &= 5 + \max\{e((x^3, y^2, z^2, xy)), e((x^2, y)) + e((y, z))\} \\ &= 5 + \max\{4, 2 + 1\} \\ &= 9. \end{aligned}$$

Another interesting result that follows from Lemma 3.1 and Theorem 3.3 is a formula of $\omega(I)$ and $\omega^\bullet(I)$ for monomial ideals of $R = k[x, y]$ where $k$ is a field and $x, y$ are indeterminates over $k$.

**Theorem 3.9.** *Let $R = k[x, y]$ and $J$ a monomial ideal of $R$. Write $J = (x^{a_1}y^{b_1}, \dots, x^{a_r}y^{b_r})$, where $\{a_i\}$ is strictly decreasing and $\{b_i\}$ is strictly increasing. Then*

$$\omega(J) = \omega^\bullet(J) = \begin{cases} a_1 + b_1 & \text{if } r = 1. \\ \max_{1 \leq i \leq r-1}\{a_i + b_{i+1}\} - 1 & \text{if } r > 1. \end{cases}$$

**Proof.** The case when $r = 1$ follows from Corollary 3.7. For $r > 1$, first observe the standard decomposition of $J$ is $J = x^{a_r}R \cap y^{b_1}R \cap (x^{a_1}, y^{b_2}) \cap (x^{a_2}, y^{b_3}) \cap \cdots \cap (x^{a_{r-1}}, y^{b_r})$ ([12, Proposition 3.2]). The case $b_1 = a_r = 0$ follows from Corollary 2.6. Suppose that at least one of $a_r$ and $b_1$ is nonzero. Then by Lemma 2.4 and Corollary 3.6,

$$\begin{aligned}\omega(J) = \omega^{\bullet}(J) &= \max\{e((x^{a_1}, y^{b_2}) \cap (x^{a_2}, y^{b_3}) \cap \cdots \cap (x^{a_{r-1}}, y^{b_r})), e(x^{a_r}R) + e(y^{b_1}R)\} \\ &= \max\{\max_{1 \le i \le r-1}\{e((x^{a_i}, y^{b_{i+1}}))\}, a_r + b_1\} \\ &= \max\{\max_{1 \le i \le r-1}\{a_i + b_{i+1} - 1\}, a_r + b_1\} \\ &= \max_{1 \le i \le r-1}\{a_i + b_{i+1}\} - 1.\end{aligned}$$

□

**Example 3.10.** If $R = k[x, y]$ and $J = (x^{11}y^4, x^8y^5, x^7y^9, x^4y^{10}, x^2y^{16})$, then by Theorem 3.9,

$$\omega(J) = \omega^{\bullet}(J) = \max\{11 + 5, 8 + 9, 7 + 10, 4 + 16\} - 1 = 19.$$

## 4. $\omega$-Linear ideals

Given an ideal $I$ of a ring $R$, we will say that $I$ is an *$\omega$-linear ideal* if $\omega(I^m) = m\omega(I)$ for each $m \in \mathbb{N}$. Perhaps the most common example of $\omega$-linear ideals can be found amongst those $P \in \text{Spec}(R)$ for which $P^n$ is $P$-primary for each $n \in \mathbb{N}$ ([1, Theorem 3.1, Theorem 5.7]). For instance,

1. $R$ is a Prüfer domain and $P^2 \neq P$.
2. $R$ is a Noetherian ring and $P$ is a maximal ideal that contains a nonzerodivisor.
3. $R = k[x_1, \ldots, x_n]$ and $P$ is a monomial ideal.

In this section, we investigate the properties of $\omega$-linear ideals. Again, we will restrict our concern to monomial ideals of a polynomial ring $R = k[x_1, \ldots, x_n]$ where $k$ is a field.

We first consider a few useful inequalities regarding monomial ideals.

**Lemma 4.1.** *Let $I$ be a monomial ideal of $R = k[x_1, \ldots, x_n]$. Then $\omega(I) \geq \max\{\deg(f) \mid f \in G(I)\}$.*

**Proof.** Let $f \in G(I)$. Then $f = \prod_{k=1}^{r} x_{i_k}^{a_k}$ for some $a_1, \ldots, a_r \in \mathbb{N}$ and $1 \leq i_1 < i_2 < \cdots < i_r \leq n$. Since $f \in I$ but $\dfrac{f}{x_{i_k}} \notin I$ for each $k \in \{1, \ldots, r\}$ by minimality of $G(I)$, we have that $I$ is not $(\deg(f) - 1)$-absorbing. Hence $\omega(I) \geq \deg(f)$, and since $f$ was chosen arbitrarily, we have the desired conclusion. □

**Lemma 4.2.** *Let $I \subseteq J$ be ideals of a ring $R$. If $\sqrt{I} = \sqrt{J}$, then $e(J) \leq \omega(I)$. In particular, if $I$ and $J$ are both $P$-primary ideals of a prime ideal $P$ of $R$, then $\omega(J) \leq \omega(I)$.*

**Proof.** Since $\sqrt{I} = \sqrt{J}$, $(\sqrt{J})^{\omega(I)} \subseteq (\sqrt{I})^{e(I)} \subseteq I \subseteq J$ by Theorem 1.2 and $e(J) \leq \omega(I)$. The second statement follows immediately since $e = \omega$ for primary ideals. □

**Lemma 4.3.** *Let $P$ be a prime monomial ideal of $R = k[x_1, \ldots, x_n]$. If $I, J$ are $P$-primary monomial ideals of $R$, then $\omega(I+J) \leq \min\{\omega(I), \omega(J)\} \leq \max\{\omega(I), \omega(J)\} = \omega(I \cap J) \leq \omega(IJ) \leq \omega(I) + \omega(J)$. Moreover, $\omega(I : J) \geq \omega(I) - \omega(J)$.*

**Proof.** Note that by Corollary 2.3, $IJ \subseteq I \cap J \subseteq I + J$ are all $P$-primary monomial ideals. Therefore $\omega(I + J) \leq \min\{\omega(I), \omega(J)\} \leq \max\{\omega(I), \omega(J)\} \leq \omega(I \cap J) \leq \omega(IJ)$ by Lemma 4.2. On the other hand, let $I = \bigcap_{i=1}^{r} Q_i$ and $J = \bigcap_{j=1}^{s} T_j$ be the standard decompositions of $I$ and $J$, respectively. Then $I \cap J = (\bigcap_{i=1}^{r} Q_i) \cap (\bigcap_{j=1}^{s} T_j)$ is an irreducible decomposition of $I \cap J$, and by throwing away any redundant components, there are $A \subseteq \{1, \ldots, r\}$ and $B \subseteq \{1, \ldots, s\}$ so that $I \cap J = (\bigcap_{i \in A} Q_i) \cap (\bigcap_{j \in B} T_j)$ is the standard decomposition of $I \cap J$. Thus by Corollary 2.6,

$$\begin{aligned} \omega(I \cap J) &= \max\{\max_{i \in A}\{e(Q_i)\}, \max_{j \in B}\{e(T_j)\}\} \\ &\leq \max\{\max_{1 \leq i \leq r}\{e(Q_i)\}, \max_{1 \leq j \leq s}\{e(T_j)\}\} \\ &= \max\{\omega(I), \omega(J)\}. \end{aligned}$$

Moreover, $(\sqrt{IJ})^{e(I)+e(J)} = P^{e(I)+e(J)} = P^{e(I)}P^{e(J)} = (\sqrt{I})^{e(I)}(\sqrt{J})^{e(J)} \subseteq IJ$, and so $e(IJ) \leq e(I)+e(J)$. Combined with Theorem 1.2, we have $\omega(IJ) \leq \omega(I)+\omega(J)$. It remains to show that $\omega(I : J) \geq \omega(I) - \omega(J)$. When $J \subseteq I$, then we have $I : J = R$ and $\omega(I : J) = 0 \geq \omega(I) - \omega(J)$ by Lemma 4.2. If $J \not\subseteq I$, then $I : J$ is $P$-primary by Corollary 2.3, and since $J(I : J) \subseteq I$, we have $\omega(I : J) + \omega(J) \geq \omega(I)$ by the first part of this lemma, hence the claim. □

As Anderson and Badawi pointed out ([1, Example 2.7]), the conclusion of Lemma 4.3 does not hold in every ring $R$. We add, that even in a polynomial ring over a field, the conclusion of the above lemma may fail if we drop any part of the hypothesis.

**Example 4.4.** Let $R = k[x, y, z]$ and $I = (x^2, xy, y^2, xz^2)$ and $J = (x^2, xy, y^2, yz^3)$, so that neither $I$ nor $J$ are primary ideals. The standard decompositions of $I, J$

and $I \cap J$ are

$$\begin{aligned} I &= (x^2, y, z^2) \cap (x, y^2) \\ J &= (x, y^2, z^3) \cap (x^2, y) \\ I \cap J &= (x, y^2) \cap (x^2, y) \\ I + J &= (x, y) \cap (x, y^2, z^3) \cap (x^2, y, z^2). \end{aligned}$$

Thus we have $\omega(I) = 3$, $\omega(J) = 4$, $\omega(I \cap J) = 2$ and $\omega(I + J) = 4$, so that $\omega(I \cap J) < \omega(I + J) = \max\{\omega(I), \omega(J)\}$.

**Example 4.5.** Let $R = k[x, y, z]$ and $I = (x, y)$ and $J = (y, z^2)$, so that $I$ and $J$ are both primary, but $\sqrt{I} \neq \sqrt{J}$. Then we have $\omega(I) = 1$, $\omega(J) = 2$ and $\omega(I \cap J) = 3$, so that $\omega(I \cap J) > \max\{\omega(I), \omega(J)\}$.

**Corollary 4.6.** *Let $I$ be a primary monomial ideal of $R = k[x_1, \dots, x_n]$. Then for each $m \in \mathbb{N}$ we have $\omega(I^m) \leq m\omega(I)$.*

**Proof.** Follows immediately by induction on $m$ and Lemma 4.3. □

Next, we derive a characterization of primary monomial $\omega$-linear ideals.

**Lemma 4.7.** *Let $R = k[x_1, \dots, x_n]$ and $Q$ a primary monomial ideal of $R$, so that $G(Q)$ consists of monomials of the ring $k[x_{i_1}, \dots, x_{i_r}]$ for some $1 \leq i_1 < i_2 < \dots < i_r \leq n$ and there exists $a_1, \dots, a_r \in \mathbb{N}$ so $\{x_{i_1}^{a_1}, \dots, x_{i_r}^{a_r}\} \subseteq G(Q)$. Choose $s \in \{1, \dots, r\}$ so $a_s = \max_{1 \leq j \leq r}\{a_j\}$.*

1. *If $G(Q) = \{x_{i_1}^{a_1}, \dots, x_{i_r}^{a_r}\}$, then $\omega(Q^m) = (m-1)a_s + \omega(Q)$ for each $m \in \mathbb{N}$.*
2. *$Q$ is $\omega$-linear if and only if $\omega(Q) = a_s$.*

**Proof.** (1) Let $Q = (x_{i_1}^{a_1}, \dots, x_{i_r}^{a_r})$. Then given $m \in \mathbb{N}$, set $S_m = \{(k_1, \dots, k_r) \in \mathbb{N}^r \mid \sum_{j=1}^{r} k_j = m + r - 1\}$ and $Q_k = (x_{i_1}^{k_1 a_1}, \dots, x_{i_r}^{k_r a_r})$ for each $k = (k_1, \dots, k_r) \in S_m$. Then $Q^m = \bigcap_{k \in S_m} Q_k$ ([14, Theorem 6.2.4]). Now by Corollary 2.6 and Lemma 2.4,

$$\omega(Q^m) = \max_{k \in S_m}\{e(Q_k)\} = \max_{k \in S_m}\{\sum_{j=1}^{r} k_j a_j\} - r + 1 = (m-1)a_s + \omega(Q).$$

(2) Fix $m \in \mathbb{N}$ and set

$$I_1 = (x_{i_1}^{a_1}, \dots, x_{i_r}^{a_r})^m, I_2 = (x_{i_1}, \dots, x_{i_{s-1}}, x_{i_s}^{ma_s}, x_{i_{s+1}}, \dots, x_{i_r}).$$

It follows that $I_1 \subseteq Q^m \subseteq I_2$ are $(x_{i_1}, \dots, x_{i_r})$-primary ideals, so we have $ma_s = \omega(I_2) \leq \omega(Q^m) \leq \omega(I_1) = (m-1)a_s + \sum_{j=1}^{r} a_j - r + 1$ by Corollary 4.2,

Lemma 2.4 and part 1 of this lemma. Therefore if $Q$ is $\omega$-linear, then $\omega(Q) = \lim_{m\to\infty} \frac{m\omega(Q)}{m} = \lim_{m\to\infty} \frac{\omega(Q^m)}{m} = a_s$. Conversely, suppose that $\omega(Q) = a_s$ and fix $m \in \mathbb{N}$. Then since $x_{i_s}^{ma_s} \in G(Q^m)$ we have $\omega(Q^m) \geq ma_s = m\omega(Q)$ by Lemma 4.1. Hence $\omega(Q^m) = m\omega(Q)$ by Corollary 4.6 and so $Q$ is $\omega$-linear. □

**Corollary 4.8.** *Let $I$ be an irreducible monomial ideal of $R = k[x_1, \ldots, x_n]$ so that $I = (x_{i_1}^{a_1}, \ldots, x_{i_r}^{a_r})$ for some $1 \leq i_1 < i_2 < \cdots < i_r \leq n$ and $a_1, \ldots, a_n \in \mathbb{N}$. Set $a_s = \max_{1\leq j\leq r}\{a_j\}$. Then the following are equivalent.*

(1) *$I$ is $\omega$-linear.*
(2) *$\omega(I^m) = m\omega(I)$ for some $m > 1$.*
(3) *$\omega(I) = a_s$.*
(4) *$a_i = 1$ for each $i \neq s$.*

**Proof.** (1) ⇒ (2) Obvious.
(2) ⇒ (3) Suppose that $\omega(I^m) = m\omega(I)$ for some $m > 1$. By Lemma 4.7(1) we have $\omega(I^m) = (m-1)a_s + \omega(I)$. Hence $\omega(I) = a_s$.
(3) ⇔ (4) Immediate consequence of Lemma 2.4.
(3) ⇔ (1) Follows from Lemma 4.7(2). □

**Lemma 4.9.** *Let $P$ be a monomial prime ideal of $R$. If $I, J$ are $P$-primary $\omega$-linear monomial ideals of $R$, then so is $I \cap J$.*

**Proof.** Without loss of generality we may assume that $\omega(I) \geq \omega(J)$. By Lemma 4.7(2), there is $j \in \{1, \ldots, r\}$ so that $x_{i_j}^{\omega(I)} \in G(I)$. There exists $a \in \mathbb{N}$ so $x_{i_j}^{a} \in G(J)$. Then again, by Lemma 4.7(2), $a \leq \omega(J)$. Now, $x_{i_j}^{\omega(I)} = lcm(x_{i_j}^{\omega(I)}, x_{i_j}^{a}) \in G(I \cap J)$. On the other hand, $\omega(I \cap J) = \omega(I)$ by Lemma 4.3. Hence $I \cap J$ is $\omega$-linear by Lemma 4.7(2). □

Given a monomial ideal $I$ of $R = k[x, y]$ we will write $I = (x^{a_1}y^{b_1}, \ldots, x^{a_r}y^{b_r})$ where $\{a_i\}$ and $\{b_i\}$ are strictly decreasing and strictly increasing sequences of non-negative integers, respectively. Similarly, if $J$ is a monomial ideal of $R$ we write $J = (x^{c_1}y^{d_1}, \ldots, x^{c_s}y^{d_s})$ where $\{c_i\}$ and $\{d_i\}$ are strictly decreasing and strictly increasing sequence of non-negative integers, respectively. Hence $b_1 = a_r = 0$ iff $I$ is $(x, y)$-primary, and $d_1 = c_s = 0$ iff $J$ is $(x, y)$-primary.

**Lemma 4.10.** *Let $R = k[x, y]$ and $I, J$ be $(x, y)$-primary monomial ideals with $\omega(I) \geq \omega(J)$. Then $\omega(IJ) \leq \omega(I) + \max\{c_1, d_s\}$.*

**Proof.** We may assume that $c_1 \geq d_s$. Then $e(I) = \omega(I) \geq \omega(J) \geq c_1$ by Lemma 4.1, so $(x,y)^{e(I)+c_1} = (x,y)^{e(I)}(x^{c_1}, y^{c_1}) = (\sqrt{I})^{e(I)}(x^{c_1}, y^{c_1}) \subseteq IJ$ are $(x,y)$-primary ideals. Therefore $\omega(IJ) \leq \omega((x,y)^{e(I)+c_1}) = e(I) + c_1 = \omega(I) + c_1$ by Lemma 4.2. □

We now classify $\omega$-linear monomial ideals $I$ in $R = k[x,y]$.

**Proposition 4.11.** *Let $R = k[x,y]$ and $I = (x^{a_1}y^{b_1}, \ldots, x^{a_r}y^{b_r})$ be a monomial ideal of $R$. Then the following are equivalent.*

(1) *$I$ is $\omega$-linear.*

(2) *$\omega(I^m) = m\omega(I)$ for some $m > 1$.*

(3) *$\omega(I) = \max\{a_1 + b_1, a_r + b_r\}$.*

**Proof.** Note that given $m \in \mathbb{N}$ and a monomial $f$ of $R$, by Lemma 3.1 we have

$$\begin{aligned} \omega(I^m) &= m\omega(I) \\ \Leftrightarrow m(deg(f)) + \omega(I^m) &= m(deg(f)) + m\omega(I) \\ \Leftrightarrow deg(f^m) + \omega(I^m) &= m(deg(f) + \omega(I)) \\ \Leftrightarrow \omega((fI)^m) &= m\omega(fI). \end{aligned}$$

Moreover, if $I$ is a principal ideal, then $I$ satisfies all of $1, 2$, and $3$ by Corollary 3.2. Hence we may assume that $I$ is a $(x,y)$-primary monomial ideal of $R$. That is, $a_r = b_1 = 0$.

(1) $\Rightarrow$ (2) is trivial.

(2) $\Rightarrow$ (3) Suppose that $\omega(I^m) = m\omega(I)$ for some $m > 1$. Note that $\omega(I^{m-1}) + \omega(I) \geq \omega(I^m) = m\omega(I)$ by Lemma 4.3 and $\omega(I^{m-1}) \leq (m-1)\omega(I)$ by Corollary 4.6, and thereby $\omega(I^{m-1}) = (m-1)\omega(I)$. Hence we must have $\omega(I^2) = 2\omega(I)$. Since $\omega(I^2) \leq \omega(I) + \max\{a_1, b_r\}$ by Lemma 4.10, $\omega(I) = \omega(I^2) - \omega(I) \leq \max\{a_1, b_r\}$. On the other hand, $\omega(I) \geq \max\{a_1, b_r\}$ by Lemma 4.1. Therefore $\omega(I) = \max\{a_1, b_r\}$.

(3) $\Rightarrow$ (1) Follows from Lemma 4.7(2). □

**Lemma 4.12.** *The set of monomial $\omega$-linear ideals of $R = k[x,y]$ is multiplicatively closed.*

**Proof.** Let $I$ and $J$ be monomial $\omega$-linear ideals of $R$. By Lemma 3.1 we may assume that $I$ and $J$ are $(x,y)$-primary ideals of $R$. Then $\omega(I) = \max\{a_1, b_r\}$, $\omega(J) = \max\{c_1, d_s\}$ by Proposition 4.11. Now, $x^{a_1+c_1}$ and $y^{b_r+d_s}$ are elements of $G(IJ)$. Hence by Lemma 4.7(2) and Lemma 4.1, to show that $IJ$ is $\omega$-linear it suffices to show that $\omega(IJ) \leq \max\{a_1 + c_1, b_r + d_s\}$. Suppose that $\omega(I) = a_1$ and

$\omega(J) = c_1$. Then all we have to show is $\omega(IJ) \leq a_1 + c_1$, which follows from Lemma 4.3. The case when $\omega(I) = b_r$ and $\omega(J) = d_s$ can be derived in the exact same manner. Therefore, without loss of generality we may assume that $\omega(I) = a_1 > b_r$ and $\omega(J) = d_s > c_1$. Observe now that $Ix^{c_1} + Jy^{b_r}$ is an $(x, y)$-primary ideal contained in $IJ$. Thus by Lemma 4.2 and Theorem 3.9 we have

$$\begin{aligned}\omega(IJ) &\leq \omega(Ix^{c_1} + Jy^{b_r}) \\ &= \max\{\max_{1\leq i\leq r-1}\{a_i + b_{i+1} + c_1\} - 1, \max_{1\leq j\leq s-1}\{c_j + d_{j+1} + b_r\} - 1\} \\ &= \max\{\omega(I) + c_1, \omega(J) + b_r\} \\ &= \max\{a_1 + c_1, b_r + d_s\}.\end{aligned}$$

□

Recall that given an ideal $I$ of a commutative ring $R$, an element $f \in R$ is said to be *integral* over $I$ if there is some $k \in \mathbb{N}$ and $c_i \in I^i$ for each $i \in \{1, \ldots, k\}$ so that

$$f^k + c_1 f^{k-1} + \cdots + c_{k-1} f + c_k = 0.$$

The set of elements of $R$ integral over $I$ is called the *integral closure* of $I$ and is denoted by $\overline{I}$. $I$ is said to be *integrally closed* if $I = \overline{I}$.

**Corollary 4.13.** *Every integrally closed monomial ideal of $R = k[x, y]$ is $\omega$-linear.*

**Proof.** Let $I$ be an integrally closed monomial ideal of $R$. It is well known that $R$ is an *integrally closed domain* (i.e., $R$ is an integral domain that contains every nonzero element of the quotient field of $R$ that is integral over $R$), and that each principal ideal of $R$ is integrally closed, and the product of an integrally closed ideal of $R$ and a nonzero element of $R$ yields another integrally closed ideal of $R$. Hence by Lemma 3.1 we may assume that $I$ is $(x, y)$-primary. Now by [16, Proposition 2.6] there are monomial ideals $I_1 = (\{x^{r-i}y^{b_i}\}_{i=0}^r)$ and $I_2 = (\{x^{a_i}y^i\}_{i=0}^r)$ of $R$ with $0 = b_0 < b_1 < \cdots < b_r$ and $a_0 > a_1 > \cdots > a_r = 0$ so $I = I_1 I_2$. Thus by Lemma 4.12, it suffices to show that $I_1$ and $I_2$ are $\omega$-linear. By Theorem 3.9, $\omega(I_1) = \max_{0\leq i\leq r-1}\{c_i\}$, where $c_i = r - i + b_{i+1} - 1$ for each $i \in \{0, 1, \ldots, r-1\}$. Since $c_{i+1} - c_i = b_{i+1} - (b_i + 1) \geq 0$ for each $i \in \{0, 1, \ldots, r-1\}$, we have $\omega(I_1) = c_{r-1} = b_r = \max\{r, b_r\}$ and $I_1$ is $\omega$-linear by Proposition 4.11. The proof that $I_2$ is $\omega$-linear follows similarly. □

**Remark 4.14.** (1) Even if $I$ and $J$ are $\omega$-linear monomial primary ideals such that $\sqrt{I} = \sqrt{J}$, we may have $\omega(I \cap J) < \omega(IJ) < \omega(I) + \omega(J)$. Indeed, set $R = k[x, y]$,

$I = (x^3, xy, y^2)$ and $J = (x^2, xy, y^3)$. Then both $I$ and $J$ are $\omega$-linear $(x, y)$-primary ideals of $R$. However, $IJ = (x^5, x^3y, x^2y^2, xy^3, y^5)$, so $\omega(IJ) = 5 < 6 = \omega(I) + \omega(J)$. On the other hand, $\omega(I \cap J) = \max\{\omega(I), \omega(J)\} = 3$ by Corollary 2.6.

(2) Not every $\omega$-linear monomial ideal of $R = k[x, y]$ is integrally closed. For example, set $I = (x^3, xy^2, y^4)$. Then $\omega(I) = 4$ by Theorem 3.9, and $I$ is $\omega$-linear by Proposition 4.11. However, $(x^2y)^2 = x^3(xy^2) \in I^2$ and $x^2y \notin I$. Thus $I$ is not integrally closed ([10, Theorem 1.4.2]).

So far, we have considered only $\omega$-linear monomial ideals of the form $fI$ where $f$ is a monomial and $I$ is a primary ideal, and most of the proof is solely based on the fact that $e(I) = \omega(I)$ when $I$ is a primary ideal. We now show that there exists a class of (integrally closed) nonprimary $\omega$-linear monomial ideals. In fact, some of the squarefree monomial ideals are $\omega$-linear. Recall that a monomial $f = x_{i_1}^{a_1} \cdots x_{i_r}^{a_r}$ is said to be *squarefree* if $a_1 = \cdots = a_r = 1$. A monomial ideal generated by squarefree monomials is said to be a *squarefree monomial ideal.*

**Lemma 4.15.** *Let $I$ be a squarefree monomial ideal. Then $\omega(I^m) \geq m\omega(I)$ for each $m \in \mathbb{N}$.*

**Proof.** Let $P_1, \ldots, P_r$ be minimal prime ideals of $I$. Then $I = \bigcap_{i=1}^r P_i$ and $\omega(I) = r$ by Proposition 3.5. Set $f_i = \sum\limits_{x_j \in G(P_i)} x_j$ for each $i \in \{1, \ldots, r\}$. Then $f := \prod_{i=1}^r f_i \in \prod_{i=1}^r P_i \subseteq I$, so $f^m \in I^m$. However, $\dfrac{f^m}{f_i} \notin P_i^m$, so $\dfrac{f^m}{f_i} \notin I^m$([10, Proposition 1.4.4]). Thus $I^m$ is not $(mr-1)$-absorbing and $\omega(I^m) \geq m\omega(I)$. □

Recall that a graph $G$ consists of a set of vertices $V = \{v_1, ..., v_n\}$ and a set of edges $E \subseteq \{v_iv_j | v_i, v_j \in V\}$, and is called *bipartite* if there exists two disjoint subsets $U_1, U_2$ of $V$ such that $E \subseteq \{v_iv_j \mid v_i \in U_1, v_j \in U_2\}$. The *edge ideal* of $G$ is defined to be the ideal $I = (\{x_ix_j | v_iv_j \in E\})$ of $R = k[x_1, ..., x_d]$, where $k$ be a field and $d$ is the number of vertices of $G$. Given a graph $G = (V, E)$, a subset $W$ of $V$ is said to be a *vertex cover* if given $v_iv_j \in E$, either $v_i \in W$ or $v_j \in W$. A vertex cover $W$ of $G$ is said to be a *minimal vertex cover* if each proper subset of $W$ is not a vertex cover of $G$.

If $I$ is an edge ideal of a graph, then it is a squarefree monomial ideal and a monomial prime ideal $P$ is a minimal ideal of $I$ if and only if the set of vertices that corresponds to $P$ is a minimal vertex cover. Also, a graph is bipartite if and only if it has no cycle of odd length as its subgraph.

Our first example of a nonprimary $\omega$-linear ideal is the edge ideal of a bipartite graph.

**Lemma 4.16.** *Let $R = k[x_1, \ldots, x_n]$. If $I$ is an ideal of $R$ that is also the edge ideal of a bipartite graph $G$, then $I$ is $\omega$-linear.*

**Proof.** Let $I$ be an edge ideal of a graph $G$ and let $P_1, \ldots, P_r$ be the set of (incomparable) minimal prime ideals of $I$. Recall that a graph $G$ is bipartite if and only if

$$I^m = \bigcap_{P \text{ is a minimal prime of } I} P^m$$

for each $m \in \mathbb{N}$ ([18, Theorem 5.9]). Hence if $G$ is bipartite, then by Proposition 3.5, $\omega(I^m) = \sum_{i=1}^{r} e(P_i^m) = \sum_{i=1}^{r} m = mr$ for each $m \in \mathbb{N}$. Therefore the conclusion follows. □

There are nonbipartite graphs whose edge ideals are $\omega$-linear.

**Theorem 4.17.** *Let $R = k[x_1, \ldots, x_n]$. Let $I = (x_1x_2, x_2x_3, \ldots, x_{n-1}x_n, x_nx_1)$ (that is, $I$ is the edge ideal of a cycle graph of length $n$). Then $I$ is $\omega$-linear.*

**Proof.** Since a cycle of even length is bipartite, by Lemma 4.16 we may assume that $n = 2l+1$ for some $l \in \mathbb{N}$. Fix $m \in \mathbb{N}$. $I$ is a squarefree monomial ideal, so $I = P_1 \cap \cdots \cap P_r$ where $P_1, \ldots, P_r$ are the minimal prime ideals of $I$ ([10, Lemma 1.3.5]). Thus by Proposition 3.5 we have $\omega(I) = \sum_{i=1}^{r} e(P_i) = r$, and we only need to show that $\omega(I^m) = mr$. Note that since $I$ is an edge ideal of a cycle of length $2l+1$, $\mathrm{Ass}(R/I^m) = \{P_1, \ldots, P_r\}$ if $m \leq l$ and $\mathrm{Ass}(R/I^m) = \{P_1, \ldots, P_r, \mathfrak{m}\}$ if $m > l$ ([5, Lemma 3.1]). Hence if $m \leq l$, then $I^m = \bigcap_{i=1}^{r} P_i^m$ and $\omega(I^m) = \sum_{i=1}^{r} e(P_i^m) = mr$ by Proposition 3.5, so we are done. Assume that $m > l$. Then $I^m = (\bigcap_{i=1}^{r} P_i^m) \cap Q$ is the canonical primary decomposition of $I^m$, where $Q$ is an $\mathfrak{m}$-primary monomial ideal of $R$ ([10, Proposition 1.4.4]). Now, $Q = (x_1^{a_1}, \ldots, x_n^{a_n}, f_1, \ldots, f_t)$ for some $a_i \in \mathbb{N}$ and monomials $f_i$. Since $I$ is a squarefree monomial ideal and $Q$ is a primary component of $I^m$, we must have $a_i \leq m$ for each $i \in \{1, \ldots, n\}$, and thus $e(Q) \leq e((x_1^{a_1}, \ldots, x_n^{a_n})) \leq mn - n + 1 \leq mr$ by Lemma 2.4 and since $n \leq r$. It follows that $\omega(I^m) = \max\{\sum_{i=1}^{r} e(P_i^m), e(Q)\} = \max\{mr, e(Q)\} = mr$ by Theorem 3.3 and Proposition 3.5. □

We close the section with the following question: Is every integrally closed monomial ideal $\omega$-linear? Integrally closed monomial ideals considered in this note (certain monomial ideals in $R = k[x, y]$, irreducible monomial ideals, or edge ideal of bipartite graphs) were all $\omega$-linear. Note also that if this question has an affirmative answer, then it follows that every edge ideal is $\omega$-linear.

**Acknowledgement.** We thank the referee whose suggestions greatly improved the presentation of this paper.

## References


[1] D. F. Anderson and A. Badawi, *On n-absorbing ideals of commutative rings*, Comm. Algebra, 39(5) (2011), 1646-1672.

[2] D. F. Anderson and S. T. Chapman, *How far is an element from being prime?*, J. Algebra Appl., 9(5) (2010), 779-789.

[3] A. Badawi, *On 2-absorbing ideals of commutative rings*, Bull. Austral. Math. Soc., 75(3) (2007), 417-429.

[4] A. Badawi, *n-Absorbing ideals of commutative rings and recent progress on three conjectures: a survey*, In: Fontana, M., Frisch, S., Glaz, S., Tartarone, F., Zanardo, P. (eds) Rings, Polynomials, and Modules, Springer, Cham, (2017), 33-52.

[5] J. Chen, S. Morey and A. Sung, *The stable set of associated primes of the ideal of a graph*, Rocky Mountain J. Math., 32(1) (2002), 71-89.

[6] H. S. Choi and A. Walker, *The radical of an n-absorbing ideal*, arXiv:1610.10077 [math.AC], (2016), J. Commutative Algebra, accepted.

[7] G. Donadze, *The Anderson-Badawi Conjecture for commutative algebras over infinite fields*, Indian J. Pure Appl. Math., 47(4) (2016), 691-696.

[8] G. Donadze, *A proof of the Anderson-Badawi* $rad(I)^n \subseteq I$ *formula for n-absorbing ideals*, Proc. Indian Acad. Sci. Math. Sci., 128(1) (2018), Art. 6 (6 pp).

[9] D. Eisenbud, Commutative Algebra, with a View Toward Algebraic Geometry, Graduate Texts in Mathematics, 150, Springer-Verlag, New York, 1995.

[10] J. Herzog and T. Hibi, Monomial Ideals, Graduate Texts in Mathematics, 260, Springer-Verlag London, Ltd., London, 2011.

[11] A. Laradji, *On n-absorbing rings and ideals*, Colloq. Math., 147(2) (2017), 265-273.

[12] E. Miller and B. Sturmfels, Combinatorial Commutative Algebra, Graduate Texts in Mathematics, 227, Springer-Verlag, New York, 2005.

[13] H. F. Moghimi and S. R. Naghani, *On n-absorbing ideals and the n-Krull dimension of a commutative ring*, J. Korean Math. Soc., 53(6) (2016), 1225-1236.

[14] W. F. Moore, M. Rogers and S. Sather-Wagstaff, Monomial Ideals and Their Decompositions, Universitext, Springer, Cham, 2018, www.ndsu.edu/pubweb/ ssatherw/DOCS/monomial.pdf.

[15] P. Nasehpour, *On the Anderson-Badawi $\omega_{R[X]}(I[X]) = \omega_R(I)$ conjecture*, Arch. Math. (Brno), 52(2) (2016), 71-78.

[16] V. C. Quiñonez, *Integral closure and other operations on monomial ideals*, J. Commut. Algebra, 2(3) (2010), 359-386.

[17] S. Sihem and H. Sana, *On Anderson-Badawi conjectures*, Beitr. Algebra Geom., 58(4) (2017), 775-785.

[18] A. Simis, W. V. Vasconcelos and R. H. Villarreal, *On the ideal theory of graphs*, J. Algebra, 167 (1994), 389-416.

**Hyun Seung Choi** (Corresponding Author)
Department of Mathematics
Glendale Community College
Glendale, California 91208, U.S.A.
e-mail: hyunc@glendale.edu

**Andrew Walker**
Department of Mathematics
Midland University
Fremont, Nebraska, 68025, U.S.A.
e-mail: ajwalk010@gmail.com

# 8

# STAR OPERATIONS ON KUNZ DOMAINS

Dario Spirito

Communicated by Abdullah Harmancı

Abstract. We study star operations on Kunz domains, a class of analytically irreducible, residually rational domains associated to pseudo-symmetric numerical semigroups, and we use them to refute a conjecture of Houston, Mimouni and Park. We also find an estimate for the number of star operations in a particular case, and a precise counting in a sub-case.

**Keywords**: Star operation, pseudo-symmetric semigroup, Kunz domain, star regular domain

## 1. Introduction

Let $D$ be an integral domain with quotient field $K$, and let $\mathcal{F}(D)$ be the set of *fractional ideals* of $D$, i.e., the set of $D$-submodules $I$ of $K$ such that $xI \subseteq D$ for some $x \in K \setminus \{0\}$.

A *star operation* on $D$ is a map $\star : \mathcal{F}(D) \longrightarrow \mathcal{F}(D)$, $I \mapsto I^\star$, such that, for every $I, J \in \mathcal{F}(D)$ and every $x \in K$:

- $I \subseteq I^\star$;
- if $I \subseteq J$, then $I^\star \subseteq J^\star$;
- $(I^\star)^\star = I^\star$;
- $x \cdot I^\star = (xI)^\star$;
- $D = D^\star$.

A fractional ideal $I$ is $\star$-*closed* if $I = I^\star$.

The easiest example of a non-trivial star operation is the *v-operation* $v : I \mapsto (D : (D : I))$, where if $I, J \in \mathcal{F}(D)$ we define $(I : J) := \{x \in K \mid xJ \subseteq I\}$. An ideal that is $v$-closed is said to be *divisorial*; if $I$ is divisorial and $\star$ is any other star operation then $I = I^\star$. We denote by $d$ the identity, which is obviously a star operation.

Recently, the cardinality of the set $\mathrm{Star}(D)$ of the star operations on $D$ has been studied, especially in the case of Noetherian [4,8] and Prüfer domains [3,5]. In particular, Houston, Mimouni and Park started studying the relationship between the cardinality of $\text{Star}(D)$ and the cardinality of $\text{Star}(T)$, where $T$ is an overring of $D$ (an *overring* of $D$ is a ring comprised between $D$ and $K$) [6,7]: they called a domain *star regular* if $|\text{Star}(D)| \geq |\text{Star}(T)|$ for every overring of $T$. While even simple domains may fail to be star regular (for example, there are domains with just one star operation having an overring with infinitely many star operations [6, Example 1.3]), they conjectured that every one-dimensional local Noetherian domain $D$ such that $1 < |\text{Star}(D)| < \infty$ is star regular, and proved it when the residue field of $D$ is infinite [6, Corollary 1.18].

In this context, a rich source of examples are *semigroup rings*, that is, subrings of the power series ring $K[[X]]$ (where $K$ is a field, usually finite) of the form $K[[S]] := K[[X^S]] := \{\sum_i a_i X^i \mid a_i = 0 \text{ for all } i \notin S\}$, where $S$ is a numerical semigroup (i.e., a submonoid $S \subseteq \mathbb{N}$ such that $\mathbb{N} \setminus S$ is finite). Star operations can also be defined on numerical semigroups [14], and there is a link between star operations on $S$ and star operations on $K[[S]]$: for example, every star operation on $S$ induces a star operation on $K[[S]]$, and $|\text{Star}(S)| = 1$ if and only if $|\text{Star}(K[[S]])| = 1$ [14, Theorem 5.3], with the latter result corresponding to the equivalence between $S$ being symmetric and $K[[S]]$ being Gorenstein [2,10]. A detailed study of star operations on some numerical semigroup rings was carried out in [15].

In this paper, we study star operations on *Kunz domains*, which are, roughly speaking, a generalization of rings in the form $K[[S]]$ where $S$ is a pseudo-symmetric semigroup (see the beginning of the next section for the definitions). We show that, if $R$ is a Kunz domain whose residue field is finite and the length of $\overline{R}/R$ is at least 4 (where $\overline{R}$ is the integral closure of $R$) then $R$ is a counterexample to Houston-Mimouni-Park's conjecture; that is, $R$ satisfies $1 < |\text{Star}(R)| < \infty$ but there is an overring $T$ of $R$ with more star operations than $R$. In Section 3, we also study more deeply one specific class of domains, linking the cardinality of $\text{Star}(R)$ with the set of vector subspaces of a vector space over the residue field of $R$, and calculate the cardinality of $\text{Star}(R)$ when the value semigroup of $R$ is $\langle 4, 5, 7\rangle$.

We refer to [13] for information about numerical semigroups, and to [1] for the passage from numerical semigroups to one-dimensional local domains.

## 2. Kunz domains

A *numerical semigroup* is a subset $S \subseteq \mathbb{N}$ such that $0 \in S$, that is closed by addition and such that $\mathbb{N} \setminus S$ is finite. If $S$ is a numerical semigroup, we let $g := g(S) := \sup(\mathbb{Z} \setminus S)$ be the *genus* of $S$ and $\mu := \mu(S) := \min(S \setminus \{0\})$ be the *multiplicity* of $S$. If $a_1, \ldots, a_n$ are coprime integers, we denote by $\langle a_1, \ldots, a_n\rangle$ the numerical semigroup generated by $a_1, \ldots, a_n$, i.e., $\langle a_1, \ldots, a_n \rangle = \{\lambda_1 a_1 + \cdots + \lambda_n a_n \mid \lambda_1, \ldots, \lambda_n \in \mathbb{N}\}$.

Let $(V, M_V)$ be a discrete valuation ring with associated valuation $\mathbf{v}$. We shall consider local subrings $(R, M_R)$ of $V$ with the following properties:

- $R$ and $V$ have the same quotient field;
- the integral closure of $R$ is $V$;
- $R$ is Noetherian;
- the conductor ideal $(R : V)$ is nonzero;
- the inclusion $R \hookrightarrow V$ induces an isomorphism of residue fields $R/M_R \longrightarrow V/M_V$.

Equivalently, $R$ is an analytically irreducible, residually rational one-dimensional Noetherian local domain having integral closure $V$. For every such $R$, the set $\mathbf{v}(R) := \{\mathbf{v}(r) \mid r \in R\}$ is a numerical semigroup. We state explicitly a property which we will be using many times.

**Proposition 2.1** ([12, Corollary to Proposition 1]). *Let $R$ be as above, and let $I \subseteq J$ be $R$-submodules of the quotient field of $R$. Then,*

$$\ell_R(J/I) = |\mathbf{v}(J) \setminus \mathbf{v}(I)|,$$

*where $\ell_R$ is the length of an $R$-module.*

We say that a numerical semigroup $S$ is *pseudo-symmetric* if $g = g(S)$ is even and, for every $a \in \mathbb{N}$, $a \neq g/2$, either $a \in S$ or $g - a \in S$. Following [1] (and using the characterization in [1, Proposition II.1.12]), we give the following definition.

**Definition 2.2.** A ring $R$ satisfying the previous conditions is a *Kunz domain* if $\mathbf{v}(R)$ is a pseudo-symmetric semigroup.

From now on, we suppose that $R$ is a Kunz domain, and we set $g := g(\mathbf{v}(R))$ and $\tau := g/2$. The hypotheses on $R$ guarantee that, if $x \in V$ is such that $\mathbf{v}(x) > g$, then $x \in R$ [10, Theorem, p.749].

Our first stage is constructing an overring $T$ of $R$ which we will use in the counterexample.

**Lemma 2.3.** *Let $y \in V$ be an element of valuation $g$, and let $T := R[y]$. Then:*

(a) *$T$ contains all elements of valuation $g$;*
(b) *$\mathbf{v}(T) = \mathbf{v}(R) \cup \{g\}$;*
(c) *$\ell_R(T/R) = 1$;*
(d) *$T = R + yR$.*

**Proof.** Let $y' \in V$ be another element of valuation $g$. Then, $\mathbf{v}(y/y') = 0$, and thus $c := y/y'$ is a unit of $V$. Hence, there is a $c' \in R$ such that the images of $c$ and $c'$ in the residue field of $V$ coincide; in particular, $c = c' + m$ for some $m \in M_V$. Hence,

$$y' = cy = (c' + m)y = c'y + my.$$

Since $c' \in R$, we have $c'y \in R[y]$; furthermore, $\mathbf{v}(my) = \mathbf{v}(m) + \mathbf{v}(y) > \mathbf{v}(y) = g$, and thus $my \in R$. Hence, $y' \in R[y]$, and thus $R[y]$ contains all elements of valuation $g$.

The fact that $\mathbf{v}(T) = \mathbf{v}(R) \cup \{g\}$ is trivial; hence, $\ell_R(T/R) = |\mathbf{v}(T) \setminus \mathbf{v}(R)| = 1$. The last point follows from the fact that $R+yR$ is an $R$-module, from $R \subsetneq R+yR \subseteq T$ and from $\ell_R(T/R) = 1$. $\square$

In particular, the previous proposition shows that $T$ is independent from the element $y$ chosen. From now on, $T$ will always denote this ring.

We denote by $\mathcal{F}_0(R)$ the set of $R$-fractional ideals $I$ such that $R \subseteq I \subseteq V$. If $I$ is any fractional ideal over $R$, and $\alpha \in I$ is an element of minimal valuation, then $\alpha^{-1}I \in \mathcal{F}_0(R)$; hence, the action of any star operation is uniquely determined by its action on $\mathcal{F}_0(R)$. Furthermore, $V^\star = V$ for all $\star \in \mathrm{Star}(R)$ (since $(R : (R : V)) = V$) and thus $I^\star \in \mathcal{F}_0(R)$ for all $I \in \mathcal{F}_0(R)$, i.e., $\star$ restricts to a map from $\mathcal{F}_0(R)$ to itself.

To analyze star operations, we want to subdivide them according to whether they close $T$ or not. One case is very simple.

**Proposition 2.4.** *If $\star \in \mathrm{Star}(R)$ is such that $T \neq T^\star$, then $\star = v$.*

**Proof.** Suppose $\star \neq v$: then, there is a fractional ideal $I \in \mathcal{F}_0(R)$ that is $\star$-closed but not divisorial. By [1, Lemma II.1.22], $\mathbf{v}(I)$ is not divisorial (in $\mathbf{v}(R)$) and thus by [1, Proposition I.1.16] there is a positive integer $n \in \mathbf{v}(I)$ such that $n+\tau \notin \mathbf{v}(I)$.

Let $x \in I$ be an element of valuation $n$, and consider the ideal $J := x^{-1}I \cap V$: being the intersection of two $\star$-closed ideals, it is itself $\star$-closed. Since $\mathbf{v}(x) > 0$, every element of valuation $g$ belongs to $J$; on the other hand, by the choice of $n$, no element of valuation $\tau$ can belong to $J$.

Consider now the ideal $L := (R : M_R)$: then, $L$ is divisorial (since $M_R$ is divisorial) and, using [1, Proposition II.1.16(1)],

$$\mathbf{v}(L) = (\mathbf{v}(R) - \mathbf{v}(M_R)) = \mathbf{v}(R) \cup \{\tau, g\}.$$

We claim that $T = J \cap L$: indeed, clearly $J \cap L$ contains $R$, and if $y$ has valuation $g$ then $y \in J \cap L$ by construction; thus $T = R + yR \subseteq J \cap L$. On the other hand, $\mathbf{v}(J \cap L) \subseteq \mathbf{v}(J) \cap \mathbf{v}(L) = \mathbf{v}(R) \cup \{g\}$, and thus $J \cap L \subseteq T$.

Hence, $T = J \cap L$; since $J$ and $L$ are both $\star$-closed, so is $T$. Therefore, if $T \neq T^\star$ then $\star$ must be the divisorial closure, as claimed. $\square$

Suppose now that $T = T^\star$. Then, $\star$ restricts to a star operation $\star_1 := \star|_{\mathcal{F}(T)}$, and the amount of information we lose in the passage from $\star$ to $\star_1$ depends on the $R$-fractional ideals that are not ideals over $T$. We can determine them explicitly.

Recall that the *canonical ideal* of a ring $R$ is a (fractional) ideal $\omega$ such that $(\omega : (\omega : I)) = I$ for every fractional ideal $I$. Not every integral domain has a canonical ideal; however, Kunz domains (or, more generally, Noetherian one-dimensional local domains whose completion is reduced [11, Korollar 2.12], so in particular domains satisfying the five properties at the beginning of this section) have a canonical ideal; furthermore, if $S$ is a Kunz domain, then an ideal $\omega \in \mathcal{F}_0(R)$ is canonical if and only if $\mathbf{v}(\omega) = S \cup \{x \in \mathbb{N} \mid g(S) - x \notin S\} = S \cup \{\tau\}$ [9, Satz 5].

**Lemma 2.5.** *Let $I \in \mathcal{F}_0(R)$, $I \neq R$. Then, the following are equivalent.*

(i) $\mathbf{v}(I) = \mathbf{v}(R) \cup \{\tau\}$*;*

(ii) *$I$ does not contain any element of valuation $g$;*

(iii) $IT \neq I$*;*

(iv) *$I$ is a canonical ideal of $R$.*

*Furthermore, in this case, there is a unit $u$ of $R$ such that $R \subseteq uI \subseteq (R : M_R)$ and $(uI)^v = (R : M_R)$.*

**Proof.** (i) $\Rightarrow$ (ii) is obvious.

(ii) $\Rightarrow$ (iii) Since $R \subseteq I$, there is an element $x$ of $I$ of valuation 0; hence, $IT$ contains an element of valuation $g$, and thus $IT \neq I$.

(iii) $\Rightarrow$ (i) Suppose there is an $x \in I$ such that $\mathbf{v}(x) \notin \mathbf{v}(R) \cup \{\tau\}$. Since $\mathbf{v}(R)$ is pseudo-symmetric, there is an $y \in R$ such that $\mathbf{v}(y) = g - \mathbf{v}(x)$; hence, $I$ contains an element (explicitly, $xy$) of valuation $g$ and, by the proof of Lemma 2.3, it follows that it contains every element of valuation $g$.

Fix now an element $y \in V$ of valuation $g$. Since $IT \neq I$, there are $i \in I$, $t \in T$ such that $it \notin I$. By Lemma 2.3(d), there are $r, r' \in R$ such that $t = r + yr'$; hence, $it = i(r + yr') = ir + iyr'$. Both $ir$ and $iyr'$ are in $I$, the former since it belongs to $IR = I$ and the latter because its valuation is at least $g$. However, this contradicts $it \notin I$; therefore, $\mathbf{v}(I) \subseteq \mathbf{v}(R) \cup \{\tau\}$.

If $\mathbf{v}(I) = \mathbf{v}(R)$, then we must have $I = R$, against our hypothesis; therefore, $\mathbf{v}(I) = \mathbf{v}(R) \cup \{\tau\}$.

(i) $\Leftrightarrow$ (iv) follows from [9, Satz 5] and the fact that $R$ is a Kunz domain.

For the last claim, we first note that $(R : M_R)$ is divisorial (since $M_R$ is divisorial). By [1, Proposition II.1.16], $\mathbf{v}((R : M_R)) = S \cup \{\tau, g\}$; if $x \in (R : M_R)$ has valuation $\tau$, then $I' := R + xR$ is a canonical ideal (since $\mathbf{v}(I') = S \cup \{\tau\}$) and is contained between $R$ and $(R : M_R)$. Since $R$ is local, and $I$ is a canonical ideal too, there is a unit $u$ of $R$ such that $I' = uI$ [11, Satz 2.8(b)]; the claim follows. □

**Proposition 2.6.** *The map*

$$\Psi\colon \mathrm{Star}(R) \setminus \{d, v\} \longrightarrow \mathrm{Star}(T)$$
$$\star \longmapsto \star|_{\mathcal{F}(T)}$$

*is well-defined and injective.*

**Proof.** By Proposition 2.4, if $\star \neq v$ then $T = T^\star$, and thus $\star|_{\mathcal{F}(T)}$ is a star operation on $T$; hence, $\Psi$ is well-defined. We claim that it is injective: suppose $\star_1 \neq \star_2$. Then, there is an $I \in \mathcal{F}_0(R)$ such that $I^{\star_1} \neq I^{\star_2}$. If $I$ is a $T$-module then $\Psi(\star_1) \neq \Psi(\star_2)$; suppose $I$ is not a $T$-module.

By Lemma 2.5, $I$ can only be $R$ or a canonical ideal of $R$. In the former case, since $\star_1$ and $\star_2$ are star operations, $R^{\star_1} = R = R^{\star_2}$, a contradiction. In the latter case, by multiplying by a unit we can suppose that $I \subsetneq (R : M_R)$. Then, $\ell((R : M_R)/I) = 1$, and thus $I^{\star_i}$ can only be $I$ or $(R : M_R)$; suppose now that $I^\star = I$ for some $\star \in \mathrm{Star}(R)$. By definition of the canonical ideal, $J = (I : (I : J))$ for every ideal $J$; since $(I : L)$ is always $\star$-closed if $I$ is $\star$-closed, it follows that $\star$ must be the identity. Since $\star_1, \star_2 \neq d$, we must have $I^{\star_1} = (R : M_R) = I^{\star_2}$, against the assumptions. Thus, $\Psi$ is injective. □

An immediate corollary of the previous proposition is that $|\mathrm{Star}(R)| \leq |\mathrm{Star}(T)| + 2$. Our counterexample thus involves finding star operations of $T$ that do not belong to the image of $\Psi$; to do so, we restrict to the case $\ell_R(V/R) \geq 4$ or, equivalently, $|\mathbb{N} \setminus \mathbf{v}(R)| \geq 4$. This excludes exactly two pseudo-symmetric numerical semigroups, namely $\langle 3, 4, 5\rangle$ and $\langle 3, 5, 7\rangle$.

**Lemma 2.7.** *Let $S$ be a pseudo-symmetric numerical semigroup, let $g := \max(\mathbb{N} \setminus S)$ and let $S' := S \cup \{g\}$. If $|\mathbb{N} \setminus S| \geq 4$, then there are $a, b \in (S' - M_{S'}) \setminus S'$, $a \neq b$, such that $2a, 2b, a + b \in S'$.*

**Proof.** Let $\mu$ be the multiplicity of $S$. We claim that $a := \tau$ and $b := g - \mu$ are the two elements we are looking for.

Since $a + M_S \subseteq S$ and $a + g > g$ (and so $a + g \in M_S$) we have $a \in (S' - M_{S'})$. Furthermore, since $|\mathbb{N} \setminus S| \geq 4$, we have $g > \mu$, and thus $b + m \geq g$ for all $m \in M_{S'}$.

By the previous point, $a+m, b+m \in S' \cup \{a,b\}$ for every $m \in M_{S'}$. Since $a = \tau$, we have $2a = g \in S'$.

If $g > 2\mu$, then $a > \mu$, and so $a+b \geq g$, which implies $a+b \in S'$; moreover, also $b > \mu$, and thus $2b = b+b \geq g - \mu + \mu = g$, so that $2b \in S'$.

If $g < 2\mu$, then $g$ must be equal to $2\mu - 2$ or to $\mu - 1$; the latter case is impossible since $|\mathbb{N} \setminus S| \geq 4$. Hence, $b = \mu - 2$ and $a = \mu - 1$. Then, $2b = 2\mu - 4$ and $a+b = 2\mu - 3$; again since $|\mathbb{N} \setminus S| \geq 4$, we must have $\mu > 3$, and thus $2b > a+b \geq \mu$. Furthermore, in this case $S' = \{0, \mu, \ldots\}$, and so $a+b, 2b \in S'$, as claimed. □

**Proposition 2.8.** *Let $K$ be the residue field of $R$, and suppose that $\ell_R(V/R) \geq 4$. There are at least $|K|+1$ star operations on $T$ that do not close $(R : M_R)$.*

**Proof.** We first note that $(R : M_R)$ is a $T$-module. Indeed, let $x \in (R : M_R)$ and $t \in T$: then, $t = r + ay$, with $r, a \in R$ and $\mathbf{v}(y) = g$, and so $xt = xr + axy$. Both $xr$ and $axy$ belong to $(R : M_R)$, the former because $(R : M_R)$ is an $R$-module and the latter since its valuation is at least $g$: hence, $xt \in (R : M_R)$. Thus, it makes sense to ask if a star operation on $T$ closes $(R : M_R)$.

Furthermore, $T \subsetneq (R : M_R) \subseteq (T : M_T)$: the first containment follows from the previous reasoning (and the fact that $(R : M_R)$ contains an element of valuation $\tau$ while $T$ does not). To see the second containment, we note that $M_T = M_R + \{x \in V \mid \mathbf{v}(x) = g\}$; thus, if $x \in (R : M_R)$ and $y \in M_T$, we can write $y = y_1 + y_2$ (with $y_1 \in M_R$ and $\mathbf{v}(y_2) = g$) and so $xy = x(y_1 + y_2) = xy_1 + xy_2$. Now $xy_1 \in R \subseteq T$, while $\mathbf{v}(xy_2) \geq g$ since $\mathbf{v}(x) \geq 0$, and thus both $xy_1$ and $xy_2$ belong to $T$. It follows that $x \in (T : M_T)$, i.e., $(R : M_R) \subseteq (T : M_T)$. Therefore, $(R : M_R)^{v_T} = (T : M_T)$, where $v_T$ is the $v$-operation on $T$.

Let $S' := \mathbf{v}(T)$: by Lemma 2.7, we can find $a, b \in (S' - M_{S'}) \setminus S'$ such that $2a, 2b, a+b \in S'$. Choose $x, y \in (T : M_T)$ such that $\mathbf{v}(x) = a$ and $\mathbf{v}(y) = b$ (and, without loss of generality, suppose $y \notin (R : M_R)$): they exist since $\mathbf{v}((T : M_T)) = (S' - M_{S'})$ [1, Proposition II.1.16].

Let $\{\alpha_1, \ldots, \alpha_q\}$ be a complete set of representatives of $R/M_R$ (or, equivalently, of $T/M_T$). Then, $x + \alpha_i y \in (T : M_T)$ for each $i$, and by the choice of $\mathbf{v}(x)$ and $\mathbf{v}(y)$ the module $T_i = T + (x + \alpha_i y)T$ is a ring, equal to $T[x + \alpha_i y]$. Define $\star_i$ as the star operation

$$I \mapsto I^{v_T} \cap IT_i.$$

We claim that $\star_i$ closes $T_i$ but not $T_j$ for $j \neq i$.

Indeed, clearly $T_i^{\star_i} = T_i$. If $j \neq i$, then $T_i T_j$ contains both $x + \alpha_i y$ and $x + \alpha_j y$, and thus it contains their difference $(\alpha_i - \alpha_j)y$. Since $\alpha_i$ and $\alpha_j$ are units corresponding to different residues, it follows that $\alpha_i - \alpha_j$ is a unit of $R$, and thus of $T$; hence, $y \in T_iT_j$. By construction, $y \in (T : M_T)$: thus, $y \in T_i^{\star_j}$. On the other hand, $y \notin T_i$, and thus $T_i^{\star_j} \neq T_i$.

Thus, $\{\star_1, \dots, \star_q\}$ are $q = |K|$ different star operations. Furthermore, none of them closes $(R : M_R)$, since

$$(R : M_R)^{\star_i} = (T : M_T) \cap (R : M_R)T[x + a_i y]$$

contains $y$, while $y \notin (R : M_R)$.

To conclude the proof, it is enough to note that none of the $\star_i$ are the divisorial closure (since they close one of the $T_i$, none of which is divisorial); thus, adding $v_T$ to the $\star_i$, we have $q+1$ star operations that do not close $(R : M_R)$. □

We are now ready to show that $R$ is the desired counterexample.

**Theorem 2.9.** *Let $R$ be a Kunz domain with finite residue field, and suppose that $\ell_R(V/R) \geq 4$. Then, $1 < |\mathrm{Star}(R)| < \infty$, but $R$ is not star regular.*

**Proof.** Since $K$ is a finite field and $R$ is not Gorenstein, by [4, Theorem 2.5] $1 < |\mathrm{Star}(R)| < \infty$, and the same for $T$.

By Proposition 2.6, we have $|\mathrm{Star}(R)| \leq 2 + |\Psi(\mathrm{Star}(R))|$; by Proposition 2.8, we have $|\Psi(\mathrm{Star}(R))| \leq |\mathrm{Star}(T)| - |K| - 1$. Hence,

$$\begin{aligned} |\mathrm{Star}(R)| &\leq 2 + |\mathrm{Star}(T)| - |K| - 1 = \\ &= |\mathrm{Star}(T)| - |K| + 1 < |\mathrm{Star}(T)| \end{aligned}$$

since $|K| \geq 2$. The claim is proved. □

## 3. The case $\mathbf{v}(R) = \langle n, n+1, \dots, 2n-3, 2n-1 \rangle$

In this section, we specialize to the case of Kunz domains $R$ such that $\mathbf{v}(R) = \langle n, n+1, \dots, 2n-1, 2n-3 \rangle = \{0, n, n+1, \dots, 2n-1, 2n-3, \dots\}$, where $n \geq 4$ is an integer. It is not hard to see that this semigroup is pseudo-symmetric, with $g = 2n-2$ and $\tau = n-1$.

We note that this semigroup is pseudo-symmetric also if $n = 3$, for which the number of star operations has been calculated in [8, Proposition 2.10]: in this case, we have $|\mathrm{Star}(R)| = 4$.

By Lemma 2.5, the only $I \in \mathcal{F}_0(R)$ such that $IT \neq I$ are $R$ and the canonical ideals. From now on, we denote by $\mathcal{G}$ the set $\{I \in \mathcal{F}_0(R) \mid IT = I\}$; we want to parametrize $\mathcal{G}$ by subspaces of a vector space.

**Lemma 3.1.** *Let $K$ be the residue field of $R$. Then, there is an order-preserving bijection between $\mathcal{G}$ and the set of vector subspaces of $K^{n-1}$.*

**Proof.** Every $I \in \mathcal{G}$ contains $T$. The quotient of $R$-modules $\pi : V \mapsto V/T$ induces a map

$$\begin{aligned} \widetilde{\pi}\colon \mathcal{G} &\longrightarrow \mathcal{P}(V/T) \\ I &\longmapsto \pi(I), \end{aligned}$$

where $\mathcal{P}(V/T)$ denotes the power set of $V/T$. It is obvious that $\widetilde{\pi}$ is injective.

The map $\pi$ induces on $V/T$ a structure of $K$-vector space of dimension $n-1$. If $I \in \mathcal{G}$, then its image along $\widetilde{\pi}$ will be a vector subspace; conversely, if $W$ is a vector subspace of $V/T$ then $\pi^{-1}(W)$ will be an ideal in $\mathcal{G}$. The claim is proved. □

For an arbitrary domain $D$ and a fractional ideal $I$ of $D$, the star operation *generated* by $I$ is the map [14, Section 5]

$$\star_I : J \mapsto (I : (I : J)) \cap J^v = J^v \cap \bigcap_{\gamma \in (I:J)\setminus\{0\}} \gamma^{-1} I;$$

this star operation has the property that, if $I$ is $\star$-closed for some $\star \in \mathrm{Star}(D)$ and $J$ is $\star_I$-closed, then $J$ is also $\star$-closed. If $\Delta \subseteq \mathcal{F}(D)$, we define $\star_\Delta$ as the map

$$\star_\Delta : J \mapsto \bigcap_{I \in \Delta} J^{\star_I}.$$

In the present case, we can characterize when an ideal is $\star_\Delta$-closed.

**Proposition 3.2.** *Let $I, J \in \mathcal{G}$ and let $\Delta \subseteq \mathcal{G}$ be a set of nondivisorial ideals.*

(a) *$I$ is divisorial if and only if $n-1 \in \mathbf{v}(I)$;*

(b) *$I^v = I \cup \{x \mid \mathbf{v}(x) \geq n-1\}$;*

(c) *if $I, J$ are nondivisorial, then $I = I^{\star_J}$ if and only if $I \subseteq \gamma^{-1}J$ for some $\gamma$ of valuation 0;*

(d) *if $I$ is nondivisorial, then $I$ is $\star_\Delta$-closed if and only if $I \subseteq \gamma^{-1}J$ for some $J \in \Delta$ and some $\gamma$ of valuation 0.*

**Proof.** (a) If $I$ is divisorial, then (since $I \neq R$) we must have $(R : M_R) \subseteq I$; in particular, $n-1 \in \mathbf{v}(I)$.

Suppose $n-1 \in \mathbf{v}(I)$; since $I$ contains every element of valuation at least $n$ (being $IT = I$), it contains also all elements of valuation $n-1$. Let $x$ be such that $\mathbf{v}(x) = n-1$: then, $\mathbf{v}(x+r) \geq n-1$ for every $r \in V$, and thus $x + I \subseteq I$. Hence, $I$ is divisorial by [1, Proposition II.1.23].

(b) Let $L := I \cup \{x \mid \mathbf{v}(x) \geq n-1\}$. If $n-1 \in \mathbf{v}(I)$, then $L = I$ and $I^v = L$ by the previous point. If $n-1 \notin \mathbf{v}(I)$, then (since $I$ contains any element of valuation at least $n$), $L$ is a fractional ideal of $R$ such that $\mathbf{v}(L) = \mathbf{v}(I) \cup \{n-1\}$; hence, it is divisorial and $\ell(L/I) = 1$. It follows that $L = I^v$, as claimed.

(c) Suppose $I \subseteq \gamma^{-1}J$, where $\mathbf{v}(\gamma) = 0$. Since $J$ is not divisorial, $n-1 \notin \mathbf{v}(J) = \mathbf{v}(\gamma^{-1}J)$; hence, using the previous point, $I = I^v \cap \gamma^{-1}J$ is closed by $\star_J$.

Conversely, suppose $I = I^{\star_J}$. Since $I$ is nondivisorial, there must be $\gamma \in (I : J)$, $\gamma \neq 0$ such that $I \subseteq \gamma^{-1}J$ and $I^v \nsubseteq \gamma^{-1}J$. If $\mathbf{v}(\gamma) > 0$, then $\gamma^{-1}J$ contains the elements of valuation $n-1$; it follows that $I^v \subseteq \gamma^{-1}J$ and thus that $I^v \subseteq I^{\star_J}$, against $I = I^{\star_J}$. Hence, $\mathbf{v}(\gamma) = 0$, as claimed.

(d) If $I \subseteq \gamma^{-1}J$ for some $J \in \Delta$ and some $\gamma$ such that $\mathbf{v}(\gamma) = 0$, then $I^{\star_\Delta} \subseteq I^{\star_J} = I$, and thus $I$ is $\star_\Delta$-closed.

Conversely, suppose $I = I^{\star_\Delta}$. For every $J \in \Delta$, the ideal $I^{\star_J}$ is contained in $I^v = I \cup \{x \mid \mathbf{v}(x) \geq n-1\}$; since $\ell(I^v/I) = 1$, it follows that $I^{\star_J}$ is either $I$ or $I^v$. Since $I = I^{\star_\Delta}$, it must be $I^{\star_J} = I$ for some $J$; by the previous point, $I \subseteq \gamma^{-1}J$ for some $\gamma$, as claimed. □

An important consequence of the previous proposition is the following: suppose that $\Delta$ is a set of nondivisorial ideals in $\mathcal{F}_0(R)$ such that, when $I \neq J$ are in $\Delta$, then $I \nsubseteq \gamma^{-1}J$ for all $\gamma$ having valuation 0. Then, for every subset $\Lambda \subseteq \Delta$, the set of ideals of $\Delta$ that are $\star_\Lambda$-closed is exactly $\Lambda$; in particular, each nonempty subset of $\Delta$ generates a different star operation.

We will use this observation to estimate the cardinality of $\mathrm{Star}(R)$ when the residue field is finite.

**Proposition 3.3.** *Let $R$ be a Kunz domain such that $\mathbf{v}(R) = \langle n, n+1, \ldots, 2n-3, 2n-1\rangle$, and suppose that the residue field of $R$ has cardinality $q < \infty$. Then,*

$$|\mathrm{Star}(R)| \geq 2^{\frac{q^{n-2}-1}{q-1}} \geq 2^{q^{n-3}}.$$

**Proof.** Let $L := \{x \in V \mid \mathbf{v}(x) \geq n\}$; then, $A := V/L$ is a $K$-algebra. Let $e_1$ be an element of valuation 1, and let $e_i := e_1^i$; then, $\{1 = e_0, e_1, \ldots, e_{n-1}\}$ projects to a $K$-basis of $A$, which for simplicity we still denote by $\{e_0, \ldots, e_{n-1}\}$. The vector subspace spanned by $e_0$ is exactly the field $K$.

Since $V$ and $L$ are stable by multiplication by every element of valuation 0, asking if $\gamma I \subseteq J$ for some $I, J \in \mathcal{F}_0(R)$ and some $\gamma$ is equivalent to asking if there is a $\overline{\gamma} \in A$ of "valuation" 0 such that $\overline{\gamma}\overline{I} \subseteq \overline{J}$, where $\overline{I}$ and $\overline{J}$ are the images of $I$ and $J$, respectively, in $A$. Hence, instead of working with ideals in $\mathcal{F}_0(R)$ we can work with vector subspaces of $A$ containing $e_0$.

Furthermore, if $V$ is a vector subspace of $A$ and $\gamma$ has valuation 0, then $\gamma V$ has the same dimension of $V$; thus, if $V$ and $W$ have the same dimension, $\gamma V \subseteq W$ if and only if $\gamma V = W$. Let $\sim$ denote the equivalence relation such that $V \sim W$ if and only if $\gamma V = W$ for some $\gamma$ of valuation 0.

Let $X$ be the set of 2-dimensional subspaces of $A$ that contain $e_0$ but not $e_{n-1}$. Then, the preimage of every element of $X$ is an element of $\mathcal{F}_0(R)$ that does not contain any element of valuation $n-1$, and thus it is nondivisorial by Proposition 3.2(b).

An element of $X$ is in the form $\langle e_0, \lambda_1 e_1 + \cdots + \lambda_{n-1}e_{n-1}\rangle$, where at least one among $\lambda_1, \ldots, \lambda_{n-2}$ is not 0; since $\langle e_0, f\rangle = \langle e_0, \lambda f\rangle$ for all $\lambda \in K$, $\lambda \neq 0$, there are exactly $(q^{n-1}-q)/(q-1)$ such subspaces.

Let $V \in X$, say $V = \langle e_0, f\rangle$, and consider the equivalence class $\Delta$ of $V$ with respect to $\sim$. Then, $W \in \Delta$ if and only if $\gamma W = V$ for some $\gamma$; since $1 \in W$, it follows that such a $\gamma$ must belong to $V$. Since $\gamma$ has valuation 0, it must be in the form $\lambda_0 e_0 + \lambda_1 f$ with $\lambda_0 \neq 0$; furthermore, if $\gamma' = \lambda\gamma$ then $\gamma^{-1}V = \gamma'^{-1}W$. Hence, the cardinality of $\Delta$ is at most $\frac{q^2-q}{q-1} = q$.

Therefore, $X$ contains elements belonging to at least

$$\frac{1}{q}\frac{q^{n-1}-q}{q-1} = \frac{q^{n-2}-1}{q-1} \geq q^{n-3}$$

equivalence classes; let $X'$ be a set of representatives of such classes, and let $Y$ be the preimage of $X'$ in the power set of $\mathcal{F}_0(R)$. Then, every subset of $Y$ generates a different star operation (with the empty set corresponding to the $v$-operation); it follows that

$$|\mathrm{Star}(R)| \geq 2^{\frac{q^{n-2}-1}{q-1}} \geq 2^{q^{n-3}},$$

as claimed. □

For $n = 4$, we can even calculate $|\mathrm{Star}(R)|$.

**Proposition 3.4.** *Let $R$ be a Kunz domain such that $\mathbf{v}(R) = \langle 4, 5, 7\rangle$, and suppose that the residue field of $R$ has cardinality $q < \infty$. Then, $|\mathrm{Star}(R)| = 2^{2q} + 3$.*

**Proof.** Consider the same setup of the previous proof. We start by claiming that two vector subspaces $W_1, W_2$ of $A$ of dimension 3 that contain $e_0$ but not $e_3$ are equivalent under $\sim$.

Indeed, any such subspace must have a basis of the form $\{e_0, e_1+\theta_1 e_3, e_2+\theta_2 e_3\}$, and different pairs $(\theta_1, \theta_2)$ induce different subspaces; let $W(\theta_1, \theta_2) := \langle e_0, e_1 + \theta_1 e_3, e_2 + \theta_2 e_3\rangle$. To show that two such subspaces are equivalent, we prove that they are all equivalent to $W(0,0)$. Let $\gamma := e_0 - \theta_2 e_1 - \theta_1 e_2$: we claim that $\gamma W(\theta_1, \theta_2) = W(0,0)$. Indeed, $\gamma e_0 = \gamma \in W(0,0)$; on the other hand,

$$\gamma(e_1 + \theta_1 e_3) = e_1 + \theta_1 e_3 - \theta_2 e_2 - \theta_1 e_3 = e_1 - \theta_2 e_2 \in W(0,0),$$

and likewise

$$\gamma(e_2 + \theta_2 e_3) = e_2 + \theta_2 e_3 - \theta_2 e_3 = e_2 \in W(0,0).$$

Hence, $W(\theta_1, \theta_2) \sim W(0,0)$.

Consider now the set $\Delta$ of nondivisorial ideals in $\mathcal{F}_0(R)$. By Lemma 2.5 and Proposition 3.2, $\Delta$ is equal to the union of the set $\mathcal{C}$ of the canonical ideals and the set $\mathcal{G}$ of the $I \in \mathcal{F}_0(R)$ such that $IT = I$. By Lemma 3.1 and Proposition 3.2, the elements of the latter correspond to the subspaces of $V/T$ containing $e_0$ but not $e_3$: hence, we can write $\mathcal{G} = \mathcal{G}_1 \cup \mathcal{G}_2 \cup \mathcal{G}_3$, where $\mathcal{G}_i$ contains the ideals of $\mathcal{G}$ corresponding to subspaces of dimension $i$.

Given $\star \in \mathrm{Star}(R)$, let $\Delta(\star) := \{I \in \Delta \mid I = I^\star\}$. We claim that $\Delta(\star)$ is one of the following:

- $\Delta$;
- $\Delta \setminus \mathcal{C}$;
- $\Lambda \cup \{T\}$ for some $\Lambda \subseteq \mathcal{G}_2$;
- the empty set.

By Proposition 2.4, if $T \neq T^\star$ (i.e., if $T \notin \Delta(\star)$) then $\star = v$, and $\Delta(\star) = \emptyset$.

If $\Delta(\star)$ contains a canonical ideal then $\star$ is the identity, and thus $\Delta(\star) = \Delta$.

If $I$ is $\star$-closed for some $I \in \mathcal{G}_3$, but no canonical ideal is $\star$-closed, then every element of $\mathcal{G}_3$ must be closed, since any other $I' \in \mathcal{G}_3$ is in the form $\gamma I$ for some $\gamma$ of valuation 0 (by the first part of the proof); furthermore, every element of $\mathcal{G}_2$ is the intersection of the elements of $\mathcal{G}_3$ containing it, and thus it is $\star$-closed. It follows that $\Delta(\star) = \Delta \setminus \mathcal{C}$; in particular, there is only one such star operation.

Let $\star$ be any star operation different from the three above. Then, $\Delta(\star)$ must contain $T$ and cannot contain any canonical ideal nor any element of $\mathcal{G}_3$. Hence, $\Delta(\star)$ must be equal to $\Lambda \cup \{T\}$ for some $\Lambda \subseteq \mathcal{G}_2$. Moreover, $\Lambda \cup \{T\}$ is equal to $\Delta(\star)$ for some $\star$ if and only if $\Lambda$ is the (possibly empty) union of equivalence classes under $\sim$. It follows that $|\mathrm{Star}(R)| = 2^x + 3$, where $x$ is the number of such equivalence classes.

By the proof of Proposition 3.3, the image of an element of $\mathcal{G}_2$ is in the form $\langle e_0, f\rangle$, where $f = \lambda_1 e_1 + \lambda_2 e_2 + \lambda_3 e_3$ with at least one between $\lambda_1$ and $\lambda_2$ nonzero. Let $V(\lambda_1, \lambda_2, \lambda_3)$ denote the subspace $\langle e_0, f\rangle$; clearly, $V(\lambda_1, \lambda_2, \lambda_3) = V(c\lambda_1, c\lambda_2, c\lambda_3)$ for every $c \in K \setminus \{0\}$. The subspaces equivalent to $V$ must have the form $(e_0+\theta f)^{-1}V$ for some $\theta \in K$, and, by using the basis $\{e_0, e_0+\theta f\}$ of $V$, we see that $(e_0+\theta f)^{-1}V(\lambda_1, \lambda_2, \lambda_3) = \langle e_0, (e_0+\theta f)^{-1}\rangle$. If $\theta = 0$, then $e_0 + \theta f = e_0$, and thus $(e_0+\theta f)^{-1}V(\lambda_1, \lambda_2, \lambda_3) = V(\lambda_1, \lambda_2, \lambda_3)$; suppose, from now on, that $\theta \neq 0$.

To calculate $(e_0 + \theta f)^{-1} = e_0 + \alpha_1 e_1 + \alpha_2 e_2 + \alpha_3 e_3$, we can simply expand the product $(e_0 + \theta f)(e_0 + \alpha_1 e_1 + \alpha_2 e_2 + \alpha_3 e_3)$, using $e_i = 0$ for $i > 3$; we obtain

$$\begin{cases} \alpha_1 = -\theta\lambda_1 \\ \alpha_2 = -\theta(\lambda_1\alpha_1 + \lambda_2) \\ \alpha_3 = -\theta(\lambda_1\alpha_2 + \lambda_2\alpha_1 + \lambda_3). \end{cases}$$

Since $\theta \neq 0$, the set $\{e_0, (e_0 + \theta f)^{-1} - e_0\}$ is a basis of $(e_0 + \theta f)^{-1}V(\lambda_1, \lambda_2, \lambda_3)$; hence, $(e_0 + \theta f)^{-1}V(\lambda_1, \lambda_2, \lambda_3) = V(\alpha_1, \alpha_2, \alpha_3)$. We distinguish two cases.

If $\lambda_1 = 0$, then $\lambda_2 \neq 0$, and so we can suppose $\lambda_2 = 1$. Then, we have

$$\begin{cases} \alpha_1 = 0 \\ \alpha_2 = -\theta \\ \alpha_3 = -\theta\lambda_3. \end{cases}$$

and so $(e_0 + \theta f)^{-1}V(0, 1, \lambda_3) = V(0, -\theta, -\theta\lambda_3) = V(0, 1, \lambda_3)$ since $\theta \neq 0$. It follows that the only subspace equivalent to $V(0, 1, \lambda_3)$ is $V(0, 1, \lambda_3)$ itself; since we have $q$ choices for $\lambda_3$, this case gives $q$ different equivalence classes.

If $\lambda_1 \neq 0$, we can suppose $\lambda_1 = 1$. Then, we get

$$\begin{cases} \alpha_1 = -\theta \\ \alpha_2 = -\theta(\alpha_1 + \lambda_2) = -\theta(-\theta + \lambda_2) \\ \alpha_3 = -\theta(-\theta(-\theta + \lambda_2) - \theta\lambda_2 + \lambda_3). \end{cases}$$

Since $\theta \neq 0$, we can divide by $-\theta$, obtaining

$$(e_0 + \theta f)^{-1}V(1, \lambda_2, \lambda_3) = V(1, -\theta + \lambda_2, \theta^2 - 2\theta\lambda_2 + \lambda_3).$$

Since $-\theta + \lambda_2 \neq -\theta' + \lambda_2$ if $\theta \neq \theta'$, we have $(e_0 + \theta f)^{-1}V(1, \lambda_2, \lambda_3) \neq (e_0 + \theta' f)^{-1}V(1, \lambda_2, \lambda_3)$ for all $\theta \neq \theta'$; thus, every equivalence class is composed by $q$ subspaces. Since there are $q^2$ such subspaces, we get $q$ additional equivalence classes.

Therefore, $\mathcal{G}_2$ is partitioned into $2q$ equivalence classes, and so $|\mathrm{Star}(R)| = 2^{2q}+3$, as claimed. $\square$

**Remark 3.5.** (1) *The estimate obtained in Proposition 3.3 grows very quickly; for example, if $q$ is fixed, it follows that the double logarithm of* $|\mathrm{Star}(R)|$ *grows (at least) linearly in $n = \ell(V/R) + 1$. This should be compared with* [8, Theorem 3.21], *where the authors analyzed a case where the growth of* $|\mathrm{Star}(R)|$ *was linear in $\ell(\overline{R}/R)$ (where $\overline{R}$ is the integral closure of $R$, which in this case is nonlocal).*

(2) *Let* $V = K[[X]]$ *be the ring of power series and consider the case* $n = 4$*. Then,* $T = K + X^4K[[X]]$*, and using Theorem 2.9 and Proposition 3.3, we have the lower bound* $|\text{Star}(T)| \geq 2^{2q}+q+2$*. This estimate is not very far from the precise counting* $|\text{Star}(T)| = 2^{2q+1} + 2^{q+1} + 2$ *obtained in* [15, Corollary 4.1.2].

## References


[1] V. Barucci, D. E. Dobbs and M. Fontana, *Maximality properties in numerical semigroups and applications to one-dimensional analytically irreducible local domains,* Mem. Amer. Math. Soc., 125(598) (1997), x+78 pp.

[2] H. Bass, *On the ubiquity of Gorenstein rings,* Math. Z., 82 (1963), 8-28.

[3] E. G. Houston, A. Mimouni and M. H. Park, *Integral domains which admit at most two star operations,* Comm. Algebra, 39(5) (2011), 1907-1921.

[4] E. G. Houston, A. Mimouni and M. H. Park, *Noetherian domains which admit only finitely many star operations,* J. Algebra, 366 (2012), 78-93.

[5] E. G. Houston, A. Mimouni and M. H. Park, *Integrally closed domains with only finitely many star operations,* Comm. Algebra, 42(12) (2014), 5264-5286.

[6] E. Houston, A. Mimouni and M. H. Park, *Star operations on overrings of Noetherian domains,* J. Pure Appl. Algebra, 220(2) (2016), 810-821.

[7] E. Houston, A. Mimouni and M. H. Park, *Star operations on overrings of Prüfer domains,* Comm. Algebra, 45(8) (2017), 3297-3309.

[8] E. G. Houston and M. H. Park, *A characterization of local Noetherian domains which admit only finitely many star operations: The infinite residue field case,* J. Algebra, 407 (2014), 105-134.

[9] J. Jäger, *Längenberechnung und kanonische Ideale in eindimensionalen Ringen,* Arch. Math. (Basel), 29(5) (1977), 504-512.

[10] E. Kunz, *The value-semigroup of a one-dimensional Gorenstein ring,* Proc. Amer. Math. Soc., 25 (1970), 748-751.

[11] E. Kunz, *Beispiel: Die kanonische Idealklasse eines eindimensionalen Cohen-Macaulay-Rings,* 103, Lecture Notes in Math., Vol. 238, Springer, Berlin, (1971), 17-24.

[12] T. Matsuoka, *On the degree of singularity of one-dimensional analytically irreducible Noetherian local rings,* J. Math. Kyoto Univ., 11 (1971), 485-494.

[13] J. C. Rosales and P. A. García-Sánchez, Numerical Semigroups, Developments in Mathematics, 20, Springer, New York, 2009.

[14] D. Spirito, *Star operations on numerical semigroups,* Comm. Algebra, 43(7) (2015), 2943-2963.

[15] B. White, *Star Operations and Numerical Semigroup Rings*, Ph.D. thesis, The University of New Mexico, 2014.

**Dario Spirito**
Dipartimento di Matematica e Fisica
Università degli Studi "Roma Tre"
Roma, Italy
email: spirito@mat.uniroma3.it

# 9

# ON 2-ABSORBING MODULES OVER NONCOMMUTATIVE RINGS

N. J. Groenewald and Bac T. Nguyen

Communicated by Sait Halıcıoğlu

ABSTRACT. Let $R$ be a noncommutative ring with identity. We define the notion of a 2-absorbing submodule and show that if the ring is commutative then the notion is the same as the original definition of that of A. Darani and F. Soheilnia. We give an example to show that in general these two notions are different. Many properties of 2-absorbing submodules are proved which are similar to the results for commutative rings.

**Keywords:** 2-Absorbing submodule, strong 2-absorbing submodule, prime submodule, completely prime submodule

## 1. Introduction

In 2007, Badawi [1] introduced the concept of 2-absorbing ideals of commutative rings with identity, which is a generalization of prime ideals, and investigated some properties. He defined a 2-absorbing ideal $P$ of a commutative ring $R$ with identity to be a proper ideal of $R$ and if whenever $a, b, c \in R$ with $abc \in P$, then $ab \in P$ or $bc \in P$ or $ac \in P$. In 2011, Darani and Soheilnia [3] introduced the concept of 2-absorbing submodules of modules over commutative rings with identities. A proper submodule $P$ of a module $M$ over a commutative ring $R$ with identity is said to be a 2-absorbing submodule of $M$ if whenever $a, b \in R$ and $m \in M$ with $abm \in P$, then $abM \subseteq P$ or $am \in P$ or $bm \in P$. One can see that 2-absorbing submodules are generalization of prime submodules. Moreover, it is obvious that 2-absorbing ideals are special cases of 2-absorbing submodules.

Throughout all rings (not necessarily commutative rings) have identities and all modules are unital left modules. In recent years the study of the absorbing property of rings, modules and related notions have been some of the topics of interest in the development of the ring and module theory. In this paper we study the notion of 2-absorbing modules over noncommutative rings. We prove basic properties of 2-absorbing submodules analogous to properties studied by Payrovi and Babaei in

[6] and [7] for 2-absorbing submodules over commutative rings. We also introduce the notion of strong 2-absorbing submodules and show that in general if $R$ is not a commutative ring then the notions of 2-absorbing and strong 2-absorbing submodules are not the same. If $R$ is a commutative ring then the notions of 2-absorbing and strong 2-absorbing submodule coincide with that of the original definition introduced by Darani and Soheilnia in [3].

## 2. 2-Absorbing submodules and ideals

Prime submodules of modules over associative rings were introduced by Dauns [4]. A proper submodule $N$ of $M$ is called prime, if $rRm \subseteq N$ implies $rM \subseteq N$ or $m \in N$ for all $r \in R$ and $m \in M$. Recently the concept of 2-absorbing submodules was introduced by Darani and Soheilnia as a generalization of 2-absorbing ideals in [3].

**Definition 2.1.** Let $P$ be a proper ideal of a ring $R$. Then $P$ is a 2*-absorbing ideal* of $R$ if $aRbRc \subseteq P$ implies $ab \in P$ or $bc \in P$ or $ac \in P$ for all $a, b, c \in R$.

**Definition 2.2.** Let $R$ be a ring and $N$ be a proper submodule of an $R$-module $M$. Then $N$ is 2*-absorbing submodule* of $M$ if $aRbRm \subseteq N$ implies $abM \subseteq N$ or $am \in N$ or $bm \in N$ for all $a, b \in R$ and $m \in M$.

**Remark 2.3.** *If $R$ is a commutative ring then this notion of a 2-absorbing submodule coincides with that of Darani and Soheilnia.*

**Definition 2.4.** Let $R$ be a noncommutative ring and $M$ an $R$-module. We define a proper submodule $P$ of $M$ to be a *strong 2-absorbing submodule* if whenever $a, b \in R$ and $m \in M$ with $abm \in P$, then $abM \subseteq P$ or $am \in P$ or $bm \in P$.

**Proposition 2.5.** *Let $R$ be a ring and $N$ be a prime submodule of an $R$-module $M$. If $aRbRm \subseteq N$ and $am \notin N$, then $bM \subseteq N$ for all $a, b \in R$ and $m \in M$.*

**Proof.** Let $a, b \in R$ and $m \in M$. Assume that $aRbRm \subseteq N$ and $am \notin N$. First, we show that $bRm \subseteq N$. Let $r$ be any element of the ring $R$. Then $aR(brm) \subseteq aR(bRm) \subseteq N$. Since $N$ is a prime submodule, $aM \subseteq N$ or $brm \in N$. Then $brm \in N$ because $am \notin N$. That is $bRm \subseteq N$. Since $N$ is a prime submodule and $am \notin N$, it follows that $m \notin N$ so that $bM \subseteq N$. □

From [5], a proper submodule $P$ of an $R$-module $M$ is called completely prime, if $am \in P$ implies $m \in P$ or $aM \subseteq P$, for each $a \in R$ and $m \in M$. An $R$-module $M$ is completely prime if the zero submodule of $M$ is a completely prime submodule of $M$. In general, an $R$-module $M/P$ is a completely prime module if and only if $P$ is a completely prime submodule of $M$.

**Proposition 2.6.** *If $N$ is a prime (completely prime) submodule of an $R$-module $M$, then $N$ is a 2-absorbing (strong 2-absorbing) submodule of $M$.*

**Proof.** Assume that $N$ is a prime (completely prime) submodule of an $R$-module $M$. Let $a, b \in R$, $m \in M$ and assume $N$ is a prime submodule such that $aRbRm \subseteq N$ but $am \notin N$. Thus $bM \subseteq N$ by Proposition 2.5. Then $bm \in N$ and $abM \subseteq aN \subseteq N$. Hence $N$ is a 2-absorbing submodule of $M$. Now, assume $N$ is a completely prime submodule and that $abm \in N$ but $bm \notin N$. Since $N$ is completely prime and $bm \notin N$, we have $aM \subseteq N$. Hence $abM \subseteq aM \subseteq N$ and we have $N$ is a strong 2-absorbing submodule of $M$. □

In general if $R$ is not a commutative ring then the notions of 2-absorbing and strong 2-absorbing submodules are not the same.

**Example 2.7.** *Let $M = \left\{ \begin{pmatrix} \bar{0} & \bar{0} \\ \bar{0} & \bar{0} \end{pmatrix}, \begin{pmatrix} \bar{1} & \bar{1} \\ \bar{0} & \bar{0} \end{pmatrix}, \begin{pmatrix} \bar{0} & \bar{0} \\ \bar{1} & \bar{1} \end{pmatrix}, \begin{pmatrix} \bar{1} & \bar{1} \\ \bar{1} & \bar{1} \end{pmatrix} \right\}$ where entries of matrices in $M$ are from $\mathbb{Z}_2 = \{\bar{0}, \bar{1}\}$ and $R = M_2(\mathbb{Z})$. ${}_RM$ is a simple module and hence also a prime module and as such 2-absorbing. However, $M$ is not a strong 2-absorbing module.*

**Proof.** It is clear that ${}_RM$ is a simple module and hence also a prime module.

Take $a = \begin{pmatrix} 2 & 1 \\ 2 & 4 \end{pmatrix}$ and $b = \begin{pmatrix} 1 & 0 \\ 3 & 5 \end{pmatrix} \in R$ and $m = \begin{pmatrix} \bar{1} & \bar{1} \\ \bar{1} & \bar{1} \end{pmatrix} \in M$.

$abm = \begin{pmatrix} 2 & 1 \\ 2 & 4 \end{pmatrix} \begin{pmatrix} 1 & 0 \\ 3 & 5 \end{pmatrix} \begin{pmatrix} \bar{1} & \bar{1} \\ \bar{1} & \bar{1} \end{pmatrix} = \begin{pmatrix} \bar{0} & \bar{0} \\ \bar{0} & \bar{0} \end{pmatrix}$ but

$am = \begin{pmatrix} 2 & 1 \\ 2 & 4 \end{pmatrix} \begin{pmatrix} \bar{1} & \bar{1} \\ \bar{1} & \bar{1} \end{pmatrix} = \begin{pmatrix} \bar{1} & \bar{1} \\ \bar{0} & \bar{0} \end{pmatrix} \neq \begin{pmatrix} \bar{0} & \bar{0} \\ \bar{0} & \bar{0} \end{pmatrix}$,

$bm = \begin{pmatrix} 1 & 0 \\ 3 & 5 \end{pmatrix} \begin{pmatrix} \bar{1} & \bar{1} \\ \bar{1} & \bar{1} \end{pmatrix} = \begin{pmatrix} \bar{1} & \bar{1} \\ \bar{0} & \bar{0} \end{pmatrix} \neq \begin{pmatrix} \bar{0} & \bar{0} \\ \bar{0} & \bar{0} \end{pmatrix}$

and $abM \neq \left\{ \begin{pmatrix} \bar{0} & \bar{0} \\ \bar{0} & \bar{0} \end{pmatrix} \right\}$ since

$ab \begin{pmatrix} \bar{1} & \bar{1} \\ \bar{0} & \bar{0} \end{pmatrix} = \begin{pmatrix} 2 & 1 \\ 2 & 4 \end{pmatrix} \begin{pmatrix} 1 & 0 \\ 3 & 5 \end{pmatrix} \begin{pmatrix} \bar{1} & \bar{1} \\ \bar{0} & \bar{0} \end{pmatrix} = \begin{pmatrix} \bar{1} & \bar{1} \\ \bar{0} & \bar{0} \end{pmatrix} \neq \begin{pmatrix} \bar{0} & \bar{0} \\ \bar{0} & \bar{0} \end{pmatrix}$. □

In [2] a module is defined to be a semi-commutative module if whenever $am = 0$ for $a \in R$ and $m \in M$, we have $aRm = 0$. A submodule $N$ of an $R$-module $M$ is a semi-commutative submodule if whenever $am \in N$ for $a \in R$ and $m \in M$, we have $aRm \subseteq N$.

**Proposition 2.8.** *Let $M$ be a left $R$ module. If $N$ is a 2-absorbing submodule which is also a semi-commutative submodule, then $N$ is a strong 2-absorbing submodule.*

**Proof.** Suppose $N$ is 2-absorbing. Let $a, b \in R$ and $m \in M$ such that $abm \in N$. Since $N$ is a semi-commutative submodule, we have $aRbRm \subseteq N$. Now $N$ 2-absorbing implies that $am \in N$ or $bm \in N$ or $abM \subseteq N$. Hence $N$ is a strong 2-absorbing submodule of $M$. □

Compare the next Theorem with Theorem 2.3 in [6].

**Theorem 2.9.** *Let $N$ be a proper submodule of an $R$-module $M$. If $N$ is a 2-absorbing submodule of $M$, then $(N :_R M) = \{r \in R : rM \subseteq N\}$ is a 2-absorbing ideal of $R$.*

**Proof.** Let $a, b, c \in R$ such that $aRbRc \subseteq (N :_R M)$ and suppose $ac \notin (N :_R M)$ and $bc \notin (N :_R M)$. We show that $ab \in (N :_R M)$. Since $ac, bc \notin (N :_R M)$, there exists $x_1, x_2 \in M$ such that $acx_1 \notin N$ and $bcx_2 \notin N$. Now, $aRbRc(x_1 + x_2) \subseteq N$. Since $N$ is a 2-absorbing submodule of $M$ we have $ab \in (N :_R M)$ or $ac(x_1+x_2) \in N$ or $bc(x_1 + x_2) \in N$. If $ac(x_1 + x_2) \in N$, then $acx_2 \notin N$ since $acx_1 \notin N$. Since $aRbRcx_2 \subseteq N$ and $bcx_2 \notin N$ and $acx_2 \notin N$ we have $ab \in (N :_R M)$. Similar to the case $bc(x_1 + x_2) \in N$, we get $ab \in (N :_R M)$. Hence $(N :_R M)$ is a 2-absorbing ideal of $R$. □

**Corollary 2.10.** [6, Theorem 2.3] *If $R$ is a commutative ring and $N$ is a 2-absorbing submodule of $M$, then $(N :_R M)$ is a 2-absorbing ideal of $R$.*

**Remark 2.11.** *The converse of the above theorem is not true in general.*

**Proof.** Let $p$ be a fixed prime integer. Then $\mathbb{Z}(p^\infty) = \{\alpha \in \mathbb{Q}/\mathbb{Z} : \alpha = r/p^n + \mathbb{Z}$ for some $r \in \mathbb{Z}$ and $n \geq 0\}$ is a non-zero submodule of $\mathbb{Q}/\mathbb{Z}$. Let $G_t = \{\alpha \in \mathbb{Q}/\mathbb{Z} : \alpha = r/p^t + \mathbb{Z}$ for some $r \in \mathbb{Z}\}$ for all $t \geq 0$. It is well known that each proper submodule of $\mathbb{Z}(p^\infty)$ is equal to $G_t$ for some $t \geq 0$. By [6, page 914], $G_t$ is not a 2-absorbing submodule of $\mathbb{Z}(p^\infty)$. We can see that $(G_t :_{\mathbb{Z}} \mathbb{Z}(p^\infty)) = 0$ is a 2-absorbing ideal of $\mathbb{Z}$ for all $t \geq 0$. □

Compare the next Proposition with Theorem 2.2 (i) in [6].

**Proposition 2.12.** *The intersection of each pair of prime (completely prime) submodules of an $R$-module $M$ is a 2-absorbing (strong 2-absorbing) submodule of $M$.*

**Proof.** Let $N$ and $K$ be two prime (completely prime) submodules of $M$. If $N = K$, then $N \cap K$ is a prime (completely prime) submodule of $M$ so that $N \cap K$ is a 2-absorbing (strong 2-absorbing) submodule of $M$. Assume that $N$ and $K$ are distinct. Since $N$ and $K$ are proper submodules of $M$, it follows that $N \cap K$ is a proper submodule of $M$. Next, let $a, b \in R$ and $m \in M$ be such that $aRbRm \subseteq N \cap K$ ($abm \in N \cap K$) but $am \notin N \cap K$ and $abM \nsubseteq N \cap K$. Then, we can conclude that

**(a):** $am \notin N$ or $am \notin K$, and
**(b):** $abM \nsubseteq N$ or $abM \nsubseteq K$.

These two conditions give 4 cases:

**(1):** $am \notin N$ and $abM \nsubseteq N$;
**(2):** $am \notin N$ and $abM \nsubseteq K$;
**(3):** $am \notin K$ and $abM \nsubseteq N$;
**(4):** $am \notin K$ and $abM \nsubseteq K$.

Let $N$ and $K$ be two prime submodules. We first consider Case(1). Since $aRbRm \subseteq N \cap K \subseteq N$ and $am \notin N$, it follows from Proposition 2.5 that $bM \subseteq N$. This is a contradiction because $abM \nsubseteq N$. Hence Case(1) does not occur. Similarly, Case(4) is not possible.

Next, Case(2) is considered. Again, we obtain that $bM \subseteq N$ and then $bm \in N$. Let $r \in R$. Since $aRbRm \subseteq N \cap K \subseteq K$, it follows that $aR(brm) \subseteq aR(bRm) \subseteq K$. Hence $aM \subseteq K$ or $brm \in K$ because $K$ is a prime submodule of $M$. If $aM \subseteq K$, then $abM \subseteq aM \subseteq K$ contradicts $abM \nsubseteq K$. Thus $brm \in K$. That is $bRm \subseteq K$. Since $K$ is a prime submodule, $bM \subseteq K$ or $m \in K$. If $bM \subseteq K$, then $abM \subseteq K$ leading to the same contradiction. Therefore, $m \in K$ and then $bm \in K$. Hence $bm \in N \cap K$. The proof of Case(3) is similar to that of Case(2).

Now, let $N$ and $K$ be completely prime submodules of $M$ and $abm \in N \cap K \subseteq N$. We consider Case(1): Since $abm \in N \cap K \subseteq N$ and $N$ completely prime we have $aM \subseteq N$ or $bm \in N$. If $aM \subseteq N$, then $abM \subseteq aM \subseteq N$ which is not possible. So, suppose $bm \in N$. Now $bM \subseteq N$ or $m \in N$. This is not possible and therefor Case (1) does not occur. Similarly, Case(4) is not possible. Next, Case(2) is considered. We have $abm \in N \cap K \subseteq K$ and since $K$ is completely prime it follows that $aM \subseteq K$ or $bm \in K$. If $aM \subseteq K$ then $abM \subseteq aM \subseteq K$ which contradicts $abM \nsubseteq K$ thus $bm \in K$. From $abm \in N \cap K \subseteq N$ we have $aM \subseteq N$ or $bm \in N$. Since $am \notin N$, $aM \subseteq N$ is not possible. Hence $bm \in N \cap K$. The proof of Case(3) is similar to that of Case(2). □

**Corollary 2.13.** [6, Theorem 2.2 (i)] *If $R$ is a commutative ring and $N$ is a prime submodule of $M$ or is an intersection of two prime submodules of $M$, then $N$ is 2-absorbing.*

Proposition 2.6 guarantees that every prime (completely prime) submodule is a 2-absorbing (strong 2-absorbing) submodule. The converse does not necessarily hold.

**Example 2.14.** *Consider $\mathbb{Z}_6$ as $\mathbb{Z}$-module, $(\bar{0})$ is not a prime (completely prime) submodule of $\mathbb{Z}_6$ since $\bar{2}.\bar{3} \in (\bar{0})$ but $\bar{3} \notin (\bar{0})$ and $2 \notin ((\bar{0})_{\mathbb{Z}} : \mathbb{Z}_6) = 6\mathbb{Z}$. By Proposition 2.12 $(\bar{0}) = (\bar{2}) \cap (\bar{3})$ is a 2-absorbing (strong 2-absorbing) submodule of $\mathbb{Z}_6$ as $\mathbb{Z}$-module.*

**Proposition 2.15.** *Let $N, W$ be two submodules of an $R$-module $M$ and $N \subseteq W$. If $N$ is a 2-absorbing (strong 2-absorbing) submodule of $M$, then $N$ is a 2-absorbing (strong 2-absorbing) submodule of $W$.*

**Proof.** If $W = M$, then there is nothing to prove. Let $aRbRx \subseteq N$ ($abx \in N$), where $a, b \in R$, $x \in W$. $N$ is a 2-absorbing (strong 2-absorbing) submodule of $M$, so either $ax \in N$ or $bx \in N$ or $ab \in (N :_R M)$. Since $N \subseteq W$ implies $(N :_R M) \subseteq (N :_R W)$, then either $ax \in N$ or $bx \in N$ or $ab \in (N :_R W)$. Hence $N$ is 2-absorbing (strong 2-absorbing) in $W$. □

**Lemma 2.16.** *Let $N$ be a proper submodule of an $R$-module $M$. Then $N$ is a 2-absorbing (strong 2-absorbing) submodule of $M$ if and only if $aRbK \subseteq N$ ($abK \subseteq N$) implies $ab \in (N : M)$, or $aK \subseteq N$ or $bK \subseteq N$ for each $a, b \in R$ and submodule $K$ of $M$.*

**Proof.** Suppose that $ab \notin (N : M)$ and $aK \nsubseteq N$ and $bK \nsubseteq N$. Then there exist $m_1, m_2$ in $K$ such that $am_1 \notin N$ and $bm_2 \notin N$. Since $aRbRm_1 \subseteq aRbK \subseteq N$ ($abm_1 \in abK \subseteq N$) and $ab \notin (N : M)$, $am_1 \notin N$, we get $bm_1 \in N$. Also, since $aRbRm_2 \subseteq aRbK \subseteq N$ ($abm_2 \in abK \subseteq N$) and $ab \notin (N : M)$, $bm_2 \notin N$, we get $am_2 \in N$. Now, since $aRbR(m_1 + m_2) \subseteq aRbK \subseteq N$ ($ab(m_1 + m_2) \in abK \subseteq N$) and $ab \notin (N : M)$ we have $a(m_1 + m_2) \in N$ or $b(m_1 + m_2) \in N$. If $a(m_1 + m_2) \in N$, i.e. $(am_1 + am_2) \in N$ then since $am_2 \in N$ we get $am_1 \in N$ which is contradiction. If $b(m_1 + m_2) \in N$, i.e. $(bm_1 + bm_2) \in N$, then since $bm_1 \in N$ we get $bm_2 \in N$ which is a contradiction. Thus either $ab \in (N : M)$ or $aK \subseteq N$ or $bK \subseteq N$. The converse is clear. □

As for the commutative case [6, Theorem 2.4] we have.

**Proposition 2.17.** *If $N$ is a 2-absorbing submodule of the $R$-module $M$ then $(N : Rm)$ is a 2-absorbing ideal of $R$ for every $m \in M \backslash N$.*

**Proof.** Let $a, b, c \in R$ and $m \in M \backslash N$ such that $aRbRc \subseteq (N : Rm)$. Hence $aRb(RcR)m \subseteq N$. Since $RcR$ is an ideal of $R$, we have $(RcR)m$ is a submodule of $M$. It now follows from Lemma 2.16 that $a(RcR)m \subseteq N$ or $b(RcR)m \subseteq N$ or $abM \subseteq N$. Hence $acRm \subseteq N$ or $bcRm \subseteq N$ or $abM \subseteq N$. Thus $ac \in (N : Rm)$ or $bc \in (N : Rm)$ or $abRm \subseteq abM \subseteq N$ i.e. $ab \in (N : Rm)$ and we are done. □

Compare the next Theorem with [7, Theorem 2.3].

**Theorem 2.18.** *Let $N$ be a proper submodule of the $R$ module $M$. If $N$ is 2-absorbing (strong 2-absorbing) submodule of $M$ and if $I$ and $J$ are ideals of $R$ and $K$ a submodule of $M$ such that $IJK \subseteq N$, then $IK \subseteq N$ or $JK \subseteq N$ or $IJ \subseteq (N :_R M)$. The converse holds for 2-absorbing submodules.*

**Proof.** Suppose $IJK \subseteq N$ and $IJ \nsubseteq (N :_R M)$. We show that $IK \subseteq N$ or $JK \subseteq N$. Suppose $IK \nsubseteq N$ and $JK \nsubseteq N$. There exists $a_1 \in I$ and $a_2 \in J$ such that $a_1K \nsubseteq N$ and $a_2K \nsubseteq N$. But $a_1Ra_2K \subseteq IJK \subseteq N$ $(a_1a_2K \subseteq IJK \subseteq N)$. Since $N$ is a 2-absorbing (strong 2-absorbing) submodule of $M$ it follows from Lemma 2.16 that $a_1a_2 \in (N :_R M)$. Since $IJ \nsubseteq (N :_R M)$. there exists $b_1 \in I$ and $b_2 \in J$ such that $b_1b_2M \nsubseteq N$. Now, since $N$ is 2-absorbing (strong 2-absorbing) and $b_1Rb_2K \subseteq IJK \subseteq N$ $(b_1b_2K \subseteq IJK \subseteq N)$ and also $b_1b_2M \nsubseteq N$ it follows from Lemma 2.16 that $b_1K \subseteq N$ or $b_2K \subseteq N$. We have the following cases:

Case (1) $b_1K \subseteq N$ and $b_2K \nsubseteq N$

Since $a_1Rb_2K \subseteq IJK \subseteq N$ $(a_1b_2K \subseteq IJK \subseteq N)$ and $a_1K \nsubseteq N$ and $b_2K \nsubseteq N$ it follows from Lemma 2.16 that $a_1b_2 \in (N :_R M)$. Since $b_1K \subseteq N$ and $a_1K \nsubseteq N$, we conclude $(a_1 + b_1)K \nsubseteq N$. On the other hand, $(a_1 + b_1)Rb_2K \subseteq N$ $((a_1 + b_1)b_2K \subseteq N)$ and neither $(a_1 + b_1)K \subseteq N$ nor $b_2K \subseteq N$, we get that $(a_1 + b_1)b_2 \in (N : M)$ by Lemma 2.16. But since $(a_1 + b_1)b_2 = (a_1b_2 + b_1b_2) \in (N : M)$ and $a_1b_2 \in (N : M)$, we get $b_1b_2 \in (N : M)$ which is a contradiction.

Case (2) $b_2K \subseteq N$ and $b_1K \nsubseteq N$

By a similar argument to case (1) we get a contradiction.

Case (3) $b_1K \subseteq N$ and $b_2K \subseteq N$

$b_2K \subseteq N$ and $a_2K \nsubseteq N$ gives $(a_2 + b_2)K \nsubseteq N$. But $a_1R(a_2 + b_2)K \subseteq N$ $(a_1(a_2 + b_2)K \subseteq N)$ and neither $a_1K \subseteq N$ nor $(a_2 + b_2)K \subseteq N$, hence $a_1(a_2 + b_2) \in (N : M)$ by Lemma 2.16. Since $a_1a_2 \in (N : M)$ and $(a_1a_2 + a_1b_2) \in (N : M)$, we have $a_1b_2 \in (N : M)$. Since $(a_1 + b_1)Ra_2K \subseteq N$ $((a_1 + b_1)a_2K \subseteq N)$ and neither $a_2K \subseteq N$ nor $(a_1 + b_1)K \subseteq N$, we conclude $(a_1 + b_1)a_2 \in (N : M)$ by

Lemma 2.16. But $(a_1 + b_1)a_2 = a_1a_2 + b_1a_2$, so $(a_1a_2 + b_1a_2) \in (N : M)$ and since $a_1a_2 \in (N : M)$, we get $b_1a_2 \in (N : M)$. Now, since $(a_1 + b_1)R(a_2 + b_2)K \subseteq N$ $((a_1 + b_1)(a_2 + b_2)K \subseteq N)$ and neither $(a_1 + b_1)K \subseteq N$ nor $(a_2 + b_2)K \subseteq N$, we have $(a_1 + b_1)(a_2 + b_2) = (a_1a_2 + a_1b_2 + b_1a_2 + b_1b_2) \in (N : M)$ by Lemma 2.16. But $a_1a_2, a_1b_2, b_1a_2 \in (N : M)$, so $b_1b_2 \in (N : M)$ which is a contradiction. Consequently $IK \subseteq N$ or $JK \subseteq N$.

The converse is clear for the 2-absorbing case since if $aRbRm \subseteq N$, then $(RaR)(RbR)Rm \subseteq N$ and we have $am \in (RaR)Rm \subseteq N$ or $bm \in (RbR)Rm \subseteq N$ or $abM \subseteq (RaR)(RbR)M \subseteq N$. □

**Corollary 2.19.** *Let $I$ and $J$ be two ideals of $R$ and $P$ a 2-absorbing submodule of $M$. If $m \in M$ such that $IJm \subseteq P$, then $Im \subseteq P$ or $Jm \subseteq P$ or $IJ \subseteq (N :_R M)$.*

**Proof.** If $IJm \subseteq P$, then $IJRm \subseteq P$ and consequently $Im \subseteq IRm \subseteq P$ or $Jm \subseteq JRm \subseteq P$ or $IJ \subseteq (P :_R M)$. □

**Lemma 2.20.** *Let $I$ be an ideal of $R$ and $N$ be a 2-absorbing (strong 2-absorbing) submodule of $M$. If $a \in R$, $m \in M$ and $IRaRm \subseteq N$ ($Iam \subseteq N$), then $am \in N$ or $Im \subseteq N$ or $Ia \subseteq (N_R : M)$.*

**Proof.** Let $am \notin N$ and $Ia \nsubseteq (N_R : M)$. Then there exists $b \in I$ such that $ba \notin (N_R : M)$. Now, $bRaRm \subseteq N$ ($bam \in N$), implies that $bm \in N$, since $N$ is a 2-absorbing (strong 2-absorbing) submodule of $M$. We have to show that $Im \subseteq N$. Let $c$ be an arbitrary element of $I$. Thus $(b + c)RaRm \subseteq IRaRm \subseteq N$ $(b + c)am \in Iam \subseteq N)$. Hence, either $(b + c)m \in N$ or $(b + c)a \in (N_R : M)$. If $(b+c)m \in N$, then by $bm \in N$ it follows that $cm \in N$. If $(b+c)a \in (N_R : M)$, then $ca \notin (N_R : M)$, but $cRaRm \subseteq N$ ($cam \in N$). Thus $cm \in N$. Hence, we conclude that $Im \subseteq N$. □

**Corollary 2.21.** *Let $N$ be a 2-absorbing (strong 2-absorbing) submodule of the $R$-module $M$. Then $(N :_M I) = \{m \in M : Im \subseteq N\}$ is a 2-absorbing (strong 2-absorbing) submodule of $M$ for every ideal $I$ of $R$.*

**Proof.** Let $I$ be an ideal of $R$ and $a, b \in R$ and $m \in M$ such that $aRbRm \subseteq (N :_M I)(abm \subseteq (N :_M I))$. Thus $IaRbRm \subseteq N(Iabm \subseteq N)$, then $(IaR)RbRm \subseteq IaRbRm \subseteq N$. Hence from Lemma 2.20 we have $(IaR)m \subseteq N$ or $(IaRb) \subseteq (N_R : M)$ or $bm \in N$. If $bm \in N$, then $Ibm \subseteq N$ and consequently $bm \in (N :_M I)$ and we are done. If $IaRb \subseteq (N_R : M)$, then $ab \in aRb \subseteq ((N :_R M) :_R I) = ((N :_M I) :_R M)$. If $(IaR)m \subseteq N$, then $am \in aRm \subseteq (N :_M I)$. Thus $bm \in (N :_M I)$

or $am \in (N :_M I)$ or $ab \in ((N :_M I) :_R M)$ which complete the proof for 2-absorbing. For $Iabm \subseteq N$ and $N$ strong 2-absorbing it follows from Lemma 2.20 that $abm \in N$ or $Im \subseteq N$ or $Iab \subseteq (N_R : M)$. If $abm \in N$, then $am \in N$ or $bm \in N$ or $ab \in (N :_R M)$. Hence for $am \in N$ it follows that $Iam \subseteq IN \subseteq N$ and we have $am \in (N :_M I)$. For $bm \in N$ it follows that $Ibm \subseteq IN \subseteq N$ and we have $bm \in (N :_M I)$. For $ab \in (N :_R M)$, we have $ab \in ((N :_R M) :_R I) = ((N :_M I) :_R M)$. For $Im \subseteq N$, we have $m \in (N :_M I)$ and thus $am \in (N :_M I)$. For $Iab \subseteq (N_R : M)$, we have $ab \in ((N :_R M) :_R I) = ((N :_M I) :_R M)$ and $(N :_M I)$ is a strong 2-absorbing submodule of $M$. □

For the next Theorem compare also [7, Theorem 2.7 and Corollary 2.8].

**Theorem 2.22.** *Let $N$ be a 2-absorbing submodule of $M$. Then $(N :_R M)$ is a prime ideal of $R$ if and only if $(N :_R P)$ is a prime ideal for every submodule $P$ of $M$ containing $N$.*

**Proof.** Let $I$ and $J$ be ideals of $R$ and $K$ a submodule of $M$ such that $N \subseteq P$ and $IJ \subseteq (N :_R P)$. Hence $IJP \subseteq N$. Since $N$ is a 2-absorbing submodule of $M$ it follows from Theorem 2.18 that $IP \subseteq N$ or $JP \subseteq N$ or $IJ \subseteq (N :_R M)$. Hence $I \subseteq (N :_R P)$ or $J \subseteq (N :_R P)$. Also for $IJ \subseteq (N :_R M)$ by the assumption that $(N :_R M)$ is a prime ideal, we get $IP \subseteq IM \subseteq N$ or $JP \subseteq JM \subseteq N$. Hence also $I \subseteq (N :_R P)$ or $J \subseteq (N :_R P)$ and we have $(N :_R P)$ is a prime ideal.

This is clear, just take $P = M$. □

**Proposition 2.23.** *Let $N$ and $K$ be submodules of an $R$-module $M$ with $K \nsubseteq N$. If $N$ is a 2-absorbing (strong 2-absorbing ) submodule of $M$, then $K \cap N$ is a 2-absorbing (strong 2-absorbing) submodule of $K$.*

**Proof.** Since $N$ and $K$ are submodules of $M$ and $K \nsubseteq N$, $K \cap N$ is a proper submodule of $K$. Assume that $N$ is a 2-absorbing (strong 2-absorbing) submodule of $M$. Let $a, b \in R$ and $x \in K$ be such that $aRbRx \subseteq N$ ($abx \in N$). Since $K$ is a submodule of $M$, $abK \subseteq K$ and $ax; bx \in K$. Moreover, since $aRbRx \subseteq K \cap N \subseteq N$ ($abx \in K \cap N \subseteq N$) and $N$ is a 2-absorbing (strong 2-absorbing ) submodule of $M$, $abM \subseteq N$ or $ax \in N$ or $bx \in N$. Thus $abK \subseteq abK \cap abM \subseteq K \cap N$ or $ax \in K \cap N$ or $bx \in K \cap N$. Therefore, $K \cap N$ is a 2-absorbing (strong 2-absorbing ) submodule of $K$. □

**Proposition 2.24.** *Let $N$ and $K$ be submodules of an $R$-module $M$ with $K \subseteq N$. Then $N$ is a 2-absorbing (strong 2-absorbing) submodule of $M$ if and only if $N/K$ is a 2-absorbing (strong 2-absorbing) submodule of $M/K$.*

**Proof.** First, assume that $N$ is a 2-absorbing (strong 2-absorbing) submodule of $M$. Then $N/K$ is a proper submodule of $M/K$. Let $a, b \in R$ and $m \in M$ be such that $aRbR(m+K) \subseteq N/K$ $(ab(m+K) \subseteq N/K)$. Let $s;t \in R$. Thus $asbtm + K = asbt(m+K) \subseteq aRbR(m+K) \subseteq N/K$. Then there exists $n \in N$ such that $asbtm + K = n + K$ so that $-n + asbtm \in K \subseteq N$ and then $asbtm \in N$. This shows that $aRbRm \subseteq N$. (For the strong 2-absorbing case, $abm + K = ab(m+K) \subseteq N/K$ and then there exists $n \in N$ such that $abm + K = n + K$ so that $-n + abm \in K \subseteq N$ and then $abm \in N$). As a result, $am \in N$ or $bm \in N$ or $abM \subseteq N$ because $N$ is a 2-absorbing (strong 2-absorbing) submodule of $M$. Therefore, $a(m+K) \in N/K$ or $b(m+K) \in N/K$ or $ab(M/K) \subseteq N/K$. Hence $N/K$ is a 2-absorbing (strong 2-absorbing) submodule of $M/K$.

Conversely, assume that $N/K$ is a 2-absorbing (strong 2-absorbing) submodule of $M/K$. Then $N$ is a proper submodule of $M$. Let $a, b \in R$ and $m \in M$ be such that $aRbRm \subseteq N (abm \in N)$. Then $aRbR(m+K) \subseteq N/K$ $(ab(m+K) \subseteq N/K)$. Since $N/K$ is a 2-absorbing (strong 2-absorbing) submodule of $M/K$, we obtain that $a(m+K) \in N/K$ or $b(m+K) \in N/K$ or $ab(M/K) \subseteq N/K$. That is $am \in N$ or $bm \in N$ or $abM \subseteq N$. This implies that $N$ is a 2-absorbing (strong 2-absorbing) submodule of $M$. □

Consider $R = R_1 \times R_2$ where each $R_i$ is an associative ring with identity, $M_i$ be an $R_i$-module where $i = 1, 2$, and $M = M_1 \times M_2$ be the $R$-module with $(r_1, r_2)(m_1, m_2) = (r_1m_1, r_2m_2)$ where $r_i \in R_i$, $m_i \in M_i, i = 1, 2$.

**Theorem 2.25.** *Let $M_1$ be an $R_1$-module, $M_2$ be an $R_2$-module, $R = R_1 \times R_2$ and $M = M_1 \times M_2$. Then*

(1) *$N_1$ is a 2-absorbing $R_1$-submodule of $M_1$ if and only if $N_1 \times M_2$ is a 2-absorbing submodule of $M$; and*

(2) *$N_2$ is a 2-absorbing $R_2$-submodule of $M_2$ if and only if $M_1 \times N_2$ is a 2-absorbing submodule of $M$.*

**Proof.** It suffices to prove only part 1. First, assume that $N_1$ is a 2-absorbing $R_1$-submodule of $M_1$. Suppose that $(a,b)R(c,d)R(m_1,m_2) \subseteq N_1 \times M_2$ where $(a,b),(c,d) \in R$ and $(m_1,m_2) \in M$.

Then $(aR_1cR_1m_1, bR_2dR_2m_2) = (a,b)R(c,d)R(m_1,m_2) \subseteq N_1 \times M_2$, i.e., $aR_1cR_1m_1 \subseteq N_1$ and $bR_2dR_2m_2 \subseteq M_2$. Since $N_1$ is a 2-absorbing $R_1$-submodule of $M_1$, $acM_1 \subseteq N_1$ or $am_1 \in N_1$ or $cm_1 \in N_1$. That is $(a,b)(c,d)M = (acM_1, bdM_2) \subseteq N_1 \times M_2$ or $(a,b)(m_1,m_2) = (am_1, bm_2) \in N_1 \times M_2$ or $(c,d)(m_1,m_2) = (cm_1, dm_2) \in N_1 \times M_2$. Therefore, $N_1 \times M_2$ is a 2-absorbing submodule of $M$. Conversely, assume that $N_1 \times M_2$ is a 2-absorbing $R$-submodule of $M$. Let $a, b \in R_1$ and $m_1 \in M_1$. Assume that $aR_1bR_1m_1 \subseteq N_1$. Let $x, y \in R_2$ and $m_2 \in M_2$. Then $(a,x)R(b,y)R(m_1,m_2) = (aR_1bR_1m_1, xR_2yR_2m_2) \subseteq N_1 \times M_2$. Since $N_1 \times M_2$ is a 2-absorbing $R$-submodule of $M$, $(a,x)(b,y)M \subseteq N_1 \times M_2$ or $(a,x)(m_1,m_2) \in N_1 \times M_2$ or $(b,y)(m_1,m_2) \in N_1 \times M_2$. Then $(abM_1, xyM_2) = (a,x)(b,y)M \subseteq N_1 \times M_2$ or $(am_1, xm_2) = (a,x)(m_1,m_2) \in N_1 \times M_2$ or $(bm_1, ym_2) = (b,y)(m_1,m_2) \in N_1 \times M_2$, i.e., $abM_1 \subseteq N_1$ or $am_1 \in N_1$ or $bm_1 \in N_1$. Therefore, $N_1$ is a 2-absorbing $R_1$-submodule of $M_1$. □

**Theorem 2.26.** *Let $M_1$ be an $R_1$-module, $M_2$ be an $R_2$-module, $R = R_1 \times R_2$ and $M = M_1 \times M_2$. Then*

(1) *$N_1$ is a strong 2-absorbing $R_1$-submodule of $M_1$ if and only if $N_1 \times M_2$ is a strong 2-absorbing submodule of $M$; and*
(2) *$N_2$ is a strong 2-absorbing $R_2$-submodule of $M_2$ if and only if $M_1 \times N_2$ is a strong 2-absorbing submodule of $M$.*

**Proof.** It suffices to prove only part 1. First, assume that $N_1$ is a strong 2-absorbing $R_1$-submodule of $M_1$. Suppose that $(a,b)(c,d)(m_1,m_2) \in N_1 \times M_2$ where $(a,b),(c,d) \in R$ and $(m_1,m_2) \in M$.
Then $(acm_1, bdm_2) = (a,b)(c,d)(m_1,m_2) \in N_1 \times M_2$, i.e., $acm_1 \in N_1$ and $bdm_2 \in M_2$. Since $N_1$ is a strong 2-absorbing $R_1$-submodule of $M_1$, it follows that $acM_1 \subseteq N_1$ or $am_1N_1$ or $cm_1 \in N_1$. That is $(a,b)(c,d)M = (acM_1, bdM_2) \subseteq N_1 \times M_2$ or $(a,b)(m_1,m_2) = (am_1, bm_2) \in N_1 \times M_2$ or $(c,d)(m_1,m_2) = (cm_1, dm_2) \in N_1 \times M_2$. Therefore, $N_1 \times M_2$ is a 2-absorbing submodule of $M$. Conversely, assume that $N_1 \times M_2$ is a strong 2-absorbing $R$-submodule of $M$. Let $a, b \in R_1$ and $m_1 \in M_1$. Assume that $abm_1 \in N_1$. Let $x, y \in R_2$ and $m_2 \in M_2$. Then $(a,x)(b,y)(m_1,m_2) = (abm_1, xym_2) \in N_1 \times M_2$. Since $N_1 \times M_2$ is a strong 2-absorbing $R$-submodule of $M$, it follows that $(a,x)(b,y)M \subseteq N_1 \times M_2$ or $(a,x)(m_1,m_2) \in N_1 \times M_2$ or $(b,y)(m_1,m_2) \in N_1 \times M_2$. Then $(abM_1, xyM_2) = (a,x)(b,y)M \subseteq N_1 \times M_2$ or $(am_1, xm_2) = (a,x)(m_1,m_2) \in N_1 \times M_2$ or $(bm_1, ym_2) = (b,y)(m_1,m_2) \in N_1 \times M_2$, i.e., $abM_1 \subseteq N_1$ or $am_1 \in N_1$ or $bm_1 \in N_1$. Therefore, $N_1$ is a 2-absorbing $R_1$-submodule of $M_1$. □

**Acknowledgement.** The authors would like to thank the referee for the valuable suggestions and comments.

## References


[1] A. Badawi, *On 2-absorbing ideals of commutative rings*, Bull. Austral. Math. Soc., 75(3) (2007), 417-429.

[2] A. M. Buhphang and M. B. Rege, *Semi-commutative modules and Armendariz modules*, Arab J. Math. Sci., 8(1) (2002), 53-65.

[3] A. Y. Darani and F. Soheilnia, *2-Absorbing and weakly 2-absorbing submodules*, Thai J. Math., 9(3) (2011), 577-584.

[4] J. Dauns, *Prime modules*, J. Reine Angew. Math., 298 (1978), 156-181.

[5] N. J. Groenewald and D. Ssevviiri, *Completely prime submodules*, Int. Electron. J. Algebra, 13 (2013), 1-14.

[6] Sh. Payrovi and S. Babaei, *On 2-absorbing submodules*, Algebra Colloq., 19 (2012) 913-920.

[7] Sh. Payrovi and S. Babaei, *On the 2-absorbing submodules*, Iran. J. Math. Sci. Inform., 10(1) (2015), 131-137.



**N. J. Groenewald** (Corresponding Author)
Department of Mathematics
Faculty of Science
Nelson Mandela University
Port Elizabetyh
South Africa
e-mail: nico.groenewald@mandela.ac.za

**Bac T. Nguyen**
Nguyen Tat Thanh University
300 A Nguyen Tat Thanh street
Ho Chi Minh city, Vietnam
and Dept. of Basic Science
Thainguyen Univ. of Economics and Business Administration
Thainguyen
Vietnam
e-mail: bacnt2008@gmail.com

# 10

# UNIQUENESS OF DECOMPOSITION, FACTORISATIONS, $G$-GROUPS AND POLYNOMIALS

Alberto Facchini and Serap Şahinkaya

Communicated by A. Çiğdem Özcan

*Dedicated to our dear friend John Clark*

ABSTRACT. In this article, we present the classical Krull-Schmidt Theorem for groups, its statement for modules due to Azumaya, and much more modern variations on the theme, like the so-called weak Krull-Schmidt Theorem, which holds for some particular classes of modules. Also, direct product of modules is considered. We present some properties of the category of $G$-groups, a category in which Remak's results about the Krull-Schmidt Theorem for groups can be better understood. In the last section, direct-sum decompositions and factorisations in other algebraic structures are considered.

**Keywords**: Krull-Schmidt Theorem, indecomposable module, direct-sum decomposition, endomorphism ring, uniserial module, artinian module

## 1. Introduction

In this paper, we mainly present the classical Krull-Schmidt Theorem for groups and modules and some of its weak versions. According to the classical Krull-Schmidt Theorem for modules, any module of finite composition length decomposes as a direct sum of indecomposable modules in an essentially unique way, that is, unique up to isomorphism of the indecomposable summands and a permutation of the summands. In Section 2, we present the historical background of that theorem.

Section 3 is devoted to commutative monoids, which provide the best algebraic tool to describe finite direct-sum decompositions of modules.

In 1975, Warfield proved that every finitely presented module over a serial ring is a finite direct sum of uniserial modules and posed a problem, essentially asking whether the Krull-Schimdt Theorem holds for finite direct sums of uniserial modules. More precisely, he asked whether the direct-sum decomposition of a finitely presented module over a serial ring into uniserial summands is unique up to isomorphism [37]. The negative answer to this question was given by the first author in 1996 [15]. He showed that even though the Krull-Schmidt Theorem does not hold for serial modules, a weak version of it holds not only for serial modules, but also for other classes of modules. Some of these classes of modules will be presented in Section 4.

It is natural to ask what happens when one considers arbitrary direct products of modules instead of arbitrary direct sums. In Section 5, we collect some results about direct product of modules and uniqueness of decomposition. In that section, we also present some properties of the category of $G$-groups, a framework in which the existence of the central automorphism in the Krull-Schmidt Theorem proved by Remak has a natural explanation.

All these results are strongly related to factorisation in a very general sense. Thus we present some results concerning factorisation into irreducible polynomials of non-commutative polynomials with coefficients in $\mathbb{Z}$, and also factorization of commutative polynomials with non-negative integral coefficients.

## 2. Historical background of the Krull-Schmidt-Remak-Azumaya Theorem

In this survey, rings will be associative rings $R$ with an identity, and modules will be unital right $R$-modules, unless otherwise stated.

In 1879, Frobenius and Stickelberger proved that any finite abelian group is a direct product of cyclic groups whose orders are powers of primes, and this powers of primes are uniquely determined by the group [23]. The "classical Krull-Schmidt Theorem for finite groups" was first stated in 1909 by Wedderburn [27], who published the following theorem.

**Theorem 2.1.** (Krull-Schmidt Theorem for finite groups) *If a finite group $G$ has two direct-product decompositions $G = G_1 \times G_2 \times \cdots \times G_t = H_1 \times H_2 \times \cdots \times H_s$ into indecomposables, then $t = s$ and there exist an automorphism $\varphi$ of $G$ and a permutation $\sigma$ of $\{1, 2, \dots, t\}$ such that $\varphi(G_i) = H_{\sigma(i)}$ for all indices $i = 1, 2, \dots, t$.*

The proof given by Wedderburn is not entirely convincing. It was Robert Erich Remak [32] who proved in his PhD dissertation (1911) that any two direct-product decompositions of a finite group into indecomposable factors are not only isomorphic, but also centrally isomorphic. More precisely, he proved the following theorem. Recall that a *central automorphism* of group $G$ is an automorphism of $G$ that induces the identity $G/\zeta(G) \to G/\zeta(G)$, where $\zeta(G)$ denotes the center of $G$.

**Theorem 2.2.** *If a finite group $G$ has two direct-product decompositions into indecomposables $G = G_1 \times G_2 \times \cdots \times G_t = H_1 \times H_2 \times \cdots \times H_s$, then $t = s$ and there exist a* central *automorphism $\varphi$ of $G$ and a permutation $\sigma$ of $\{1, 2, \ldots, t\}$ such that $\varphi(G_i) = H_{\sigma(i)}$ for all indices $i = 1, 2, \ldots, t$.*

Otto Yulyevich Schmidt [33] then gave a simplified proof of Remak's results. These results were transfered to modules of finite length by Krull and Schimdt, getting a theorem that, in modern terminology, can be stated as follows.

**Theorem 2.3.** (The Krull-Schmidt Theorem) *Let $R$ be a ring and $M$ be a module of finite length. Then there exists a decomposition*

$$M = M_1 \oplus M_2 \oplus \cdots M_r$$

*into indecomposable submodules $M_i$ of $M$. Moreover, if $M = N_1 \oplus N_2 \oplus \cdots N_s$ is another decomposition of $M$ into indecomposable modules, then $r = s$ and there exists a permutation $\sigma$ of $\{1, \cdots, r\}$ such that $M_i \cong N_{\sigma(i)}$ for every $i \in \{1, \cdots, r\}$.*

Arbitrary modules do not decompose in an "essentially unique" way in general. Here is an example.

**Example 2.4.** [17] *Let $R$ be a commutative integral domain with at least two distinct maximal ideals $M$ and $N$ that are not principal ideals. Then the morphism $M \oplus N \to R$, defined by $(x, y) \to x + y$, is an $R$-module epimorphism, which necessarily splits because $R$ is a projective $R$-module. The kernel of this morphism is isomorphic to $M \cap N$, so that there is a splitting short exact sequence $0 \to M \cap N \to M \oplus N \to R \to 0$. Thus $M \oplus N$ is isomorphic to $R \oplus (M \cap N)$. But $M$ and $N$ are not principal ideals, so that they are not isomorphic to $R$. Therefore the two direct-sum decompositions are essentially different. Notice that $R$ is an $R$-module of Goldie dimension 1 since $R$ is a commutative integral domain. Thus $R$ and its submodules $M$, $N$ and $M \cap N$ are all modules of Goldie dimension 1, which proves that the two essentially different direct-sum decompositions are direct-sum decompositions into indecomposables.*

Let $\Omega$ be a set. Recall that an $\Omega$-*group* is a pair $(H, \varphi)$, where $H$ is a group and $\varphi \colon \Omega \to \mathrm{End}(H)$ is a mapping. Krull [25] extended the results known for groups to the case of abelian operator groups with the ascending and descending chain conditions (operator groups = $\Omega$-groups). Notice that if $H$ is abelian, then $\mathrm{End}(H)$ is a ring, so that the mapping $\varphi\colon \Omega \to \mathrm{End}(H)$ extends uniquely to a ring homomorphism $\mathbb{Z}\langle\Omega\rangle \to \mathrm{End}(H)$, where $\mathbb{Z}\langle\Omega\rangle$ denotes the free ring with free set $\Omega$ of generators (=ring of non-commutative polynomials with coefficients in $\mathbb{Z}$ in the set of non-commuting indeterminates $\Omega$). Thus abelian $\Omega$-groups are exactly left $\mathbb{Z}\langle\Omega\rangle$-modules. The theory was subsequently further deepened by Schmidt [34].

Øystein Ore (Oslo, 1899-1968) unified the proofs from various categories: groups, abelian operator groups, rings and algebras. He showed that the theorem of Wedderburn holds for modular lattices with descending and ascending chain conditions.

The Krull-Schmidt Theorem was extended to the case of possibly infinite direct sums of modules with local endomorphism rings by Azumaya [6]. Notice that any indecomposable module of finite composition length has a local endomorphism ring (this is the so-called Fitting Lemma) and any module with a local endomorphism ring is necessarily indecomposable [4, p. 144].

**Theorem 2.5.** (Krull-Schmidt-Remak-Azumaya Theorem) *Let $M$ be a module that is a direct sum of modules with local endomorphism rings. Then $M$ is a direct sum of indecomposable modules in an essentially unique way in the following sense. If*

$$M = \bigoplus_{i\in I} M_i = \bigoplus_{j\in J} N_j,$$

*where all the submodules $M_i$ $(i \in I)$ and $N_j$ $(j \in J)$ are indecomposable, then there exists a bijection $\varphi\colon I \to J$ such that $M_i \cong N_{\varphi(i)}$ for every $i \in I$.*

After this brief historical introduction, our aim now is to describe direct-sum decompositions of a module $M_R$ as a finite direct sum $M_R = M_1 \oplus \cdots \oplus M_n$ of direct summands $M_i$. Several behaviours are possible. For instance:

- There can be uniqueness of direct-sum decomposition into indecomposables. This is the case described, for example, for modules that are direct sums of modules with local endomorphism rings (Krull-Schmidt-Remak-Azumaya).
- There are modules with a direct-sum decomposition into indecomposables, such that this decomposition is not unique in the sense of the Krull-Schmidt-Remak-Azumaya Theorem, but there are only finitely many such direct-sum decompositions up to isomorphism. This happens, for instance, for torsion-free abelian groups of finite rank [26].
- For some classes of modules, direct-sum decompositions into indecomposables are not unique, but they enjoy some kind of regularity.

- But, in general, there is no direct-sum decomposition into indecomposables, and no uniqueness as in Example 2.4.

## 3. The reduced monoid $V(\mathcal{C})$

The best way to describe finite direct-sum decompositions of a module $M_R$ is by making use of commutative monoids. A *commutative monoid* is a semigroup with a binary operation that is associative, commutative and has an identity element. All the monoids in this paper will be commutative and additive, that is, their operation will be denoted as an addition $+$, and their identity element will be denoted by 0. For a commutative monoid $M$, let $U(M)$ denote the group of all $a \in M$ with an opposite $-a$ in $M$. A commutative additive monoid $M$ is *reduced* if $x, t \in M$ and $x + t = 0$ implies $x = t = 0$. That is, if $U(M) = 0$. For every monoid $M$, the monoid $M_{red} := M/U(M)$, whose elements are the cosets $m + U(M)$, is reduced.

We will look at classes of right $R$-modules as full subcategories of the category Mod-$R$ of all right $R$-modules. Let $\mathcal{C}$ be a category and let $V(\mathcal{C})$ denote a *skeleton* of $\mathcal{C}$, that is, a class of representatives of the objects of $\mathcal{C}$ modulo isomorphism. In order to avoid set-theoretical problems, in this survey we could consider only *skeletally small* categories, that is, categories $\mathcal{C}$ in which the class $\mathrm{Ob}(\mathcal{C})$ contains a set of representatives of the objects up to isomorphism. Equivalently, a category $\mathcal{C}$ is skeletally small if it has a skeleton whose class of objects is a set. For every object $A$ in $\mathcal{C}$, there is a unique object $\langle A\rangle$ in $V(\mathcal{C})$ isomorphic to $A$. Thus there is a mapping $\mathrm{Ob}(\mathcal{C}) \to V(\mathcal{C})$, $A \mapsto \langle A\rangle$, that associates to every object $A$ of $\mathcal{C}$ the unique object $\langle A\rangle$ in $V(\mathcal{C})$ isomorphic to $A$. Assume that a product $A \times B$ exists in $\mathcal{C}$ for every pair $A, B$ of objects of $\mathcal{C}$. Define an addition $+$ in $V(\mathcal{C})$ by $A + B := \langle A \times B\rangle$ for every $A, B \in V(\mathcal{C})$. If the category $\mathcal{C}$ is not skeletally small, we have that its skeleton $V(\mathcal{C})$ is a class that is not a set. This class becomes therefore a *large* monoid, that is, a monoid that is not a set but a class, like in the next lemma.

**Lemma 3.1.** *Let $\mathcal{C}$ be a category with a terminal object and in which a product $A \times B$ exists for every pair $A, B$ of objects of $\mathcal{C}$. Then $V(\mathcal{C})$ is a large reduced commutative monoid.*

It is easy to prove that the Krull-Schmidt property holds in the additive category $\mathcal{C}$ if and only if the monoid $V(\mathcal{C})$ is a *free monoid*, that is, isomorphic to the direct sum $\mathbb{N}^{(I)}$ for some class $I$.

An element $u$ of a commutative monoid $M$ is called an *order-unit* in $M$ if, for every $x \in M$, there exists an integer $n \geq 0$ such that $x \leq nu$. Here $\leq$ denotes the *algebraic preorder* on the commutative monoid $M$, that is, the reflexive and transitive relation $\leq$ on $M$ defined, for every $x, y \in M$, by $x \leq y$ if there exists $z \in M$ with $x + z = y$. Hence $u$ is an order-unit in $M$ if and only if, for every $x \in M$, there exist an integer $n \geq 0$ and an element $y \in M$ such that $x + y = nu$. It is now easy to define the category of commutative monoids with order-unit. Its objects are the pairs $(M, u)$, where $M$ is any commutative monoid and $u \in M$ is an order-unit. Its morphisms $f\colon (M, u) \to (M', u')$ are the monoid morphisms $f\colon M \to M'$ such that $f(u) = u'$.

For any ring $R$, let $\mathcal{C}$ be the class of all finitely generated projective right $R$-modules. We will denote by $V(R)$ the monoid $V(\mathcal{C})$. Then $V(R)$ is a commutative reduced monoid with order-unit $\langle R_R \rangle$.

The following theorem was first proved by Bergman for finitely generated monoids with order-unit [7, Theorems 6.2 and 6.4]. Then it was extended by Bergman and Dicks to arbitrary monoids with order-unit [8, p. 315]. Recall that a ring $R$ is *hereditary* if all its right ideals and all its left ideals are projective modules.

**Theorem 3.2.** [7,8] *Let $k$ be a field and let $(M, u)$ be a commutative reduced monoid with order-unit. Then there exists a hereditary $k$-algebra $R$ such that $(M, u)$ and $(V(R), \langle R_R \rangle)$ are isomorphic monoids with order-unit.*

**Corollary 3.3.** *Let $k$ be a field and let $M$ be a commutative reduced monoid. Then there exists a class $\mathcal{C}$ of finitely generated projective right modules over a hereditary $k$-algebra $R$ such that $M \cong V(\mathcal{C})$.*

## 4. Weak Krull-Schmidt Theorem

Let $R$ be any ring. In this section, we will see what happens for some very special classes of right $R$-modules that are not direct sums of modules $M_i$ with local endomorphism ring $\mathrm{End}(M_i)$. A right $R$-module $M_R$ is *uniserial* if its lattice of submodules is linearly ordered, that is, if for any submodules $A, B$ of $M_R$ either $A \subseteq B$ or $B \subseteq A$, and it is *serial* if it is a direct sum of uniserial modules.

The endomorphism ring of a uniserial module has at most two maximal right (left) ideals, as the following theorem shows.

**Theorem 4.1.** [15] *Let $U_R$ be a non-zero uniserial module over a ring $R$, $E := \mathrm{End}(U_R)$ its endomorphism ring, $I := \{ f \in E \mid f$ is not injective$\}$ and $K := \{ f \in E \mid f$ is not surjective$\}$. Then $I$ and $K$ are two two-sided completely prime ideals of $E$, and every proper right ideal of $E$ and every proper left ideal of $E$ is contained either in $I$ or in $K$. Moreover,*

(a) *either $E$ is a local ring with maximal ideal $I \cup K$, or*
(b) *$E/I$ and $E/K$ are division rings, and $E/J(E) \cong E/I \times E/K$.*

In 1975, Warfield proved that every finitely presented module over a serial ring is a finite direct sum of uniserial modules, and asked whether the direct decomposition of a finitely presented module into uniserial summands is unique up to isomorphism [37]. Warfield's question was answered completely in [15], by showing that, although there exist serial rings for which the Krull-Schmidt Theorem does not hold for finitely presented modules, it is possible to prove a weak form of the Krull-Schmidt Theorem. Facchini's weak form of the Krull-Schmidt Theorem holds only for finite direct sums of uniserial modules, not for infinite ones. In order to state the theorem, we need the concepts of monogeny class and epigeny class of a module.

Two right $R$-modules $U$ and $V$ are said to have

(1) *the same monogeny class*, denoted $[U]_m = [V]_m$, if there exist a monomorphism $U \to V$ and a monomorphism $V \to U$;
(2) *the same epigeny class*, denoted $[U]_e = [V]_e$, if there exist an epimorphism $U \to V$ and an epimorphism $V \to U$.

The weak Krull-Schmidt Theorem for finite families of uniserial modules can be formulated as follows:

**Theorem 4.2.** (Weak Krull-Schmidt Theorem [15]) *Let $U_1, \dots, U_n, V_1, \dots, V_t$ be $n+t$ non-zero uniserial right modules over a ring $R$. Then the direct sums $U_1 \oplus \cdots \oplus U_n$ and $V_1 \oplus \cdots \oplus V_t$ are isomorphic $R$-modules if and only if $n = t$ and there exist two permutations $\sigma$ and $\tau$ of $\{1, 2, \dots, n\}$ such that $[U_i]_m = [V_{\sigma(i)}]_m$ and $[U_i]_e = [V_{\tau(i)}]_e$ for every $i = 1, 2, \dots, n$.*

**Remark 4.3.** *In [21], Zahra Nazemian and the first author have studied the factorisations $A = A_1 \dots A_n$ of a right ideal $A$ of a not-necessarily commutative ring $R$ as a product of right ideals $A_1, \dots, A_n$, with $R/A \cong R/A_1 \oplus \cdots \oplus R/A_n$ canonically and the right modules $R/A_1, \dots, R/A_n$ uniserial. The main example of such a factorisation is the factorization into powers of prime ideals for non-zero ideals of a Dedekind domain $R$.*

Theorem 4.2 must be modified if we want it to hold for infinite families of uniserial modules [13, Theorem 4.9]. This was done by Příhoda, who proved the converse of [13, Theorem 4.7] in [29]. First recall a definition. A right $R$-module $U_R$ is said to be *quasismall* if, for every family $\{ M_i \mid i \in I \}$ of $R$-modules such that $U$ is isomorphic to a direct summand of $\bigoplus_{i\in I} M_i$, there is a finite set $I' \subseteq I$ such that $U_R$ is isomorphic to a direct summand of $\bigoplus_{i\in I'} M_i$.

**Theorem 4.4.** [29, Theorem 2.6] *Let $\{U_i \mid i \in I\}$ and $\{V_j \mid j \in J\}$ be nonempty families of non-zero uniserial modules. Let $I' = \{i \in I \mid U_i$ is quasismall$\}$ and $J' = \{j \in J \mid V_j$ is quasismall$\}$. Then $\bigoplus_{i\in I} U_i \cong \bigoplus_{j\in J} V_j$ if and only if there exist a bijection $\sigma\colon I \to J$ and a bijection $\tau\colon I' \to J'$ such that $[U_i]_m = [V_{\sigma(i)}]_m$ for every $i \in I$ and $[U_i]_e = [V_{\tau(i)}]_e$ for every $i \in I'$.*

If a uniserial module $U$ has local endomorphism ring, then any direct summand of a direct sum $U^{(I)}$ of copies of $U$ is a direct sum of copies of $U$, because any uniserial module is $\sigma$-small (a module is *$\sigma$-small* if it is countable ascending union of small submodules) and one can use [16, Theorem 2.52].

**Theorem 4.5.** [30, Theorem 1.1] *Let $U$ be a non-zero uniserial right module over a ring $R$. Then:*

(1) *If $gf \neq 0$ for every monomorphism $f\colon U \to U$ and every epimorphism $g\colon U \to U$, then every direct summand of a direct sum $U^{(I)}$ of copies of $U$ is a direct sum of copies of $U$.*
(2) *If $U$ is quasismall and there exist a monomorphism $f\colon U \to U$ and an epimorphism $g\colon U \to U$ such that $gf = 0$, then every direct summand of a direct sum $U^{(I)}$ of copies of $U$ is isomorphic to $U^{(J)} \oplus V^{(K)}$, where $J$ and $K$ are suitable sets and $V$ is the unique uniserial module in the same monogeny class of $U$ that is not quasismall.*
(3) *If $U$ is not quasismall, then every direct summand of a direct sum $U^{(I)}$ of copies of $U$ is a direct sum of copies of $U$.*

Recall that a right module over a ring $R$ is *cyclically presented* if it is isomorphic to $R/aR$ for some element $a \in R$. For any ring $R$, the endomorphism ring $\mathrm{End}_R(R/aR)$ of a non-zero cyclically presented module $R/aR$ is isomorphic to $E/aR$, where $E := \{r \in R \mid ra \in aR\}$ is the *idealizer* of $aR$. The following theorem is proved in [2, Theorem 2.1].

**Theorem 4.6.** *Let $a$ be a non-zero non-invertible element of an arbitrary local ring $R$, let $E$ be the idealizer of $aR$, and let $E/aR$ be the endomorphism ring of the cyclically presented right $R$-module $R/aR$. Set $I := \{r \in R \mid ra \in aJ(R)\}$ and $K := J(R) \cap E$. Then $I$ and $K$ are two two-sided completely prime ideals of $E$ containing $aR$, the union $(I/aR) \cup (K/aR)$ is the set of all non-invertible elements of $E/aR$, and every proper right ideal of $E/aR$ and every proper left ideal of $E/aR$*

*is contained either in $I/aR$ or in $K/aR$. Moreover, exactly one of the following two conditions holds:*

(a) *Either $I$ and $K$ are comparable (that is, $I \subseteq K$ or $K \subseteq I$), in which case $E/aR$ is a local ring, or*

(b) *$I$ and $K$ are not comparable, and in this case $E/I$ and $E/K$ are division rings, $J(E/aR) = (I \cap K)/aR$, and $(E/aR)/J(E/aR)$ is canonically isomorphic to the direct product $E/I \times E/K$.*

For any ring $R$, let $U(R)$ denote the group of all invertible elements of $R$. If $R/aR$ and $R/bR$ are cyclically presented modules over a local ring $R$, we say that $R/aR$ and $R/bR$ *have the same lower part*, and write $[R/aR]_l = [R/bR]_l$, if there exist $u, v \in U(R)$ and $r, s \in R$ with $au = rb$ and $bv = sa$. (It is possible to prove that two cyclically presented modules over a local ring have the same lower part if and only if their Auslander-Bridger transposes have the same epigeny class.)

**Theorem 4.7.** (Weak Krull-Schmidt Theorem for Cyclically Presented Modules) *Let $a_1, \dots, a_n, b_1, \dots, b_t$ be $n + t$ non-invertible elements of a local ring $R$. Then the direct sums $R/a_1R \oplus \cdots \oplus R/a_nR$ and $R/b_1R \oplus \cdots \oplus R/b_tR$ are isomorphic right $R$-modules if and only if $n = t$ and there exist two permutations $\sigma, \tau$ of $\{1, 2, \dots, n\}$ such that $[R/a_iR]_l = [R/b_{\sigma(i)}R]_l$ and $[R/a_iR]_e = [R/b_{\tau(i)}R]_e$ for every $i = 1, 2, \dots, n$.*

The Weak Krull-Schmidt Theorem for cyclically presented modules has an immediate consequence as far as equivalence of matrices is concerned. Recall that two $m \times n$ matrices $A$ and $B$ with entries in a ring $R$ are said to be *equivalent* matrices, denoted $A \sim B$, if there exist an $m \times m$ invertible matrix $P$ and an $n \times n$ invertible matrix $Q$ with entries in $R$ (that is, matrices invertible in the rings $M_m(R)$ and $M_n(R)$, respectively) such that $B = PAQ$. We denote by $\mathrm{diag}(a_1, \dots, a_n)$ the $n \times n$ diagonal matrix whose $(i, i)$ entry is $a_i$ and whose other entries are zero.

**Remark 4.8.** *If $R$ is a* commutative *local ring and $a_1, \dots, a_n, b_1, \dots, b_n$ are elements of $R$, then* $\mathrm{diag}(a_1, \dots, a_n) \sim \mathrm{diag}(b_1, \dots, b_n)$ *if and only if there exists a permutation $\sigma$ of $\{1, 2, \dots, n\}$ with $a_i$ and $b_{\sigma(i)}$ associates for every $i = 1, 2, \dots, n$. Here $a, b \in R$ are* associates *if they generate the same principal ideal of $R$. Let's prove this. Assume $R$ commutative and local, and* $\mathrm{diag}(a_1, \dots, a_n) \sim \mathrm{diag}(b_1, \dots, b_n)$. *Then $R/a_1R \oplus \cdots \oplus R/a_nR \cong R/b_1R \oplus \cdots \oplus R/b_tR$. Now the endomorphism ring of $R/a_iR$ is the ring $R/a_iR$, because $R$ is commutative, and therefore the endomorphism ring of each $R/a_iR$ is local, because $R$ is local. Similarly for the modules $R/b_jR$. Hence the Krull-Schmidt-Remak-Azumaya Theorem* *implies that there exists a permutation* $\sigma$ *of* $\{1, 2, \ldots, n\}$ *with* $R/a_iR \cong R/b_{\sigma(i)}R$ *for every* $i = 1, 2, \ldots, n$. *Taking the annihilators of these isomorphic cyclic modules, we find that* $a_iR = b_{\sigma(i)}R$ *for every* $i = 1, 2, \ldots, n$. *Hence* $a_i$ *and* $b_{\sigma(i)}$ *are associates.*

If the ring $R$ is local, but not-necessarily commutative, we have the following result.

**Proposition 4.9.** *Let* $a_1, \ldots, a_n, b_1, \ldots, b_n$ *be elements of a local ring* $R$. *Then* $\operatorname{diag}(a_1, \ldots, a_n) \sim \operatorname{diag}(b_1, \ldots, b_n)$ *if and only if there exist two permutations* $\sigma, \tau$ *of* $\{1, 2, \ldots, n\}$ *with*

$$[R/a_iR]_l = [R/b_{\sigma(i)}R]_l \quad \textit{and} \quad [R/a_iR]_e = [R/b_{\tau(i)}R]_e$$

*for every* $i = 1, 2, \ldots, n$.

As far as infinite direct sums are concerned, the case of cyclically presented modules over local rings is much simpler than that of uniserial modules [3]. The reason for this is that cyclically presented modules are finitely generated, hence small, so the pathology of non-quasismall modules can not appear in this setting.

Thus we have seen that, similarly to the endomorphism ring of a uniserial module over an arbitrary ring, the endomorphism ring of a cyclically presented module over a local ring also has at most two maximal right ideals (Theorem 4.6). By this fact, it is not surprising that there is an analogy between the behaviour of a direct sum of uniserial modules over arbitrary rings and the behaviour of a direct sum of cyclically presented modules over local rings. Therefore it makes sense to try to see which further parts of the theory of uniserial modules also hold for cyclically presented modules over local rings. Here are some very natural questions.

(1) Does the weak Krull-Schmidt Theorem hold for a direct sum of infinitely many cyclically presented modules over local rings?
(2) Is every direct summand of a direct sum of cyclically presented modules over a local ring a direct sum of cyclically presented modules?
(3) Is every direct summand of a direct sum of finitely many cyclically presented modules over a local ring a direct sum of cyclically presented modules?

The answer to the first question was given in [3, Theorem 3.1]. The answer to the second question is negative. To this end, there is an example given by Puninski in [31, Proposition 8.1]. For the third question, there exist countably generated indecomposable relatively divisible projective modules over local rings that are not direct sums of cyclically presented modules, and relatively divisible projective modules over local rings that are not direct sums of indecomposable modules. But, according to [36, Corollary 2], every relatively divisible projective module over a commutative local ring is a direct sum of cyclically presented modules.

Let us pass to consider another class of modules. For a right module $A_R$ over a ring $R$, let $E(A_R)$ denote the injective envelope of $A_R$. We say that two modules $A_R$ and $B_R$ *have the same upper part*, and write $[A_R]_u = [B_R]_u$, if there exist a homomorphism $\varphi\colon E(A_R) \to E(B_R)$ and a homomorphism $\psi\colon E(B_R) \to E(A_R)$ such that $\varphi^{-1}(B_R) = A_R$ and $\psi^{-1}(A_R) = B_R$.

Now we need a very standard technique of homological algebra that allows to extend a morphism between two modules to their injective resolutions. Let us present it. Assume that $E_0, E_1, E_0', E_1'$ are indecomposable injective right modules over a ring $R$, and that $\varphi\colon E_0 \to E_1, \varphi'\colon E_0' \to E_1'$ are two right $R$-module morphisms. A morphism $f\colon \ker\varphi \to \ker\varphi'$ extends to a morphism $f_0\colon E_0 \to E_0'$. Now $f_0$ induces a morphism $\widetilde{f_0}\colon E_0/\ker\varphi \to E_0'/\ker\varphi'$, which extends to a morphism $f_1\colon E_1 \to E_1'$. Thus we get a commutative diagram with exact rows

$$\begin{array}{ccccccc} 0 & \longrightarrow & \ker\varphi & \longrightarrow & E_0 & \xrightarrow{\varphi} & E_1 \\ & & \downarrow{\scriptstyle f} & & \downarrow{\scriptstyle f_0} & & \downarrow{\scriptstyle f_1} \\ 0 & \longrightarrow & \ker\varphi' & \longrightarrow & E_0' & \xrightarrow[\varphi']{} & E_1'. \end{array} \tag{1}$$

The morphisms $f_0$ and $f_1$ are not uniquely determined by $f$.

**Theorem 4.10.** *Let $E_0$ and $E_1$ be indecomposable injective right modules over a ring $R$, and let $\varphi\colon E_0 \to E_1$ be a non-zero non-injective morphism. Let $S :=$ $\mathrm{End}_R(\ker\varphi)$ denote the endomorphism ring of $\ker\varphi$. Set $I := \{\, f \in S \mid$ the endomorphism $f$ of $\ker\varphi$ is not a monomorphism$\}$ and $K := \{\, f \in S \mid$ the endomorphism $f_1$ of $E_1$ is not a monomorphism$\} = \{\, f \in S \mid \ker\varphi \subset f_0^{-1}(\ker\varphi)\,\}$. Then $I$ and $K$ are two two-sided completely prime ideals of $S$, and every proper right ideal of $S$ and every proper left ideal of $S$ is contained either in $I$ or in $K$. Moreover, exactly one of the following two conditions holds:*

(a) *Either $I$ and $K$ are comparable (that is, $I \subseteq K$ or $K \subseteq I$), in which case $S$ is a local ring with maximal ideal $I \cup K$, or*

(b) *$I$ and $K$ are not comparable, and in this case $S/I$ and $S/K$ are division rings and $S/J(S) \cong S/I \times S/K$.*

The endomorphism ring of the kernel of a morphism between indecomposable injective modules has the same structure as the endomorphism ring of a uniserial module or a cyclically presented module over a local ring. More precisely, the endomorphism ring of the kernel of a morphism between two indecomposable injective modules is either local or has two maximal ideals, the kernel is determined up to isomorphism by its monogeny class and its upper part, and a weak form of the Krull-Schmidt Theorem also holds for direct sums of these kernels, as the following theorem shows.

**Theorem 4.11.** (Weak Krull-Schmidt Theorem for kernels of morphisms between indecomposable injective modules, [19, Theorem 2.7] and [14]) *Let $\varphi_i \colon E_{i,0} \to E_{i,1}$ $(i = 1, 2, \dots, n)$ and $\varphi'_j \colon E'_{j,0} \to E'_{j,1}$ $(j = 1, 2, \dots, t)$ be $n + t$ non-injective morphisms between indecomposable injective right modules $E_{i,0}, E_{i,1}, E'_{j,0}, E'_{j,1}$ over an arbitrary ring $R$. Then the direct sums $\oplus_{i=0}^{n} \ker \varphi_i$ and $\oplus_{j=0}^{t} \ker \varphi'_j$ are isomorphic $R$-modules if and only if $n = t$ and there exist two permutations $\sigma, \tau$ of $\{1, 2, \dots, n\}$ such that $[\ker \varphi_i]_m = [\ker \varphi'_{\sigma(i)}]_m$ and $[\ker \varphi_i]_u = [\ker \varphi'_{\tau(i)}]_u$ for every $i = 1, 2, \dots, n$.*

There are also some other classes of modules that satisfy the Weak Krull-Schmidt Theorem. One is the class of couniformly presented modules [20]. It generalizes the class of cyclically presented modules over a local ring. An $R$-module $M$ is said to be *couniform* if it has dual Goldie dimension 1, that is, it is nonzero and the sum of any two proper submodules of $M_R$ is a proper submodule of $M_R$. An $R$-module $M$ is *couniformly presented* if it is non-zero and there exists an exact sequence

$$0 \to C_R \to P_R \to M_R \to 0 \tag{2}$$

with both $C_R$ and $P_R$ couniform and $P_R$ projective. Under these hypotheses, the exact sequence (2) is called a *couniform presentation* of the couniformly presented module $M_R$.

**Theorem 4.12.** (Weak Krull-Schmidt Theorem for couniformly presented modules [20]) *Let $M_1, \dots, M_n, N_1, \dots, N_t$ be $n + t$ couniformly presented right $R$-modules. Then the direct sums $M_1 \oplus \cdots \oplus M_n$ and $N_1 \oplus \cdots \oplus N_t$ are isomorphic $R$-modules if and only if $n = t$ and there exist two permutations $\sigma$ and $\tau$ of $\{1, 2, \dots, n\}$ such that $[M_i]_l = [N_{\sigma(i)}]_l$ and $[M_i]_e = [N_{\tau(i)}]_e$ for every $i = 1, 2, \dots, n$.*

A similar behaviour, as far as direct-sum decompositions are concerned, takes place for the short exact sequences

$$0 \to A_R \to B_R \to C_R \to 0, \tag{3}$$

with $A_R$ and $C_R$ uniserial modules. The endomorphism ring of such a sequence in the category of all short exact sequences has at most four maximal ideals, and the isomorphism types of these sequences (3) are described by four invariants $[B]_{m,l}$, $[B]_{e,l}$, $[B]_{m,u}$, $[B]_{e,u}$ [10,11].

We conclude this section by describing one of the general patterns that allow to treat all the previous examples at the same time. Let $\mathcal{C}$ be a full subcategory of the category Mod-$R$ for some ring $R$ and assume that every object of $\mathcal{C}$ is an indecomposable right $R$-module. Define a *completely prime ideal* $\mathcal{P}$ of $\mathcal{C}$ as an assignment of a subgroup $\mathcal{P}(A,B)$ of the additive abelian group $\mathrm{Hom}_R(A,B)$ to every pair $(A,B)$ of objects of $\mathcal{C}$, with the following two properties: (1) for every $A,B,C \in \mathrm{Ob}(\mathcal{C})$, every $f\colon A \to B$ and every $g\colon B \to C$, one has that $gf \in \mathcal{P}(A,C)$ if and only if either $f \in \mathcal{P}(A,B)$ or $g \in \mathcal{P}(B,C)$; (2) $\mathcal{P}(A,A)$ is a proper subgroup of $\mathrm{Hom}_R(A,A)$ for every object $A \in \mathrm{Ob}(\mathcal{C})$. Let $\mathcal{P}$ be a completely prime ideal of $\mathcal{C}$. If $A,B$ are objects of $\mathcal{C}$, we say that $A$ and $B$ *have the same* $\mathcal{P}$ *class*, and write $[A]_{\mathcal{P}} = [B]_{\mathcal{P}}$, if $\mathcal{P}(A,B) \neq \mathrm{Hom}_R(A,B)$ and $\mathcal{P}(B,A) \neq \mathrm{Hom}_R(B,A)$.

**Theorem 4.13.** [22] *Let $\mathcal{C}$ be a full subcategory of* Mod-$R$ *and $\mathcal{P},\mathcal{Q}$ be two completely prime ideals of $\mathcal{C}$. Assume that all objects of $\mathcal{C}$ are indecomposable right $R$-modules and that, for every $A \in \mathrm{Ob}(\mathcal{C})$, $f\colon A \to A$ is an automorphism of $A$ if and only if $f \notin \mathcal{P}(A,A) \cup \mathcal{Q}(A,A)$. Then, for every $A_1,\dots,A_n,B_1,\dots,B_t \in \mathrm{Ob}(\mathcal{C})$, the modules $A_1 \oplus \cdots \oplus A_n$ and $B_1 \oplus \cdots \oplus B_t$ are isomorphic if and only if $n = t$ and there exist two permutations $\sigma,\tau$ of $\{1,2,\dots,n\}$ such that $[A_i]_{\mathcal{P}} = [B_{\sigma(i)}]_{\mathcal{P}}$ and $[A_i]_{\mathcal{Q}} = [B_{\tau(i)}]_{\mathcal{Q}}$ for all $i = 1,\dots,n$.*

A further remark: for the classes $\mathcal{C}$ of modules described so far, the fact that the weak form of the Krull-Schmidt Theorem holds can be described by saying that the corresponding monoid $V(\mathcal{C})$ is a subdirect product of two free monoids.

## 5. Direct products of modules whose endomorphism rings have at most two maximal ideals

In the previous section, we have seen that the Weak Krull-Schmidt Theorem holds not only for uniserial modules, but also for cyclically presented modules over a local ring $R$, for kernels of morphisms between indecomposable injective modules, for couniformly presented modules, and more generally, for a number of classes of modules with at most two maximal right ideals. In this section, we will see that a similar result can hold not only for direct sums, but also for direct products of modules.

In order to present the main result in the most general setting, that of modules whose endomorphism rings have at most two maximal right ideals, we begin from the Weak Krull-Schimidt Theorem for direct products of uniserial modules.

**Theorem 5.1.** [1] *Let* $\{ U_i \mid i \in I \}$ *and* $\{ V_j \mid j \in J \}$ *be two families of uniserial modules over an arbitrary ring* $R$. *Assume that there exist two bijections* $\sigma, \tau \colon I \to J$ *such that* $[U_i]_m = [V_{\sigma(i)}]_m$ *and* $[U_i]_e = [V_{\tau(i)}]_e$ *for every* $i \in I$. *Then* $\prod_{i\in I} U_i \cong \prod_{j\in J} V_j$.

A full subcategory $\mathcal{C}$ of Mod-$R$ is said to satisfy Condition (DSP) (*direct summand property*) if whenever $A$, $B$, $C$, $D$ are right $R$-modules with $A \oplus B \cong C \oplus D$ and $A$, $B$, $C \in \mathrm{Ob}(\mathcal{C})$, then also $D \in \mathrm{Ob}(\mathcal{C})$.

**Theorem 5.2.** [1] *Let* $\mathcal{C}$ *be a full subcategory of* Mod-$R$ *in which all objects are indecomposable right* $R$*-modules and let* $\mathcal{P}, \mathcal{Q}$ *be two completely prime ideals of* $\mathcal{C}$ *with the property that, for every* $A \in \mathrm{Ob}(\mathcal{C})$, $f \colon A \to A$ *is an automorphism if and only if* $f \notin \mathcal{P}(A,A) \cup \mathcal{Q}(A,A)$. *Assume that* $\mathcal{C}$ *satisfies Condition* (DSP). *Let* $\{ A_i \mid i \in I \}$ *and* $\{ B_j \mid j \in J \}$ *be two families of objects of* $\mathcal{C}$. *Suppose that there exist two bijections* $\sigma, \tau \colon I \to J$ *such that* $[A_i]_{\mathcal{P}} = [B_{\sigma(i)}]_{\mathcal{P}}$ *and* $[A_i]_{\mathcal{Q}} = [B_{\tau(i)}]_{\mathcal{Q}}$ *for every* $i \in I$. *Then the* $R$*-modules* $\prod_{i\in I} A_i$ *and* $\prod_{j\in J} B_j$ *are isomorphic.*

## 6. The category of $G$-groups

We can also find the same behaviour in other algebraic structures, not only in modules. We can find it, for instance, in groups, Lie algebras, $G$-groups, and so on. We begin with $G$-groups. The category of $G$-groups shows a behaviour that is a very pleasant combination of the behaviour of groups and that of left modules over a ring.

Let $G$ be a group. A *(left) $G$-group* is a pair $(H, \varphi)$, where $H$ is a group and $\varphi \colon G \to \mathrm{Aut}(H)$ is a group homomorphism. Equivalently, a $G$-group is a group $H$ endowed with a mapping $\cdot \colon G \times H \to H$, $(g,h) \mapsto gh$, called *left scalar multiplication*, such that

(a) $g(hh') = (gh)(gh')$

(b) $(gg')h = g(g'h)$

(c) $1_G h = h$

for every $g, g' \in G$ and every $h, h' \in H$.

The notion of a $G$-group is classical. Sometimes $G$ is called an *operator group* on $H$ [35, Definition 8.1].

As an example of a $G$-group, let $\alpha_g \colon G \to G$, $\alpha_g \colon h \mapsto ghg^{-1}$, be the *inner automorphism* of $G$ determined by $g$, for every $g \in G$. Then there is a canonical group morphism $\alpha \colon G \to \mathrm{Aut}(G)$, $g \mapsto \alpha_g$, which makes $G$ a $G$-group. It is called the *regular* $G$-group, exactly how the left $R$-module ${}_RR$ is called the regular left $R$-module.

The category $G$-**Grp** of $G$-groups has as objects all pairs $(H, \varphi)$, where $H$ is any group and $\varphi\colon G \to \mathrm{Aut}(H)$ is a group homomorphism. The morphisms in the category $G$-**Grp** are the group morphisms $f\colon H \to H'$ such that $f(gh) = gf(h)$ for every $g \in G$, $h \in H$. The category $G$-**Grp** is a semiabelian category, in which the abelian objects are the objects $(H, \varphi)$ with $H$ an abelian group. We will see at the end of this section that the category $G$-**Grp** provides the proper setting to study direct product decompositions of a group $G$ as a direct product of finitely many indecomposable groups.

**Remark 6.1.** *Instead of considering left $G$-groups, one can study right $G$-groups. A right $G$-group is defined as a pair $(H, \varphi')$, where $H$ is a group and $\varphi'\colon G \to \mathrm{Aut}(H)$ is a group antihomomorphism. Equivalently, a right $G$-group is a group $H$ endowed with a mapping $\cdot\colon H \times G \to H$, $(h, g) \mapsto hg$, called right scalar multiplication, such that* (a) *becomes $(hh')g = (hg)(h'g)$, and so on. We can thus construct the category of* **Grp**-*$G$ of all right $G$-groups. But if $G^{\mathrm{op}}$ denotes the* opposite group *of the group $G$, that is, the set $G$ with multiplication $(x, y) \mapsto yx$, then $G$ is isomorphic to $G^{\mathrm{op}}$ via the isomorphism $G \to G^{\mathrm{op}}$, $g \mapsto g^{-1}$. Now it is easily seen that the category $G$-**Grp** of left $G$-groups is isomorphic to the category* **Grp**-*$G^{\mathrm{op}}$ of right $G^{\mathrm{op}}$-groups, and that* **Grp**-*$G$ is isomorphic to $G^{\mathrm{op}}$-***Grp**. *But from the group isomorphism $G \to G^{\mathrm{op}}$, we also have that the categories $G$-***Grp** *and* **Grp**-*$G$ are isomorphic. So there is no need to introduce both left $G$-groups and right $G$-groups, because they form isomorphic categories, and therefore corresponding objects in this category isomorphism are indistinguishable from an algebraic point of view.*

Notice the strict analogy between the categories $R$-Mod and $G$-**Grp**. Objects of $R$-Mod are all pairs $(H, \varphi)$, where $H$ is any abelian group and $\varphi\colon R \to \mathrm{End}(H)$ is a ring homomorphism. As we have already said, the regular $G$-group $(G, \alpha)$ plays, in the category $G$-**Grp**, a role pretty much similar to the role of the regular module ${}_RR$ in the category $R$-Mod.

The automorphisms in $G$-**Grp** of the regular object $G$ are exactly the central automorphisms of $G$, that is, the automorphisms of $G$ that belong to the kernel of the canonical mapping $\mathrm{Aut}(G) \to \mathrm{Aut}(G/\zeta(G))$. Thus the automorphisms of $G$ in $G$-**Grp** are exactly the automorphisms of the group $G$ of the form $1 + \varphi$ for some endomorphism $\varphi$ of $G$ with $\varphi(G) \subseteq \zeta(G)$. Here the sum of two endomorphisms $\varphi, \psi$ of a group $G$ is defined by $(\varphi + \psi)(g) = \varphi(g)\psi(g)$. The sum $\varphi - \psi$ is an endomorphism of $G$ only when $[\varphi(G), \psi(G)] = 1$. It is easy to see that if $\varphi$ is a group endomorphism of $G$ with $\varphi(G) \subseteq \zeta(G)$, then $1 + \varphi$ is an automorphism of $G$ if and only if the restriction of $1+\varphi$ to $\zeta(G)$ is an automorphism of the abelian group $\zeta(G)$.

The central automorphism $\varphi$ in the statement of Theorem 2.2, whose existence was proved by Remak, corresponds exactly to the fact that Remark, in the study of direct-product decompositions of a group $G$, was studying the decompositions of the regular object $G$ in the category $G$-**Grp** and not the decompositions of the group $G$ in the category **Grp**. Of course, he didn't know what a category is.

An interesting role in this setting is played by the full subcategory $\mathcal{C}_G$ of $G$-**Grp** consisting of all the objects $(H, \varphi)$ of $G$-**Grp** for which the image of the group homomorphism $\varphi\colon G \to \mathrm{Aut}(H)$ contains the group $\mathrm{Inn}(H)$ of all inner automorphisms of $H$ [18].

Another setting in which the Weak Krull-Schmidt Theorem holds in Group Theory is in the study of abelian normal subgroups of a group $G$ [5]. If $G$ is an arbitrary group and $H$ is an abelian normal subgroup of $G$, then the conjugation $\alpha\colon G \to \mathrm{Aut}(G)$ induces an action $\alpha\colon G \to \mathrm{Aut}(H)$ because $H$ is normal in $G$. If $\mathbb{Z}[G]$ is the group algebra of $G$ with integer coefficients, then the conjugation $\alpha\colon G \to \mathrm{Aut}(H)$ extends to a ring morphism $\mathbb{Z}[G] \to \mathrm{End}(H)$, so that $H$ turns out to be a left module over the ring $\mathbb{Z}[G]$. Hence Theorem 4.2 can be applied to this left module whenever $H$ is a direct sum of uniserial left $\mathbb{Z}[G]$-modules.

## 7. Factorisation

Everything we've seen until now corresponds, in a broad sense, to study factorisations of elements in suitable monoids, for instance in the commutative monoid $V(R)$. Let's see how what we have learned about direct-sum decompositions (direct-product decompositions) also holds for factorisations in other classes of monoids. For instance, it can be applied to factorisations of elements in a commutative domain $R$, because clearly factorisations of elements in an integral domain $R$ are exactly factorisations in the multiplicative monoid of $R$. We all know that a *unique factorisation domain* (UFD) is a commutative integral domain $R$ such that:

(i) $R$ is *atomic*, that is, every element $a \in R$, $a \neq 0$ and $a$ non-invertible, is a product of finitely many irreducible elements of $R$.

(ii) If $p_1, \dots, p_n, q_1, \dots, q_m$ are irreducible elements of $R$ and $p_1 \dots p_n = q_1 \dots q_m$, then $n = m$ and there exists a permutation $\sigma$ of $\{1, 2, \dots, n\}$ such that $p_i$ and $q_{\sigma(i)}$ are associates for every $i = 1, 2, \dots, n$.

Thus an integral domain $R$ is a unique factorisation domain if and only if the multiplicative monoid $R \setminus \{0\}$ is isomorphic to the direct product of the abelian group $U(R)$ and a free commutative monoid $F$.

Notice that, according to a very nice result by J. Coykendall and W. W. Smith [12]:

**Theorem 7.1.** *A commutative integral domain $R$ is a* UFD *if and only if*

(i) *$R$ is atomic.*

(ii) *If $n \geq 2$ and $p_1, \dots, p_n, q_1, \dots, q_n$ are irreducible elements of $R$ and $p_1 \dots p_n = q_1 \dots q_n$, then there exists a permutation $\sigma$ of $\{1, 2, \dots, n\}$ such that $p_i$ and $q_{\sigma(i)}$ are associates for every $i = 1, 2, \dots, n$.*

In an integral domain $R$, every prime element is irreducible. If $R$ is a UFD, the converse holds. More precisely, an integral domain $R$ is a UFD if and only if every irreducible is prime and $R$ satisfies the ascending chain condition on principal ideals, if and only if $R$ is atomic and every irreducible is prime.

**Proposition 7.2.** *The following conditions are equivalent for two prime elements $a, b$ of a commutative integral domain $R$:*

(i) *$a = bu$ for some invertible element $u \in R$.*

(ii) *$aR = bR$.*

(iii) *$R/aR \cong R/bR$.*

(iv) *$[R/aR]_m = [R/bR]_m$.*

(v) *$[R/aR]_e = [R/bR]_e$.*

(vi) *$[R/aR]_l = [R/bR]_l$.*

Let's pass to consider, now, factorisation of polynomials into irreducible polynomials.

The case of commutative polynomials in commuting indeterminates is well known: the ring $\mathbb{Z}[x_1, \dots, x_n]$ of all polynomials whose coefficients are in the ring of integers $\mathbb{Z}$ and with $x_1, \dots, x_n$ commuting indeterminates is a UFD.

Let's see how the situation changes when we pass to the ring $\mathbb{Z}\langle x_1, \dots, x_n\rangle$, the free ring on $n$ objects. The elements of $\mathbb{Z}\langle x_1, \dots, x_n\rangle$ are non-commutative polynomials with coefficients in $\mathbb{Z}$ and with $x_1, \dots, x_n$ non-commuting indeterminates. The ring $\mathbb{Z}\langle x_1, \dots, x_n\rangle$ is atomic, in the sense that polynomials do factorise as a product of irreducible polynomials. The invertible elements in $\mathbb{Z}\langle x_1, \dots, x_n\rangle$ are only 1 and $-1$. But the factorisation $x(yx - 2) = (xy - 2)x$ in the ring $\mathbb{Z}\langle x, y\rangle$ shows that a polynomial in $\mathbb{Z}\langle x_1, \dots, x_n\rangle$ does not necessarily factorise as a product of irreducible polynomials in a unique way up to the sign of the irreducible factors. Nevertheless the following theorem holds:

**Theorem 7.3.** (Brungs' Theorem [9]) *Every polynomial in $R := \mathbb{Z}\langle x_1, \dots, x_n\rangle$ factorises as a product of irreducible polynomials. Moreover, if $p_1, \dots, p_n, q_1, \dots, q_m$*

*are irreducible polynomials in* $R$ *and* $p_1 \dots p_n = q_1 \dots q_m$*, then* $n = m$ *and there exists a permutation* $\sigma$ *of* $\{1, 2, \dots, n\}$ *such that* $[R/p_iR]_m = [R/q_{\sigma(i)}R]_m$.

For instance, consider the factorisation $x(yx - 2) = (xy - 2)x$ in the ring $R = \mathbb{Z}\langle x, y\rangle$. We have that $[R/(xy-2)R]_m = [R/(yx-2)R]_m$, because left multiplications

$$\lambda_y \colon R/(xy-2)R \to R/(yx-2)R \quad \text{and} \quad \lambda_x \colon R/(yx-2)R \to R/(xy-2)R$$

are right $R$-module monomorphisms.

Let's pass to consider another interesting type of polynomials, those in the multiplicative monoid $\mathbb{N}_0[x]$, consisting of all polynomials in $\mathbb{Z}[x]$ whose integer coefficients are all $\geq 0$.

The multiplicative monoid $\mathbb{N}_0[x]$ is not a ring, it is a commutative semiring. The multiplicative monoid $\mathbb{N}_0[x] \setminus \{0\}$ is atomic, in the sense that all its elements are finite products of atoms, that is, polynomials irreducible in $\mathbb{N}_0[x] \setminus \{0\}$. The unique invertible element in $\mathbb{N}_0[x]$ is 1. The next example shows that a polynomial in $\mathbb{N}_0[x]$ does not necessarily factorise as a product of irreducible polynomials in a unique way up to the order of the factors:

**Example 7.4.** *From the theory of cyclotomic polynomials, we know that the factorisation of* $x^n - 1$ *in the* UFD $\mathbb{Q}[x]$ *is* $x^n - 1 = \prod_{d|n} \Phi_d(x)$*, where* $\Phi_d(x)$ *is the* $d$*-th cyclotomic polynomial. Here*

$\Phi_1(x) = x - 1$, $\Phi_2(x) = x + 1$, $\Phi_3(x) = x^2 + x + 1$, $\Phi_4(x) = x^2 + 1$, $\Phi_5(x) = x^4 + x^3 + x^2 + x + 1$, $\Phi_6(x) = x^2 - x + 1$. *Thus* $x^6 - 1 = \Phi_1(x)\Phi_2(x)\Phi_3(x)\Phi_6(x) = (x-1)(x+1)(x^2+x+1)(x^2-x+1)$*, so we have the factorisation*

$$x^5 + x^4 + x^3 + x^2 + x + 1 = (x+1)(x^2+x+1)(x^2-x+1) \tag{4}$$

*into irreducibles in* $\mathbb{Q}[x]$*. Multiplying the first factor and the last one in the factorisation (4), we get that* $(x+1)(x^2-x+1) = x^3+1 \in \mathbb{N}_0[x]$ *and, multiplying the last two factors, we get that* $(x^2+x+1)(x^2-x+1) = x^4+x^2+1 \in \mathbb{N}_0[x]$*. Thus we get two essentially different factorisations*

$$(x^3+1)(x^2+x+1) = (x+1)(x^4+x^2+1) \tag{5}$$

*of* $x^5 + x^4 + x^3 + x^2 + x + 1$ *into irreducibles of* $\mathbb{N}_0[x]$*.*

*Thus factorisations into irreducibles in* $\mathbb{N}_0[x]$ *are not unique (but every polynomial in* $\mathbb{N}_0[x]$ *has only finitely many distinct factorisations into irreducibles).*

Example 7.4 can be applied to show that Krull-Schmidt fails in other categories. For instance, consider the category of finite partially ordered sets. This category has coproducts (disjoint unions), products (direct products with the component-wise order) and a terminal object 1 (the partially ordered set with one element), which is the identity with respect to product. Let $L = \{0, 1\}$ denote the partially ordered set with two elements $0 < 1$. Then, for every $n \geq 0$, the direct product $L^n$ is a connected partially ordered set with $2^n$ elements, and its automorphism group is the symmetric group $S_n$. If we compute the identity (5), which is an identity in the semiring $\mathbb{N}_0[x]$, replacing $x$ with $L$ (and the natural number 1 with the partially ordered set 1), we get two essentially different direct-product decompositions of the partially ordered set $1\dot{\cup}L\dot{\cup}L^2\dot{\cup}L^3\dot{\cup}L^4\dot{\cup}L^5$ into indecomposable partially ordered sets, that is, we get that

$$(L^3\dot{\cup}1) \times (L^2\dot{\cup}L\dot{\cup}1) \cong (L\dot{\cup}1) \times (L^4\dot{\cup}L^2\dot{\cup}1).$$

This example, due to Nakayama and Hashimoto [24,28], shows that Krull-Schmidt fails in the category of finite partially ordered sets.

The possibility of applying the identity (4) in $\mathbb{N}_0[x]$ to get an isomorphism in the category of finite partially ordered sets is due to the fact that distributivity holds for partially ordered sets: $X \times (Y\dot{\cup}Z) \cong (X \times Y)\dot{\cup}(X \times Z)$. More generally, recall that a category $\mathcal{C}$ with finite products $(-) \times (-)$ and coproducts $(-) + (-)$ is called *(finitary) distributive* if, for any objects $X, Y, Z$ of $\mathcal{C}$, the canonical morphism

$$X \times Y + X \times Z \to X \times (Y + Z)$$

is an isomorphism. It is now easily seen that Nakayama and Hashimoto's technique can be applied to any distributive category.

## References


[1] A. Alahmadi and A. Facchini, *Direct products of modules whose endomorphism rings have at most two maximal ideals*, J. Algebra, 435 (2015), 204-222.

[2] B. Amini, A. Amini and A. Facchini, *Equivalence of diagonal matrices over local rings*, J. Algebra, 320(3) (2008), 1288-1310.

[3] A. Amini, B. Amini and A. Facchini, *Weak Krull-Schmidt for infinite direct sums of cyclically presented modules over local rings*, Rend. Semin. Mat. Univ. Padova, 122 (2009), 39-54.

[4] F. W. Anderson and K. R. Fuller, Rings and Categories of Modules, New York, Springer-Verlag, 1974.

[5] M. J. Arroyo Paniagua and A. Facchini, *G-groups and biuniform abelian normal subgroups*, Adv. Group Theory Appl., 2 (2016), 79-111.

[6] G. Azumaya, *Corrections and supplementaries to my paper concerning Krull-Remak-Schmidt's theorem*, Nagoya Math. J., 1 (1950), 117-124.

[7] G. M. Bergman, *Coproducts and some universal ring constructions*, Trans. Amer. Math. Soc., 200 (1974), 33-88.

[8] G. M. Bergman and W. Dicks, *Universal derivations and universal ring constructions*, Pac. J. Math., 79 (1978), 293-337.

[9] H.-H. Brungs, *Ringe mit eindeutiger Faktorzerlegung,* J. Reine Angew. Math., 236 (1969), 43-66.

[10] F. Campanini, *On a category of chains of modules whose endomorphism rings have at most* $2n$ *maximal right ideals*, Comm. Algebra, 46(5) (2018), 1971-1982.

[11] F. Campanini and A. Facchini, *On a category of extensions whose endomorphism rings have at most four maximal ideals*, to appear in "Advances in Rings and Modules, S. López-Permouth, J. K. Park, C. Roman and S. T. Rizvi Eds, Contemp. Math., 2018.

[12] J. Coykendall and W. W. Smith, *On unique factorisation domains*, J. Algebra, 332 (2011), 62-70.

[13] N. V. Dung and A. Facchini, *Weak Krull-Schmidt for infinite direct sums of uniserial modules*, J. Algebra, 193 (1997), 102-121.

[14] Ş. Ecevit, A. Facchini, and M. T. Koşan, *Direct sums of infinitely many kernels*, J. Aust. Math. Soc., 89(2) (2010), 199-214.

[15] A. Facchini, *Krull-Schmidt fails for serial modules*, Trans. Amer. Math. Soc., 348(11) (1996), 4561-4575.

[16] A. Facchini, Module Theory. Endomorphism Rings and Direct Sum Decompositions in Some Classes of Modules, Birkhäuser Verlag, Basel, 1998.

[17] A. Facchini, *Direct-sum decompositions of modules with semilocal endomorphism rings*, Bull. Math. Sci., 2(2) (2012), 225-279.

[18] A. Facchini and M. Altun-Özarslan, *The Krull-Schmidt-Remak-Azumaya Theorem for G-groups*, to appear in the proceedings of the Conference "Noncommutative rings and their applications, V", Lens 12-15 June 2017, Contemp. Math., 2018.

[19] A. Facchini, Ş. Ecevit and M. T. Koşan, *Kernels of morphisms between indecomposable injective modules*, Glasg. Math. J., 52(A) (2010), 69-82.

[20] A. Facchini and N. Girardi, *Couniformly presented modules and dualities*, Advances in ring theory, Trends Math., Birkhâuser/Springer Basel AG, Basel, (2010), 149-164.

[21] A. Facchini and Z. Nazemian, *Equivalence of some homological conditions for ring epimorphisms*, to appear in J. Pure Appl. Algebra, 2018.

[22] A. Facchini and P. Příhoda, *The Krull-Schmidt theorem in the case two*, Algebr. Represent. Theory, 14(3) (2011), 545-570.

[23] H. Frobenius and H. Stickelberger, *Über Gruppen von vertauschbaren Elementen*, J. Reine Angew. Math., 86 (1879), 217-262.

[24] J. Hashimoto, *On direct product decomposition of partially ordered sets*, Ann. of Math., 54(2) (1951), 315-318.

[25] W. Krull, *Über verallgemeinerte endliche Abelsche Gruppen*, Math. Z., 23(1) (1925), 161-196.

[26] E. L. Lady, *Summands of finite rank torsion-free abelian groups*, J. Algebra, 32 (1974), 51-52.

[27] J. H. Maclagan-Wedderburn, *On the direct product in the theory of finite groups*, Ann. of Math., 10(4) (1909), 173-176.

[28] T. Nakayama, and J. Hashimoto, *On a problem of G. Birkhoff*, Proc. Amer. Math. Soc., 1 (1950), 141-142.

[29] P. Příhoda, *A version of the weak Krull-Schmidt theorem for infinite direct sums of uniserial modules*, Comm. Algebra, 34(4) (2006), 1479-1487.

[30] P. Příhoda, Add($U$) *of a uniserial module*, Comment. Math. Univ. Carolin., 47(3) (2006), 391-398.

[31] G. Puninski, *Some model theory over a nearly simple uniserial domain and decompositions of serial modules*, J. Pure Appl. Algebra, 163(3) (2001), 319-337.

[32] R. E. Remak, *Über die Zerlegung der endlichen Gruppen in indirekte unzerlegbare faktoren*, dissertation, 1911.

[33] O. Y. Schmidt, *Sur les produits directs*, Bull. Soc. Math. France, 41 (1913), 161-164.

[34] O. Y. Schmidt, *Über unendliche Gruppen mit endlicher Kette*, Math. Z., 29 (1929), 34-41.

[35] M. Suzuki, Group Theory I, Springer-Verlag, Berlin-New York, 1982.

[36] R. B. Warfield, *Purity and algebraic compactness for modules*, Pacific J. Math., 28 (1969), 699-719.

[37] R. B. Warfield, *Serial rings and finitely presented modules*, J. Algebra, 37 (1975), 187-222.

**Alberto Facchini** (Corresponding Author)
Dipartimento di Matematica Pura e Applicata
Università di Padova
35121 Padova, Italy
e-mail: facchini@math.unipd.it

**Serap Şahinkaya**
Department of Mathematics
Faculty of Science
Gebze Technical University
41400 Gebze, Kocaeli, Turkey
e-mail: ssahinkaya@gtu.edu.tr

# 11

# EXTENDING TABLE ALGEBRAS TO HOPF ALGEBRAS

Allen Herman and Gurmail Singh

Communicated by Abdullah Harmancı

Abstract. Let $A$ be a table algebra with standard basis $\mathbf{B}$, multiplication $\mu$, unit map $\eta$, skew-linear involution $*$, and degree map $\delta$. In this article we study the possible coalgebra structures $(A, \Delta, \delta)$ on $A$ for which $(A, \mu, \eta, \Delta, \delta)$ becomes a Hopf algebra with respect to some antipode. We show that such Hopf algebra structures are not always available for noncommutative table algebras. On the other hand, commutative table algebras will always have a Hopf algebra structure induced from an algebra-isomorphic group algebra. To illustrate our approach, we derive Hopf algebra comultiplications on table algebras of dimension 2 and 3.

**Keywords**: Table algebra, Hopf algebra, fusion ring, association scheme

## 1. Introduction

A *bialgebra* $(A, \mu, \eta, \Delta, \delta)$ is both an associative algebra and a coassociative coalgebra $A$ over a commutative ring $K$ with multiplication $\mu : A \otimes A \to A$, unit map $\eta : K \to A$, comultiplication $\Delta : A \to A \otimes A$, and counit $\delta : A \to K$, that satisfies extra compatibility conditions that $\Delta : A \to A \otimes A$ and $\delta$ are algebra homomorphisms and $\delta \circ \eta = id_K$. For non-experts, the algebra condition on $(A, \mu, \eta)$ is

$$\mu \circ (id \otimes \eta) = \mu \circ (\eta \otimes id) = id$$

as maps on $A \otimes K \simeq K \otimes A \simeq A$, and the coalgebra condition on $(A, \Delta, \delta)$ is obtained by reversing arrows:

$$(id \otimes \delta) \circ \Delta = (\delta \otimes id) \circ \Delta = id.$$

An algebra anti-automorphism $S : A \to A$ of a bialgebra $(A, \mu, \eta, \Delta, \delta)$ is called an *antipode* when

$$\eta \circ \delta = \mu \circ (\mathcal{S} \otimes id) \circ \Delta = \mu \circ (id \otimes \mathcal{S}) \circ \Delta.$$

A *Hopf algebra* is a bialgebra with antipode. Hopf algebras are a class of bialgebras that includes group algebras, since it is well-known that a group algebra $KG$ of a group $G$ over $K$ becomes a Hopf algebra over $K$ when we define comultiplication by $\Delta(g) = g \otimes g$ and counit by $\delta(g) = 1$, and the antipode by $\mathcal{S}(g) = g^{-1}$, for all $g \in G$.

In the case where the base field is the field of complex numbers $\mathbb{C}$ (which we will assume from now on), table algebras are associative algebras that also generalize group algebras. (For an extensive history on the evolution and scope of this and similar notions, we refer the reader to Blau's survey article [1].) A table algebra of finite rank $r$ is an $r$-dimensional associative algebra $A$ with skew-linear involution $*$ whose distinguished basis $\mathbf{B} = \{1 = b_0, b_1, \ldots, b_{r-1}\}$ satisfies the following three properties:

TA1: The structure constants relative to $\mathbf{B}$ are nonnegative real numbers. We denote these by $\lambda_{ijk}$ for $i, j, k \in \{0, 1, \ldots, r-1\}$, where $b_i b_j = \sum_k \lambda_{ijk} b_k$, for all $b_i, b_j \in \mathbf{B}$.

TA2: $\mathbf{B}^* = \mathbf{B}$; in particular, $b_0^* = b_0 = 1$ is the multiplicative identity of the algebra, and for all $b_i \in \mathbf{B}$ there is a unique $i^* \in \{0, 1, \ldots, r-1\}$ such that $(b_i)^* = b_{i^*}$.

TA3: For all $i \in \{0, 1, \ldots, r-1\}$, $b_{i^*}$ is the unique pseudo-inverse of $b_i$; this means for all $i, j \in \{0, 1, \ldots, r-1\}$, $\lambda_{ij0} > 0 \iff j = i^*$, and for all $i \in \{0, 1, \ldots, r-1\}$, $\lambda_{ii^*0} = \lambda_{i^*i0}$.

The complex group algebra $\mathbb{C}G$ of a finite group $G$ is a table algebra with distinguished basis $G$. Its skew-linear involution is given by $(\alpha g)^* = \bar{\alpha} g^{-1}$, for all $g \in G$, $\alpha \in \mathbb{C}$, and the pseudo-inverse of $g \in G$ is its group inverse $g^{-1}$. The adjacency algebras of association schemes are another important example of table algebras, in which the roles of involution and pseudo-inverse are played by conjugate transpose and transpose, respectively. Note that table algebras, being finite-dimensional algebras with involution over $\mathbb{C}$, are always semisimple algebras. For further overview of the theory of table algebras, see Blau's survey [1].

As is customary we will write $b_0$ for the multiplicative identity of a table algebra, so the unit map of a table algebra is given by $\eta(\alpha) = \alpha b_0$, for $\alpha \in \mathbb{C}$, and multiplication $\mu$ is as defined by property TA1. Since table algebras are a class of associative algebras that includes group algebras, and group algebras become Hopf algebras in a natural way, it makes sense to ask:

*Is there a way to define a counit $\delta$, comultiplication $\Delta$, and antipode $\mathcal{S}$ on a table algebra $(A, \mu, \eta)$ so that $(A, \mu, \eta, \Delta, \delta)$ becomes a Hopf algebra with antipode $\mathcal{S}$?*

The main results of this article show that the answer to this question is NO for some noncommutative table algebras, but YES for all commutative table algebras. Since fusion rule algebras can be characterized as commutative table algebras after a basis renormalization (see [1]), the answer is also YES for fusion rule algebras. The negative answer is a consequence of the existence of noncommutative table algebras of dimensions 5 and 7, combined with the fact that Hopf algebras of prime dimension must be isomorphic to group algebras. Since commutative table algebras of dimension $r$ are algebra isomorphic to the group algebra of any abelian group of order $r$, for each distinct isomorphism type of these groups there will be a distinct Hopf algebra structure on the table algebra. To Hopf algebraists, the possible comultiplications on commutative semisimple algebras are well-known to be the coalgebra structures for duals of group algebras. We offer a self-contained proof of this fact that also produces the values of the comultiplication on the basis of primitive idempotents. By interpolating between the character tables of the table algebra and the group we can produce formulas for the values of the Hopf algebra comultiplication map on elements of $\mathbf{B}$ for the most natural choices of counits. We will apply this approach to give explicit Hopf algebra comultiplications for table algebras in dimensions 2 and 3.

## 2. Choosing the counit and antipode of a table algebra

Let $A$ be a table algebra with distinguished basis $\mathbf{B}$. Since each distinguished basis element $b_i$ can be realized as a nonnegative matrix in its regular representation, there is a unique algebra homomorphism $\delta : \mathbb{C}\mathbf{B} \to \mathbb{C}$ called the *degree map* for which $\delta(b_i) > 0$ for all $b_i \in \mathbf{B}$ [1]. Since $\delta$ is an algebra homomorphism, it is compatible with $\mu$, and $\delta(b_0) = 1$ implies $\delta(\eta(z)) = z$, for all $z \in \mathbb{C}$, so it is also compatible with our unit map. In the case $\mathbf{B}$ is a finite group the degree map is the augmentation map, which is the counit for the usual Hopf algebra structure on the group algebra. So the degree map of the table algebra is a good candidate for the counit.

In choosing the degree map for the counit, we can scale the distinguished basis elements by positive constants to arrange that our renormalized basis satisfies $\delta(b_i) = \lambda_{ii^*0}$, for all $b_i \in \mathbf{B}$ (see [1]). We call such a table algebra basis *standard*, and from now on assume our table algebra bases to be standard ones.

When $A$ is a table algebra, it has a natural involution $*$ that is skew-linear relative to its defining basis $\mathbf{B}$. This means the conjugate involution $\bar{*}$ given by $(\alpha b_i)^{\bar{*}} = \bar{\bar{\alpha}} b_{i^*} = \alpha b_{i^*}$ for all $\alpha \in \mathbb{C}$ and $b_i \in \mathbf{B}$ will be linear relative to the basis $\mathbf{B}$

of $A$. In the case of group algebras $\bar{*}$ corresponds to the linear map that restricts to inversion on the group basis, which gives the antipode for the Hopf algebra structure on the group algebra. So $\bar{*}$ is one candidate for our antipode. However, the next example shows it will not always be the correct one.

**Remark 2.1.** For group algebras $\mathbb{C}G$ of finite groups $G$, it is in fact the case that any representation of degree one can be the counit of a Hopf algebra structure on $\mathbb{C}G$. When $\chi \in Irr(G)$ is a nontrivial representation $\chi : G \to \mathbb{C}^{\times}$, then it is straightforward to show that the set $G_{\chi} = \{g_{\chi} := \chi(g)^{-1}g : g \in G\}$ is a group, isomorphic to $G$, that will be another group basis of $\mathbb{C}G$. As a representation of the algebra $\mathbb{C}G$, $\chi$ restricts to the trivial character of $G_{\chi}$. For the antipode of $\mathbb{C}G_{\chi}$ we need to consider a different involution $\bar{*}'$ of $\mathbb{C}G$, which is given by $(\alpha g_{\chi})^{\bar{*}'} = (\bar{\alpha} g_{\chi})^{\bar{*}} = \alpha g_{\chi}^{-1}$, for all $\alpha \in \mathbb{C}$ and $g_{\chi} \in G_{\chi}$. So we can define a new Hopf algebra comultiplication on $\mathbb{C}G$ by setting $\Delta_{\chi}(g_{\chi}) = g_{\chi} \otimes g_{\chi}$, and this Hopf algebra structure on $\mathbb{C}G$ has counit $\chi$ and antipode $\bar{*}'$.

It remains to find good candidates for the comultiplication map $\Delta$ that will make $(A, \mu, \eta, \Delta, \delta)$ into a Hopf algebra with respect to some suitable antipode. There is, however, no obvious candidate for $\Delta$. One table algebra comultiplication given by $\Delta(b_i) = \frac{1}{\delta(b_i)}(b_i \otimes b_i)$, for all $b_i \in \mathbf{B}$ was considered earlier by Hanaki in [2] in the case where the table algebra arises from an association scheme. He observed it does not produce a Hopf algebra. While this $\Delta$ is coassociative, it is not an algebra homomorphism, so it will not produce a bialgebra structure. Hanaki's motivation for considering bialgebra structures that extend the algebra structure of a table algebra is that the tensor product of two representations of $A$ becomes a representation of $A$ on pre-composing with algebra homomorphism $\Delta$. So even if we cannot find a Hopf algebra structure extending the table algebra structure, it would still be useful to find a bialgebra structure.

## 3. Cases of non-existence

Let $A$ be a table algebra with standard basis $\mathbf{B}$ and degree map $\delta$. A simple strategy to find a comultiplication $\Delta$ that makes $(A, \mu, \eta, \Delta, \delta)$ into a bialgebra is to produce an algebra isomorphism of $A$ with a known bialgebra $H$, and borrow the necessary coalgebra structure from $H$. If, in addition, this $H$ is a Hopf algebra, then its antipode will also induce one on $A$. Since our table algebras are semisimple algebras, we can focus on finite-dimensional Hopf algebras that are semisimple as algebras. These are known to be isomorphic as Hopf algebras to group algebras or their duals in the following cases:

(i) $H$ is commutative (as this implies it is cocommutative and cosemisimple) [6] (see also the general prime and prime-square dimension cases treated in [7] and [4]);

(ii) $H$ has dimension divisible by two distinct primes [5].

In the next section we will show that when table algebras are algebra isomorphic to group algebras, our table algebra will have a Hopf algebra structure whose counit is our degree map $\delta$.

On the other hand, we know from results of Masuoka in [4] and [5] that the smallest dimension for which there are noncommutative semisimple Hopf algebras not isomorphic to group algebras is 8 (see [6, pg. 528-529]). Since noncommutative table algebras of dimension 5 and 7 have been shown to exist in [3], we can conclude these table algebras cannot possess a Hopf algebra structure at all.

**Theorem 3.1.** *Noncommutative table algebras exist for which there is no possible Hopf algebra comultiplication.*

**Remark 3.2.** Noncommutative table algebras of dimension 5 are not realized as the adjacency algebras of association schemes, but there are noncommutative table algebras of dimension 7 that are realized as adjacency algebras of association schemes, the smallest of which occurs for an association scheme of order 12.

**Remark 3.3.** The results of [3] are given in the slightly more general setting of reality-based algebras with a positive degree map. These are semisimple algebras with a distinguished basis satisfying most of the table algebra axioms (TA1)-(TA3) except the real strucuture constants $\lambda_{ijk}$ are allowed to be negative when $k \neq 0$ or $j \neq i^*$, and the positive degree map is an extra assumption. All of the results of this article apply immediately in this more general setting. In particular our results below for commutative table algebras generalize immediately to commutative reality based algebras with positive degree map, which are commonly known as $C$-algebras.

## 4. Hopf algebra structures induced by group algebra isomorphism

**Theorem 4.1.** *Let $A$ be an $r$-dimensional table algebra with degree map $\delta$. If $A$ is algebra isomorphic to a group algebra $\mathbb{C}G$ for a finite group $G$ of order $r$, then the algebra structure of $A$ can be extended to a Hopf algebra structure on $A$ whose counit is $\delta$.*

**Proof.** Let $(A, \mu, \eta)$ be the algebra structure on $A$. Suppose $\Theta : A \to \mathbb{C}G$ is an algebra isomorphism. Then $\chi := \delta \circ \Theta^{-1}$ is a degree one representation, so as in Remark 2.1 we can adjust the group basis of $\mathbb{C}G$ if necessary to arrange that coalgebra structure on $\mathbb{C}G$ has comultiplication given by $\Delta_G(g) = g \otimes g$, for all $g \in G$, counit $\chi$, and antipode $S(g) = g^{-1}$, for all $g \in G$. We can then see that $A$ has the Hopf algebra structure whose bialgebra is $(A, \mu, \eta, (\Theta^{-1} \otimes \Theta^{-1}) \circ \Delta \circ \Theta, \delta)$ and whose antipode is $\Theta^{-1} \circ S \circ \Theta$. □

**Remark 4.2.** In the case of commutative table algebras of dimension $r$, the calculation of $\Delta$ in Theorem 4.1 can done by interpolating between the character tables of the table algebra and the group.

Let $N$ be the character table of the abelian group $G$ and let $E = \{e_\chi : \chi \in Irr(G)\}$ be the basis of primitive idempotents for $\mathbb{C}G$. Then the formula $g = \sum_\chi \chi(g) e_\chi$ for all $g \in G$ tells us that $N$ is the change of basis matrix from the basis $E$ to the basis $G$.

On the other hand, if $P$ is the character table (first eigenmatrix) of the commutative table algebra $A$ with respect to the basis $\mathbf{B}$, then the primitive idempotent formula

$$e_\chi = \frac{m_\chi}{n} \sum_i \frac{\chi(b_i^*)}{\delta(b_i)} b_i$$

indicates that the change of basis matrix from $E$ to the standard basis $\mathbf{B}$ is the matrix $M$ obtained by dividing each column of $P$ by the degree of the corresponding element $b_i$ and multiplying each row of $P$ by the multiplicity of the corresponding character $\chi \in Irr(A)$. The composition of these change of basis matrices realizes a $*$-algebra isomorphism between $A$ and $\mathbb{C}G$ because the elements of $E$ are $*$-invariant; i.e. $e_\chi^* = e_\chi$ for all $e_\chi \in E$ [1, Proposition 2.9]. Furthermore, the degree map and augmentation map both correspond to the first rows of $P$ and $N$, so they are identified and so the counit is preserved. Using the change of basis matrix $MN^{-1}$ allows us to write elements of $\mathbf{B}$ in terms of the basis $G$, then calculate the value of the comultiplication using $\Delta_G$, and the inverse change of basis $NM^{-1}$ allows us to write the elements of $G$ in terms of the basis $\mathbf{B}$. So this gives an explicit formula for a Hopf algebra comultiplication that comes directly from the entries of $MN^{-1}$ and $NM^{-1}$. This technique will be used in the next section to generate explicit Hopf algebra comultiplication formulas for table algebras of small rank.

## 5. Values of Hopf algebra comultiplications on primitive idempotents

Experts familiar with the theory of finite-dimensional Hopf algebras are aware that finite-dimensional commutative semisimple Hopf algebras are cocommutative and cosemisimple, and thus isomorphic to group algebras or their duals. These results are normally established using properties of integrals and the convolution algebra. In this section we give a self-contained proof of this that has the side-benefit of producing a useful formula for the values of the Hopf algebra comultiplication on the basis of primitive idempotents. The proof given here is due to the referee of an earlier version of this article, it has the benefit of showing $S^2 = 1$.

**Theorem 5.1.** *Let $A$ be a commutative semisimple algebra over $\mathbb{C}$, and let $E$ be its basis of primitive idempotents $\{e_\chi : \chi \in Irr(A)\}$. Suppose the algebra structure on $A$ extends to a Hopf algebra structure $(A, \mu, \eta, \Delta, \delta)$ with antipode $S$. Then there is an abelian group structure on $Irr(A)$ such that for all $e_\chi$ in $E$,*

$$\Delta(e_\chi) = \sum_{\psi\star\theta=\chi} e_\psi \otimes e_\theta.$$

**Proof.** Since the antipode $S$ is an anti-automorphism of $A$, $S$ acts as a permutation on the set $E$. Let $S(e_\chi) = e_{S(\chi)}$ for all $\chi \in Irr(A)$. Also, $\Delta : A \to A \otimes A$ is an algebra homomorphism, so $\Delta(e_\chi)$ is an idempotent of $A \otimes A$. Since $A$ is a finite-dimensional commutative semisimple algebra over an algebraically closed field, there is a subset $C_\chi$ of $Irr(A) \times Irr(A)$ (possibly empty) such that

$$\Delta(e_\chi) = \sum_{(\theta,\phi)\in C_\chi} (e_\theta \otimes e_\phi).$$

Now,

$$1 \otimes 1 = \Delta(1) = \sum_{\chi\in Irr(A)} \Delta(e_\chi) = \sum_{\chi\in Irr(A)} \sum_{(\theta,\phi)\in C_\chi} (e_\theta \otimes e_\phi)$$

so every $(\theta, \phi) \in Irr(A) \times Irr(A)$ is an element of $C_\chi$ for some $\chi \in Irr(A)$. Since $1 \otimes 1 = \sum_{\theta\in Irr(A)} \sum_{\phi\in Irr(A)} (e_\theta \otimes e_\phi)$, it follows that for every $\theta, \phi \in Irr(A)$, there is a unique $\chi \in Irr(A)$ such that $(\theta, \phi) \in C_\chi$. Therefore, setting $\theta \star \phi = \chi$ iff $(\theta, \phi) \in C_\chi$, for all $\theta, \phi \in Irr(A)$ gives a well-defined binary operation on $Irr(A)$.

We have that the counit $\delta \in Irr(A)$ satisfies the condition that $(1 \otimes \delta) \circ \Delta$ is equal to the natural isomorphism $(1 \otimes 1)^{-1} : A \to A \otimes \mathbb{C}$. So whenever $\chi \star \delta = \psi$, we have

$$e_\psi \otimes 1 = (1 \otimes \delta)\Delta(e_\psi) = \sum_{(\theta,\phi)\in C_\psi} (e_\theta \otimes \delta(e_\phi)) = \sum_{\theta:(\theta,\delta)\in C_\psi} (e_\theta \otimes 1).$$

This implies that $(\theta, \delta) \in C_\psi$ iff $\theta = \psi$. Since $(\chi, \delta) \in C_\psi$ we also have $\chi = \psi$. Therefore, for each $\chi \in Irr(A)$ we have $\chi \star \delta = \chi$. Similarly we can show $\delta \star \chi = \chi$, and so $\delta$ is a $\star$-identity in $Irr(A)$.

Finally,

$$\delta_{\chi\delta} 1_A = \eta(\chi(e_\delta)) = \mu(S \otimes 1)\Delta(e_\delta) = \sum_{(\theta,\phi)\in C_\delta} e_{S(\theta)} e_\phi = \sum_{(\theta,S(\theta))\in C_\delta} e_{S(\theta)},$$

so $C_\delta = \{(\theta, S(\theta)) : \theta \in Irr(A)\}$. Therefore, for all $\theta \in Irr(A)$, we have $\theta \star S(\theta) = \delta$. So for all $\theta \in Irr(A)$, $S(\theta)$ is a right inverse for $\theta$, and we can similarly show it will be a left inverse. It follows then that $S^2$ is the identity permutation on $Irr(A)$, and hence $S^2$ acts as the identity operation on $A$.

Finally, from coassociativity we have $(1 \otimes \Delta) \circ \Delta = (\Delta \otimes 1) \otimes \Delta$, applied to $e_\chi$ for $\chi \in Irr(A)$ this gives

$$\sum_{\theta\star\phi=\chi} \sum_{\alpha\star\beta=\theta} (e_\alpha \otimes e_\beta \otimes e_\phi) = \sum_{\alpha\star\psi=\chi} \sum_{\beta\star\phi=\psi} (e_\alpha \otimes e_\beta \otimes e_\phi).$$

Now, $e_\alpha \otimes e_\beta \otimes e_\phi$ appears on the left when $(\alpha \star \beta) \star \phi = \chi$, and it appears on the right when $\alpha \star (\beta \star \phi) = \chi$. Since this holds for every $\chi \in Irr(A)$, $\star$ is an associative operation on $Irr(A)$. Therefore, $\star$ is a group operation.

We now have that the coalgebra structure of $(A, \Delta, \delta)$ is the dual of the group algebra $\mathbb{C}[Irr(A)]$ for the group $(Irr(A), \star)$. Since this has the same dimension as $A$, the group $(Irr(A), \star)$ must be abelian. This proves the theorem. □

**Remark 5.2.** The coalgebra structure on $A = \mathbb{C}E$ is determined up to coalgebra isomorphism by the isomorphism type of the group $(Irr(A), \star)$. If two of these comultiplications defined on the same basis set $E$ produce isomorphic coalgebra structures, then there would be a permutation $\mathcal{P}$ of $E$ for which

$$\Delta(\mathcal{P}(e_\chi)) = (\mathcal{P} \otimes \mathcal{P}) \sum_{\psi\phi=\chi} (e_\psi \otimes e_\phi)$$

for all $\chi \in \hat{G}$. But $\mathcal{P}$ induces a permutation $\sigma$ on $\hat{G}$ satisfying

$$\Delta(e_{\sigma(\chi)}) = \sum_{\psi\phi=\chi} (e_{\sigma(\psi)} \otimes e_{\sigma(\phi)})$$

for all $\chi \in \hat{G}$, so the two corresponding group multiplication tables would have to be permutation isomorphic. But this implies the two groups would be isomorphic. So the group algebras of two nonisomorphic abelian groups of order $r$ are never isomorphic as Hopf algebras.

## 6. Table algebras of dimensions 2 and 3

In this section we apply the character table approach indicated in Remark 4.2 to produce explicit formulas for Hopf algebra comultiplications for table algebras of ranks 2 and 3. Since these are prime dimension, these Hopf algebra structures will be unique up to isomorphism.

**Example 6.1.** Let $\mathbf{B} = \{1 = b_0, b_1\}$ be the standard basis for a table algebra $A$ of dimension 2. Since the involution must fix the multiplicative identity, it also sends

$b_1$ to $b_1$. Let $\delta : A \to \mathbb{C}$ be the degree map of our table algebra. If $k$ is the positive real number for which $\delta(b_1) = k$, then the assumption $\mathbf{B}$ is a standard basis implies $\delta(b_1) = \lambda_{110} = k$. Since $\delta$ is a $\mathbb{C}$-linear map for which $\delta(b_1^2) = \delta(b_1)^2$, we must have

$$\mu(b_1 \otimes b_1) = b_1^2 = kb_0 + (k-1)b_1.$$

As $b_0$ is the multiplicative identity, this completely determines the multiplication map $\mu$ and the unit map $\eta$ of $A$. Our antipode $\bar{*}$ is the identity map on $A$ in this case, because $*$ fixes $b_0$ and hence also $b_1$.

To use our approach, we will generate $*$-algebra isomorphisms between $A$ and $\mathbb{C}C_2$ using their character tables. The character table of $\mathbb{C}C_2$ relative to the group basis $C_2$ is $N = \begin{bmatrix} 1 & 1 \\ 1 & -1 \end{bmatrix}$, and $N$ is precisely the change of basis matrix from the basis of primitive idempotents $E$ to the group basis $C_2$.

On the other hand, the character table of the 2-dimensional table algebra determined by the parameter $k$ relative to its standard basis $\mathbf{B}$ is

| | $b_0$ | $b_1$ | $m_\chi$ |
|---|---|---|---|
| $\delta$ | 1 | $k$ | 1 |
| $\phi$ | 1 | $-1$ | $k$ |

The last column gives the multiplicity of the irreducible character. To obtain the change of basis matrix that converts from coefficients in the basis of primitive idempotents $E$ of $\mathbb{C}\mathbf{B}$ to the table algebra basis $\mathbf{B}$, we divide the columns corresponding to the elements $b_i \in \mathbf{B}$ by the degree $\delta(b_i)$, and the row corresponding to each $\chi \in Irr(A)$ by the multiplicity $m_\chi$. So the change of basis matrix form $E$ to $\mathbf{B}$ is

$$M = \begin{bmatrix} 1 & 1 \\ k & -1 \end{bmatrix}, \text{ and its inverse is } M^{-1} = \frac{1}{k+1}\begin{bmatrix} 1 & 1 \\ k & -1 \end{bmatrix}.$$

The change of basis matrix from the basis $C_2$ to $\mathbf{B}$ is thus $MN^{-1}$, and from $\mathbf{B}$ to $C_2$ is $NM^{-1}$.

Therefore, the element $g = e_\delta - e_\chi = \frac{1-k}{k+1}b_0 + \frac{2}{k+1}b_1$ is an element with $g^2 = b_0$ for which $\{b_0, g\} \simeq C_2$ is a group basis for $A$. Note that $\delta(g) = 1$ and $g^* = g$, so the induced Hopf algebra structure on $A$ has counit $\delta$ and antipode $\bar{*}$.

Let $\Delta$ be the Hopf algebra comultiplication on the group algebra $\mathbb{C}\langle g \rangle$ with $\Delta(b_0) = b_0$ and $\Delta(g) = g \otimes g$. We wish to compute a formula for $\Delta(b_i)$ in terms of the basis of elementary tensors $b_{ij} := b_i \otimes b_j$ for $b_i, b_j \in \mathbf{B}$. Since our change of basis matrices also tell us $b_1 = \frac{k-1}{2}b_0 + \frac{k+1}{2}g$, we can compute

$$\begin{aligned}
\Delta(b_1) &= \tfrac{k-1}{2}\Delta(b_0) + \tfrac{k+1}{2}\Delta(g) \\
&= \tfrac{k-1}{2}(b_0 \otimes b_0) + \tfrac{k+1}{2}(g \otimes g) \\
&= \tfrac{k-1}{2}(b_0 \otimes b_0) + \tfrac{k+1}{2}[(\tfrac{1-k}{k+1}b_0 + \tfrac{2}{k+1}b_1) \otimes (\tfrac{1-k}{k+1}b_0 + \tfrac{2}{k+1}b_1)] \\
&= [\tfrac{k-1}{2} + \tfrac{(1-k)^2}{2(k+1)}]b_{00} + \tfrac{(k+1)(2)(1-k)}{2(k+1)^2}[b_{01} + b_{10}] + \tfrac{(2)(2)(k+1)}{2(k+1)^2}b_{11}
\end{aligned}$$

$$\implies \Delta(b_1) = \frac{k(k-1)}{k+1}b_{00} + \frac{1-k}{k+1}[b_{01} + b_{10}] + \frac{2}{k+1}b_{11}.$$

**Example 6.2.** *The Hopf algebra structure of a triangle.* When $k$ is a positive integer larger than 1, the adjacency algebra of a complete graph on $k+1$ vertices is the table algebra of dimension 2 in Example 6.1 determined by the parameter $k$. If we let $b_0$ be the $(k+1) \times (k+1)$ identity matrix and $b_1$ be the adjacency matrix of the complete graph on $k+1$ vertices, then the formula in Example 6.1 defines the Hopf algebra structure of a complete graph on $k+1$ vertices. Its comultiplication takes values in the space $(k+1)^2 \times (k+1)^2$-matrices spanned by the $b_{ij}$'s, where $b_{ij} = b_i \otimes b_j$. For example, when $b_1$ is the adjacency matrix of a triangle (the case $k = 2$), then the Hopf algebra structure is determined by the $9 \times 9$ matrix

$$\Delta(b_1) = \frac{1}{3}(2b_{00} - b_{01} - b_{10} + 2b_{11}).$$

**Example 6.3.** *Nonreal table algebras of dimension* 3. Suppose $A$ is a 3-dimensional nonreal table algebra with standard basis $\mathbf{B} = \{b_0, b_1, b_2\}$, with $b_1^* = b_2$. The table algebra conditions imply that there is a unique $u \geq 0$ such that $b_1 b_2 = (2u+1)b_0 + ub_1 + ub_2$, and $b_1^2 = ub_1 + (u+1)b_2$. So $\delta(b_1) = \delta(b_2) = 2u+1$.

For $G = C_3$ the character table is

$$N = \begin{bmatrix} 1 & 1 & 1 \\ 1 & \zeta & \zeta^2 \\ 1 & \zeta^2 & \zeta \end{bmatrix},$$

for a primitive cube root of unity $\zeta$. We again view $N$ as the change of basis matrix from the basis of primitive idempotents $E$ to the basis $G$. It is important for our approach that the first row of $N$ corresponds to the trivial character, the ordering of the second and third columns is reversed it results in a permutation of some of our structure constants for the Hopf algebra comultiplication.

The character table of the 3-dimensional nonreal table algebra determined by the parameter $u$ can be calculated from the eigenvalues of the regular matrices of

$\{b_0, b_1, b_2\}$:

| | $b_0$ | $b_1$ | $b_1^*$ | $m_\chi$ |
|---|---|---|---|---|
| $\delta$ | 1 | $2u+1$ | $2u+1$ | 1 |
| $\phi$ | 1 | $\frac{1}{2}(-1+\sqrt{-3-4u})$ | $\frac{1}{2}(-1-\sqrt{-3-4u})$ | $2u+1$ |
| $\bar{\phi}$ | 1 | $\frac{1}{2}(-1-\sqrt{-3-4u})$ | $\frac{1}{2}(-1+\sqrt{-3-4u})$ | $2u+1$ |

To obtain the change of basis matrix $M$ that changes the basis from $E$ to $\mathbf{B}$, we multiply each row by the multiplicity of the corresponding irreducible character and divide each column by the degree of the corresponding element. So

$$M = \begin{bmatrix} 1 & 1 & 1 \\ 2u+1 & \frac{1}{2}(-1+\sqrt{-3-4u}) & \frac{1}{2}(-1-\sqrt{-3-4u}) \\ 2u+1 & \frac{1}{2}(-1-\sqrt{-3-4u}) & \frac{1}{2}(-1+\sqrt{-3-4u}) \end{bmatrix}.$$

Again it is important here that the first row of $M$ corresponds to the degree map. Now we use the change of basis matrices $MN^{-1}$ and $NM^{-1}$ to express $b_1 = \alpha e + \beta g + \gamma g^{-1}$ and $g = rb_0 + sb_1 + tb_2$, where $\alpha, \beta, \gamma$ are the entries in the second row of $MN^{-1}$ and $r, s, t$ are the entries in the second row of $NM^{-1}$. From these we can obtain a formula for $\Delta(b_1)$:

$$\begin{aligned} \Delta(b_1) &= \alpha(e\otimes e) + \beta(g\otimes g) + \gamma(g^{-1}\otimes g^{-1}) \\ &= \alpha b_{00} + \beta[(rb_0 + sb_1 + tb_2)\otimes(rb_0 + sb_1 + tb_2)] \\ &\quad + \gamma[(\bar{r}b_0 + \bar{t}b_1 + \bar{s}b_2)\otimes(\bar{r}b_0 + \bar{t}b_1 + \bar{s}b_2)] \\ &= (\alpha + \beta r^2 + \gamma\bar{r}^2)b_{00} + (\beta rs + \gamma\bar{r}\bar{t})b_{01} + (\beta rt + \gamma\bar{r}\bar{s})b_{02} \\ &\quad + (\beta sr + \gamma\bar{t}\bar{r})b_{10} + (\beta s^2 + \gamma\bar{t}^2)b_{11} + (\beta st + \gamma\bar{t}\bar{s})b_{12} \\ &\quad + (\beta tr + \gamma\bar{s}\bar{r})b_{20} + (\beta ts + \gamma\bar{s}\bar{t})b_{21} + (\beta t^2 + \gamma\bar{s}^2)b_{22}. \end{aligned}$$

Our formula for $\Delta(b_1)$ comes from the calculations of

$$MN^{-1} = \begin{bmatrix} 1 & 0 & 0 \\ \frac{2u}{3} & \frac{(3+4u)-\sqrt{9+12u}}{6} & \frac{(3+4u)+\sqrt{9+12u}}{6} \\ \frac{2u}{3} & \frac{(3+4u)+\sqrt{9+12u}}{6} & \frac{(3+4u)-\sqrt{9+12u}}{6} \end{bmatrix}$$

and

$$NM^{-1} = \begin{bmatrix} 1 & 0 & 0 \\ \frac{-2u}{3+4u} & \frac{\sqrt{3}(\sqrt{9+12u}-(3+4u))}{2(3+4u)\sqrt{3+4u}} & \frac{\sqrt{3}(\sqrt{9+12u}+(3-4u))}{2(3+4u)\sqrt{3+4u}} \\ \frac{-2u}{3+4u} & \frac{\sqrt{3}(\sqrt{9+12u}+(3+4u))}{2(3+4u)\sqrt{3+4u}} & \frac{\sqrt{3}(\sqrt{9+12u}-(3-4u))}{2(3+4u)\sqrt{3+4u}} \end{bmatrix}.$$

The second rows of these matrices determine $\alpha, \beta, \gamma$ and $r, s, t$. Now we leave it to the reader to verify the formulas $\Delta(b_0) = b_{00}$,

$$\Delta(b_1) = \frac{1}{4u+3}((4u^2+2u)b_{00} - 2ub_{01} + 0b_{02} - 2ub_{10} + (u+3)b_{11} - ub_{12} + 0b_{20} - ub_{21} + ub_{22}),$$

and

$$\Delta(b_2) = \frac{1}{4u+3}((4u^2+2u)b_{00}+0b_{01}-2ub_{02}+0b_{10}+ub_{11}-ub_{12}-2ub_{20}-ub_{21}+(u+3)b_{22}).$$

**Example 6.4.** *Real Table Algebras of Dimension* 3. Let $A$ be a real table algebra whose standard basis $\mathbf{B} = \{b_0, b_1, b_2\}$ satisfies $b_i^* = b_i$ for $i = 0, 1, 2$. Let $\delta(b_1) = k$ and $\delta(b_2) = \ell$ be the degrees of the nonidentity basis elements, and let $u, v$ be the structure constants determined by $b_1 b_2 = ub_1 + vb_2$. Since the degree map is an algebra homomorphism these parameters satisfy the necessary condition $k\ell = ku + \ell v$. The remaining structure constants are determined by the identities

$$b_1^2 = kb_0 + (k-u-1)b_1 + (k-v)b_2, \text{ and } b_2^2 = \ell b_0 + (\ell - u)b_1 + (\ell - v - 1)b_2.$$

From the regular matrices of $b_1$ and $b_2$ we can determine the character table of $A$ relative to $\mathbf{B}$ in terms of the parameters $k$, $\ell$, $u$, and $v$.

| | $1$ | $b_1$ | $b_2$ | $m_\chi$ |
|---|---|---|---|---|
| $\delta$ | $1$ | $k$ | $\ell$ | $1$ |
| $\phi$ | $1$ | $\frac{1}{2}((v-u-1)+\sqrt{\alpha})$ | $\frac{1}{2}((u-v-1)-\sqrt{\alpha})$ | $m_\phi$ |
| $\psi$ | $1$ | $\frac{1}{2}((v-u-1)-\sqrt{\alpha})$ | $\frac{1}{2}((u-v-1)+\sqrt{\alpha})$ | $m_\psi$ |

where $\alpha = (v-u-1)^2 + 4v$. The multiplicities can be calculated from the orthogonality relations, which lead to the equation $\frac{1+k+\ell}{m_\phi} = 1 + \frac{\phi(b_1)^2}{k} + \frac{\phi(b_2)^2}{\ell}$ and a similar one for $m_\psi$. These reduce to

$$m_\phi = \frac{k\ell + k^2\ell + k\ell^2}{k\ell + \ell\phi_1^2 + k\phi_2^2} \text{ and } m_\psi = \frac{k\ell + k^2\ell + k\ell^2}{k\ell + \ell\psi_1^2 + k\psi_2^2},$$

where $\phi_1 = \phi(b_1)$, $\phi_2 = \phi(b_2)$, $\psi_1 = \psi(b_1)$, and $\psi_2 = \psi(b_2)$ are the corresponding character table entries.

Therefore, the change of basis matrix from the basis of primitive idempotents $E$ to $\mathbf{B}$ is

$$M := \begin{bmatrix} 1 & 1 & 1 \\ m_\phi & \frac{m_\phi}{k}\phi_1 & \frac{m_\phi}{\ell}\phi_2 \\ m_\psi & \frac{m_\psi}{k}\psi_1 & \frac{m_\psi}{\ell}\psi_2 \end{bmatrix}.$$

We know $A \simeq \mathbb{C}C_3$ as $*$-algebras. Let $N$ be the character table of $C_3$ given in the previous example. The change of basis matrix from the basis $C_3$ to our table algebra basis $\mathbf{B}$ will be

$$MN^{-1} = \begin{bmatrix} 1 & 0 & 0 \\ \frac{m_\phi}{3}(1+\frac{\phi_1}{k}+\frac{\phi_2}{\ell}) & \frac{m_\phi}{3}(1+\frac{\phi_1}{k}\zeta^2+\frac{\phi_2}{\ell}\zeta) & \frac{m_\phi}{3}(1+\frac{\phi_1}{k}\zeta+\frac{\phi_2}{\ell}\zeta^2) \\ \frac{m_\psi}{3}(1+\frac{\psi_1}{k}+\frac{\psi_2}{\ell}) & \frac{m_\psi}{3}(1+\frac{\psi_1}{k}\zeta^2+\frac{\psi_2}{\ell}\zeta) & \frac{m_\psi}{3}(1+\frac{\psi_1}{k}\zeta+\frac{\psi_2}{\ell}\zeta^2) \end{bmatrix}$$

$$:= \begin{bmatrix} 1 & 0 & 0 \\ \alpha_1 & \beta_1 & \bar{\beta}_1 \\ \alpha_2 & \beta_2 & \bar{\beta}_2 \end{bmatrix},$$

and its inverse (obtained using Wolfram Math Alpha©) is the change of basis matrix from $\mathbf{B}$ to $C_3$:

$$NM^{-1} = \begin{bmatrix} 1 & 0 & 0 \\ r & s & t \\ \bar{r} & \bar{s} & \bar{t} \end{bmatrix},$$

where

$$r = \frac{\phi_1\psi_2 - \phi_2\psi_1 + k(\phi_2-\psi_2)\zeta + \ell(\psi_1-\phi_1)\zeta^2}{\phi_1(\psi_2-\ell)+\phi_2(k-\psi_1)+\psi_1\ell-\psi_2 k}$$

$$s = \frac{\psi_1\ell - \psi_2 k + k(\psi_2-\ell)\zeta + \ell(k-\psi_1)\zeta^2}{m_\phi(\phi_1(\psi_2-\ell)+\phi_2(k-\psi_1)+\psi_1\ell-\psi_2 k)}, \text{ and}$$

$$t = \frac{\phi_1 k - \phi_2\ell + k(\ell-\phi_2)\zeta + \ell(\phi_1-k)\zeta^2}{m_\psi(\phi_1(\psi_2-\ell)+\phi_2(k-\psi_1)+\psi_1\ell-\psi_2 k)}.$$

As before we can use $MN^{-1}$ and its inverse to derive a formula for the Hopf algebra comultiplication that is induced by the Hopf algebra comultiplication on the group algebra. We have that $b_1 = \alpha_1 b_0 + \beta_1 g + \bar{\beta}_1 g^{-1}$, $b_2 = \alpha_2 b_0 + \beta_2 g + \bar{\beta}_2 g^{-1}$, $g = rb_0 + sb_1 + tb_2$, and $g^{-1} = \bar{r}b_0 + \bar{s}b_1 + \bar{t}b_2$. Note that $\alpha_1$ and $\alpha_2$ are real follows from $b_1$ and $b_2$ being $*$-invariant. Again the ordering of the last two idempotents in the basis $E$ is not fixed, but switching the order only interchanges the second and third rows of these change of basis matrices. We will have

$$\begin{aligned}
\Delta(b_1) &= \alpha_1\Delta(b_0) + \beta_1\Delta(g) + \bar{\beta}_1\Delta(g^{-1}) \\
&= \alpha_1 b_{00} + \beta_1(g\otimes g) + \bar{\beta}_1(g^{-1}\otimes g^{-1}) \\
&= \alpha_1 b_{00} + \beta_1[(rb_0+sb_1+tb_2)\otimes(rb_0+sb_1+tb_2)] \\
&\quad + \bar{\beta}_1[(\bar{r}b_0+\bar{s}b_1+\bar{t}b_2)\otimes(\bar{r}b_0+\bar{s}b_1+\bar{t}b_2)] \\
&= (\alpha_1+\beta_1 r^2+\bar{\beta}_1\bar{r}^2)b_{00} + (\beta_1 rs+\bar{\beta}_1\bar{r}\bar{s})b_{01} + (\beta_1 rt+\bar{\beta}_1\bar{r}\bar{t})b_{02} \\
&\quad + (\beta_1 sr+\bar{\beta}_1\bar{s}\bar{r})b_{10} + (\beta_1 s^2+\bar{\beta}_1\bar{s}^2)b_{11} + (\beta_1 st+\bar{\beta}_1\bar{s}\bar{t})b_{12} \\
&\quad + (\beta_1 tr+\bar{\beta}_1\bar{t}\bar{r})b_{20} + (\beta_1 ts+\bar{\beta}_1\bar{t}\bar{s})b_{21} + (\beta_1 t^2+\bar{\beta}_1\bar{t}^2)b_{22},
\end{aligned}$$

and the formula for $\Delta(b_2)$ has the same pattern with $\alpha_2$ and $\beta_2$ replacing $\alpha_1$ and $\beta_1$. Note that the coefficients of the $b_{ij}$ occurring in $\Delta(b_1)$ and $\Delta(b_2)$ are guaranteed to be real numbers.

The explicit formulas in the 3-dimensional table algebra case are too long to express concisely here, but become easier in specific cases. We will demonstrate this with some actual calculations for table algebras corresponding to familiar strongly-regular graphs.

**Example 6.5.** *The Hopf algebra structure of a rectangle.* A rectangle is a strongly-regular graph on 4 vertices, which means that it generates an association scheme whose adjacency matrices form a table algebra basis. If $b_2$ is the $4 \times 4$ adjacency matrix of the rectangle, then the parameters of Example 6.4 for this table algebra are $k = 1$, $\ell = 2$, $u = 0$, and $v = 1$. Therefore, the character table of $\mathbb{C}\mathbf{B}$ is

| | $1$ | $b_1$ | $b_2$ | $m_\chi$ |
|---|---|---|---|---|
| $\delta$ | $1$ | $1$ | $2$ | $1$ |
| $\phi$ | $1$ | $1$ | $-2$ | $1$ |
| $\psi$ | $1$ | $-1$ | $0$ | $2$ |

and the change of basis matrix from the group basis $C_3$ to the table algebra basis $\{b_0, b_1, b_2\}$ is

$$MN^{-1} = \begin{bmatrix} 1 & 1 & 1 \\ 1 & 1 & -1 \\ 2 & -2 & 0 \end{bmatrix} \cdot \frac{1}{3} \begin{bmatrix} 1 & 1 & 1 \\ 1 & \zeta^2 & \zeta \\ 1 & \zeta & \zeta^2 \end{bmatrix} = \begin{bmatrix} 1 & 0 & 0 \\ \frac{1}{3} & \frac{-2}{3}\zeta & \frac{-2}{3}\zeta^2 \\ 0 & -\frac{2}{3}\zeta - \frac{4}{3}\zeta^2 & -\frac{4}{3}\zeta - \frac{2}{3}\zeta^2 \end{bmatrix}.$$

Using this matrix and its inverse, we have $b_1 = \frac{1}{3}e - \frac{2}{3}\zeta g - \frac{2}{3}\zeta^2$, $\quad b_2 = (-\frac{2}{3}\zeta - \frac{4}{3}\zeta^2)g + (-\frac{4}{3}\zeta - \frac{2}{3}\zeta^2)g^2$ and $g = \frac{1}{4}\zeta^2 b_0 - \frac{3}{4}\zeta^2 b_1 - \frac{1}{4}(2\zeta + \zeta^2)b_2$, and $g^2 = \frac{1}{4}\zeta b_0 - \frac{3}{4}\zeta b_1 - \frac{1}{4}(\zeta + 2\zeta^2)b_2$. From these we can obtain formulas for $\Delta(b_1)$ and $\Delta(b_2)$ as before:

$$\Delta(b_1) = \frac{1}{8}(3b_{00} - b_{01} - b_{02} - b_{10} + 3b_{11} + 3b_{12} - b_{20} + 3b_{21} - b_{22})$$

and

$$\Delta(b_2) = \frac{1}{8}(-b_{00} + 3b_{01} - b_{02} + 3b_{10} - 9b_{11} + 3b_{12} - b_{20} + 3b_{21} + 3b_{22}).$$

**Example 6.6.** *The Hopf algebra structure of a pentagon.* A pentagon is a strongly regular graph on 5 vertices whose adjacency matrix generates the real table algebra of dimension 3 with parameters $k = \ell = 2$ and $u = v = 1$. Its character table is

| | $1$ | $b_1$ | $b_2$ | $m_\chi$ |
|---|---|---|---|---|
| $\delta$ | $1$ | $2$ | $2$ | $1$ |
| $\phi$ | $1$ | $\frac{-1+\sqrt{5}}{2}$ | $\frac{-1-\sqrt{5}}{2}$ | $2$ |
| $\psi$ | $1$ | $\frac{-1-\sqrt{5}}{2}$ | $\frac{-1+\sqrt{5}}{2}$ | $2$ |

Therefore, the change of basis matrices between the bases $C_3$ and $\mathbf{B}$ are

$$MN^{-1} = \begin{bmatrix} 1 & 0 & 0 \\ \frac{1}{3} & \frac{5+\sqrt{-15}}{6} & \frac{5-\sqrt{-15}}{6} \\ \frac{1}{3} & \frac{5-\sqrt{-15}}{6} & \frac{5+\sqrt{-15}}{6} \end{bmatrix} \text{ and } NM^{-1} = \begin{bmatrix} 1 & 0 & 0 \\ \frac{-1}{5} & \frac{3-\sqrt{-15}}{10} & \frac{3+\sqrt{-15}}{10} \\ \frac{-1}{5} & \frac{3+\sqrt{-15}}{10} & \frac{3-\sqrt{-15}}{10} \end{bmatrix}.$$

So the unique Hopf algebra structure on the adjacency algebra of a pentagon has its comultiplication given by

$$\Delta(b_1) = \frac{1}{5}(2b_{00} - b_{01} + 0b_{02} - b_{10} + b_{11} + 2b_{12} + 0b_{20} + 2b_{21} - 2b_{22})$$

and

$$\Delta(b_2) = \frac{1}{5}(2b_{00} + 0b_{01} - b_{02} + 0b_{10} - 2b_{11} + 2b_{12} - b_{20} + 2b_{21} + b_{22}).$$

**Example 6.7.** The *Petersen graph* is a strongly regular graph whose adjacency matrix generate the real table algebra of dimension 3 with parameters $k = 3$, $\ell = 6$, $u = 2$, and $v = 2$. Substituting these into our formulas, the character table of this table algebra is

| | $1$ | $b_1$ | $b_2$ | $m_\chi$ |
|---|---|---|---|---|
| $\delta$ | $1$ | $3$ | $6$ | $1$ |
| $\phi$ | $1$ | $-2$ | $1$ | $4$ |
| $\psi$ | $1$ | $1$ | $-2$ | $5$ |

so our change of basis matrix is

$$M = \begin{bmatrix} 1 & 1 & 1 \\ 4 & -\frac{8}{3} & \frac{2}{3} \\ 5 & \frac{5}{3} & -\frac{5}{3} \end{bmatrix}.$$

From this matrix and the character table $N$ of the group of order 3 we can produce our formula for the Hopf algebra comultiplication on the adjacency algebra of the Petersen graph:

$$\Delta(b_1) = \frac{1}{10}(5b_{00} - 5b_{01} + 3b_{02} - 5b_{10} + 0b_{11} + 3b_{12} + 3b_{20} + 3b_{21} - 3b_{22})$$

and

$$\Delta(b_2) = \frac{1}{10}(22b_{00} - 3b_{01} - 2b_{02} - 3b_{10} - 3b_{11} + 3b_{12} - 2b_{20} + 3b_{21} + 0b_{22}).$$

**Acknowledgement.** The authors would like to express their gratitude to the referee for many helpful comments, and in particular for suggesting a better proof of Theorem 5.1.

## References

[1] H. I. Blau, *Table algebras*, European J. Combin., 30(6) (2009), 1426-1455.

[2] A. Hanaki, *Character products of association schemes*, J. Algebra, 283(2) (2005), 596-603.

[3] A. Herman, M. Muzychuk and B. Xu, *The recognition problem for table algebras and reality-based algebras*, J. Algebra, 479 (2017), 173-191.

[4] A. Masuoka, *Semisimple Hopf algebras of dimension 6,8*, Israel J. Math., 92 (1995), 361-373.

[5] A. Masuoka, *The $p^n$ theorem for semisimple Hopf algebras*, Proc. Amer. Math. Soc., 124 (1996), 735-737.

[6] D. E. Radford, Hopf Algebras, Series on Knots and Everything, 49, World Scientific Publishing Co. Pte. Ltd., Hackensack, NJ, 2012.

[7] Y. Zhu, *Hopf algebras of prime dimension*, Internat. Math. Res. Notices, 1 (1994), 53-59.

**Allen Herman** (Corresponding Author) and **Gurmail Singh**
Department of Mathematics and Statistics
University of Regina
Regina, Canada, S4S 0A2
e-mails: Allen.Herman@uregina.ca (A. Herman)
singh28g@uregina.ca (G. Singh)

# 12

# ON THE ASSOCIATED PRIME IDEALS AND THE DEPTH OF POWERS OF SQUAREFREE PRINCIPAL BOREL IDEALS

Jürgen Herzog, Bahareh Lajmiri and Farhad Rahmati

Communicated by Abdullah Harmancı

Abstract. We study algebraic properties of powers of squarefree principal Borel ideals $I$, and show that $\operatorname{astab}(I) = \operatorname{dstab}(I)$. Furthermore, the behaviour of the depth function $\operatorname{depth} S/I^k$ is considered.

**Keywords**: Squarefree principal Borel ideal, socle, astab, dstab

## 1. Introduction

Borel ideals appear in characteristic zero as generic initial ideals. By applying the stretching operator of Kalai [17] to a Borel ideal, one obtains a squarefree Borel ideal, also called squarefree strongly stable ideal. It has the same graded Betti numbers as the original ideal, see for example [14, Lemma 11.2.6]. This class of ideals was introduced by Hibi and the first author of this paper in [2]. In the sequel, algebraic properties of squarefree strongly stable ideals and their powers have been studied by many authors, see for example [1], [3], [11], [12] and [18].

Among the squarefree strongly stable ideals, the squarefree principal Borel ideals and their powers are best understood. The squarefree principal ideal with Borel generator $u$ is denoted by $B_S(u)$. We use some of the known results from Aslam [3], Francisco, Mermin and Schweig [12] and De Negri [7], to get some additional information about the algebraic and homological properties of powers of $B_S(u)$. These known facts are recalled in Section 3. In this section we also show that $B_S(u)$ is normally torsion free if and only if it is almost normally torsion free, and that this happens if and only if the Borel generator $u$ of the ideal has a specific form, see Corollary 3.4. In Corollary 3.13 we determine the height and bigheight of $B_S(u)$, and characterize those squarefree principal Borel ideals which are Cohen-Macaulay. Corollary 3.12 makes more explicit the set $\operatorname{Ass}^{\infty}(B_S(u))$, as it is described by Aslam in Theorem 3.10.

For our proofs we essentially use monomial localization. Some of the basic facts about monomial localization are recalled in Section 2. We close this section by characterizing in Corollary 2.5 $k$-strongly stable ideals. This type of ideals play an important role in Section 4, where the socle of the powers of $B_S(u)$ is determined. It turns out, see Corollary 4.2, that the socle of $B_S(u)^k$ is a $(k-1)$-stable set of monomials. By using this fact we exhibit in Corollary 4.5 for each $k$ a monomial $w_k$ with the property that $\operatorname{depth} S/B_S(u)^k = 0$, if and only if $w_k x_n \in B_S(u)^k$. Here $u$ is a squarefree monomial in $S = K[x_1, \dots, x_n]$ with $x_1 \nmid u$ and $x_n \mid u$. These conditions on $u$ are not restrictive, because one can always reduce to this case, see Lemma 5.1 and Proposition 5.2.

The invariant $\operatorname{astab}(u)$ is the smallest number $k$ for which $\operatorname{Ass}(B_S(u)^k$ stabilizes, and $\operatorname{dstab}(B_S(u))$ is the smallest number $k$ for which $\operatorname{depth} S/B_S(u)^k$ stabilizes. As the main result of Section 5 we show in Theorem 5.5 that for any $u$, $\operatorname{astab}(B_S(u)) = \operatorname{dstab}(B_S(u))$. For general monomial ideals $I$, the the Ass-stability and the depth stability are usually unrelated.

For the proof of Theorem 5.5 we use the interesting fact, shown in Theorem 5.4, that the depth stability of monomial localizations of $B_S(u)$ are bounded by the depth stability of $B_S(u)$.

In the last section we study the depth of $S/B_S(u)^k$ as a function of $k$. For short we set $f(k) = \operatorname{depth} S/B_S(u)^k$. Since all powers of $B_S(u)^k$ have a linear resolution it follows from [13, Proposition 2.1] that $f(k) \leq f(k-1)$ for $k \geq 1$. We also know that the depth function $f(k)$ becomes constant for $k \geq \deg u$, see Proposition5.3. By considering many example it seems to be the case that $f(k) < f(k-1)$, before $f(k)$ becomes constant. This is true if $\deg u \leq 3$, and follows from Proposition 6.5, where we compute $f(k)$ explicitly for $\deg u \leq 3$. In Proposition 6.2 we show that $f(1) = \deg u - 1$, and in Proposition 6.3 we determine all squarefree monomials $u$ for which $f(2) = 0$. Finally in Theorem 6.6 we show that $f(2) < f(1)$, as expected, unless $f(1) = f(k)$ for some $k \geq 2$, and in this case $f(k)$ is a constant function.

## 2. Preliminaries

In this section we introduce some concepts and results which are important in this paper.

*Monomial localizations.* In this paper, monomial localizations are an important tool. Let $K$ be a field and let $S = K[x_1, \dots, x_n]$ be the polynomial ring in $n$ variables. Let $T \subset [n]$. We define the monomial prime ideal $P_T$ to be the ideal $(x_j \colon j \in T^c)$. Here $T^c = [n] \setminus T$ and $[n] = \{1, \dots, n\}$. We also set $[0] = \emptyset$. Notice that $P_{[0]} = (x_1, \dots, x_n)$ and $P_{[n]} = (0)$.

**Definition 2.1.** Let $I \subset K[x_1, \ldots, x_n]$ be a monomial ideal, and let $T \subset [n]$. The *monomial localization* of $I$ with respect to $P_T$ is the monomial ideal $I(P_T) \subset S(P_T)$, where $S(P_T) = K[x_j : j \in T^c]$ and $I(P_T) = \varphi(I)S(P_T)$, where $\varphi : S \to S(P_T)$ is the $K$-algebra homomorphism with $x_j \mapsto x_j$ if $j \in T^c$, and $x_j \mapsto 1$ if $j \in T$.

Monomial localization with respect to $P_T$ and the usual localization with respect to the prime ideal $P_T$ are related as follows:

$$IS_{P_T} = I(P_T)S_{P_T}.$$

This identity justifies the name "monomial localization".

For example, if $I = (x_1x_3, x_1x_4, x_2x_3x_4, x_3x_5x_7x_8) \subset S = K[x_1, \ldots, x_8]$ and $T = \{3, 5\}$, then $S(P_T) = K[x_1, x_2, x_4, x_6, x_7, x_8]$ and $I(P_T) = (x_1, x_2x_4, x_7x_8)$.

If $I = P_T$, then $I(P_T)$ is the graded maximal ideal of $S(P_T)$ which we denote by $\mathfrak{m}_{S(P_T)}$.

To simplify notation, we set $I_{(j)} = I(P_{\{j\}})$ and $S_{(j)}$ for $S(P_{\{j\}})$. Note that $S_{(j)} = K[x_i : i \in [n] \setminus \{j\}]$.

*Strongly stable ideals.* Let $u = x_1^{a_1} \cdots x_n^{a_n}$ be a monomial in $S = K[x_1, \ldots, x_n]$. We set $\nu_i(u) = a_i$ for $i = 1, \ldots, n$. Now let $I \subset S = K[x_1, \ldots, x_n]$ be a monomial ideal. We denote by $G(I)$ the unique set of monomial generators of $I$. For a given integer $k \geq 1$, we let $I^{\leq k}$ be the ideal generated by all $u \in G(I)$ with $\nu_i(u) \leq k$ for $i = 1, \ldots, n$.

A monomial $u \in S$ be can be written as $u = x_{i_1}x_{i_2} \cdots x_{i_d}$ with $i_1 \leq i_2 \leq \ldots \leq i_d$. The monomial $u$ is called *squarefree* if $i_1 < i_2 < \ldots < i_d$. A monomial ideal $I$ is called a *squarefree monomial ideal* if all monomials in $G(I)$ are squarefree. Note that for any monomial ideal, the ideal $I^{\leq 1}$ is a squarefree monomial ideal. Moreover, the ideal $I$ is squarefree if and only if $I = I^{\leq 1}$.

**Definition 2.2.** Let $I \subset S = K[x_1, \ldots, x_n]$ be a monomial ideal, and let $k \geq 1$ be an integer, or $k = \infty$. Then $I$ is called *$k$-strongly stable*, if

(i) $I = I^{\leq k}$;
(ii) for all $u \in G(I)$ and all integers $1 \leq i < j \leq n$ with $\nu_j(u) > 0$ and $\nu_i(u) < k$ it follows that $x_i(u/x_j) \in I$.

The following special cases are of particular interest: let $I$ be a monomial ideal.

($\alpha$) If $k = \infty$, then there is no bound on the exponents and $\infty$-strongly stable is simply called *strongly stable*. In other words, $I$ is strongly stable, if for $u \in G(I)$ and all $j$ such that $x_j$ divides $u$, it follows that $x_i(u/x_j) \in I$ for all $i \leq j$.

($\beta$) $I$ is called *squarefree strongly stable*, if it is 1-strongly stable.

Let $u_1, \ldots, u_m$ be monomials in $S$ with $\nu_i(u_j) \leq k$ for $i = 1, \ldots, n$ and $j = 1, \ldots, m$. There exists a unique smallest $k$-strongly stable ideal containing $u_1, \ldots, u_m$ which we denote by $B_S^k(u_1, \ldots, u_m)$. The monomials $u_1, \ldots u_m$ are called *Borel generators* of $B_S^k(u_1, \ldots, u_m)$. A monomial ideal $I$ is called *$k$-principal Borel* if $I = B_S^k(u)$ for some monomial $u$ with $\nu_i(u) \leq k$ for $i = 1, \ldots, n$. We call 1-principal Borel ideals also squarefree principal Borel. The $k$-principal Borel ideals appear as powers of squarefree principal Borel ideals, see Theorem 4.1.

To simplify notation, we write $B_S(u_1, \ldots, u_m)$ for $B_S^1(u_1, \ldots, u_m)$ when $u_1, \ldots, u_m$ are squarefree monomials. The unique strongly stable ideal containing the monomials $u_1, \ldots, u_m$ will be denoted by $B_S^\infty(u_1, \ldots, u_m)$.

For example, let $I = B_S(x_2x_4, x_1x_3)$. Then $G(I) = \{x_1x_2, x_1x_3, x_1x_4, x_2x_3, x_2x_4\}$.

Let $u, v$ be monomials of same degree, and assume that $\nu_i(u) \leq k$ for $i = 1, \ldots, n$. Then we write $v \preceq_k u$ if $v \in B_S^k(u)$, and $v \prec_k u$ if $v \in B_S^k(u)$ and $v \neq u$. For $\preceq_\infty$ we simply write $\preceq$.

For each $d$, $\preceq_k$ defines a partial order on the set of monomials of degree $d$ whose exponents are bounded by $k$. For example, one has $x_1^2x_2^3x_3 \prec_3 x_2^3x_3^3$. A set $\mathcal{S}$ of monomials of degree $d$ whose exponents are bounded by $k$ is called a *$k$-stable set*, if for $u \in \mathcal{S}$ and $v \preceq_k u$ it follows that $v \in \mathcal{S}$.

Note that $B_S^k(u_1, \ldots, u_m)$ is generated by $\bigcup_{i=1}^m \{v\colon\ v \preceq_k u_i\}$. Therefore,

$$B_S^k(u_1, \ldots, u_m) = \sum_{i=1}^m B_S^k(u_i).$$

Let $v, u \in S$ be monomials of degree $d$, where $v = x_{i_1} \cdots x_{i_d}$ with $i_1 \leq i_2 \leq \cdots \leq i_d$ and $u = x_{j_1} \cdots x_{j_d}$ with $j_1 \leq j_2 \leq \cdots \leq j_d$. By [14, Lemma 4.2.4] one has

$$v \preceq u \text{ if and only if } i_k \leq j_k \text{ for all } k = 1, \ldots, d. \qquad (1)$$

**Remark 2.3.** Let $u, v \in S$ be monomials. It is clear that if $v \preceq_k u$, then $v \preceq u$. On the other hand, if the exponents of $u$ and $v$ are bounded by $k$ and $v \preceq u$, then $v \preceq_k u$. This follows from the next lemma.

**Lemma 2.4.** *Let $u \in S$ be a monomial with $\nu_i(u) \leq k$ for $i = 1, \ldots, n$. Then*

$$B_S^k(u) = B_S^\infty(u)^{\leq k}.$$

**Proof.** Let $v \in B_S^k(u)$. Then $v \preceq_k u$, and hence $v \preceq u$. Therefore, $v \in B_S^\infty(u)$. Since the exponent of $v$ is bounded by $k$ it follows that $v \in B_S^\infty(u)^{\leq k}$.

Conversely, let $v \in B_S^\infty(u)^{\leq k}$, and let $v = x_{i_1} \cdots x_{i_d}$ with $i_1 \leq i_2 \leq \cdots \leq i_d$ and $u = x_{j_1} \cdots x_{j_d}$ with $j_1 \leq j_2 \leq \cdots \leq j_d$.

Let $w$ be a monomial of degree $d$ with $v \preceq w$, and let $w = x_{s_1} \cdots x_{s_d}$ with $s_1 \leq s_2 \leq \cdots \leq s_d$. We define

$$\delta(w, v) = \sum_{l=1}^{d} (s_l - i_l).$$

Since $v \preceq w$, it follows from (1) that $\delta(w, v) \geq 0$, and we have $\delta(w, v) = 0$ if and only if $w = v$.

Let $l+1$ be the smallest integer such that $j_{l+1} > i_{l+1}$. Let $u_1 = x_{j_{l-1}-1} u / x_{j_{l+1}}$. Then the exponents of $u_1$ are bounded by $k$, unless $j_{l+1} - 1 = j_l = j_{l-1} = \cdots = j_{l-k+1}$. Assume $j_{l+1} - 1 = j_l$. By the choice of $l$ it follows that $i_{l-k+1} = \cdots i_l$. Since the exponents of $v$ are bounded by $k$, it follows that $i_{l+1} > i_l$ Moreover, $j_{l+1} > i_{l+1}$. Then $j_{l+1} \geq i_l + 2 = j_l + 2$, a contradiction. This shows that indeed the exponents of $u_1$ are bounded by $k$, and $v \preceq u_1$, and hence $v \in B_S^\infty(u_1)^{\leq k}$. Since $\delta(u_1, v) < \delta(u, v)$, we may assume by induction that $v \in B_S^k(u_1) \subset B_S^k(u)$, because $u_1 \in B_S^k(u)$. □

An immediate consequence of Lemma 2.4 is

**Corollary 2.5.** *Let $u, v$ be monomials of degree $d$ and assume that $\nu_i(u), \nu_i(v) \leq k$ for $i = 1, \ldots, n$. Let $u = x_{j_1} \cdots x_{j_d}$ with $j_1 \leq j_2 \leq \ldots \leq j_d$ and $v = x_{i_1} \cdots x_{i_d}$ with $i_1 \leq i_2 \leq \cdots \leq i_d$. Then $v \prec_k u$ if and only if $i_r \leq j_r$ for $r = 1, \ldots, d$.*

If $L \subset [n]$, $S' = K[x_i \colon i \in L]$ and $u_1, \ldots, u_m \in S'$ are monomials with $\nu_i(u_j) \leq k$ for $i \in L$ and $j = 1, \ldots, m$. Then $B_{S'}^k(u_1, \ldots, u_m)$ is defined in a similar way as in the case $S' = S$. For example, if $n = 6$ and $L = \{2, 4, 6\}$, then $S' = K[x_2, x_4, x_6]$. Let $u = x_4^2 x_6 \in S'$ and $k = 2$, then $B_{S'}^2(u) = \{x_2^2 x_4, x_2^2 x_6, x_2 x_4 x_6, x_2 x_4^2, x_4^2 x_6\}$.

## 3. Associated prime ideals of $B_S(u)^k$

In this paper we mainly study squarefree principal Borel ideals. This class of ideals behave well under localization, see [3, Theorem 1.2].

**Theorem 3.1.** *Let $u \in S$ be a squarefree monomial of degree $d$, $u = x_{i_1} x_{i_2} \cdots x_{i_d}$ with $i_1 < i_2 < \ldots < i_d$, and let $j$ be an integer with $i_{k-1} < j \leq i_k$, where we set $i_0 = 0$. Then $B_S(u)_{(j)} = B_{S_{(j)}}(v)$, where $v = u / x_{i_k}$.*

**Corollary 3.2.** *Let $u \in S$ be a squarefree monomial, $I = B_S(u)$ and $P \subset S$ a monomial prime ideal. Then $I(P)$ is a squarefree principal Borel ideal in $S(P)$.*

**Proof.** Let $P = P_T$. Then $I(P_T)$ is the ideal in $S(P_T)$ which is obtained from $I$ by the substitutions $x_j \mapsto 1$ for $j \in T$. Thus the result follows by repeated application of Theorem 3.1. □

A graded ideal $I \subset S$ is called *normally torsionfree*, if $\mathrm{Ass}(I) = \mathrm{Ass}(I^k)$ for all $k \geq 1$. Moreover, we call $I$ *almost normally torsionfree*, if $\mathrm{Ass}(I^k) \subseteq \mathrm{Ass}(I) \cup \{\mathfrak{m}\}$ for all $k$, where $\mathfrak{m} = (x_1, x_2, \ldots, x_n)$ is the graded maximal ideal of $S$.

**Corollary 3.3.** *Let $u = x_i x_n$, Then $B_S(u)$ is almost normally torsionfree. Moreover, $B_S(u)$ is normally torsionfree, if and only if $i = 1$.*

**Corollary 3.4.** *Let $u = x_{i_1} \cdots x_{i_{d-1}} x_n$, and let $d \geq 3$. Then the following conditions are equivalent:*

(a) *$B_S(u)$ is normally torsionfree.*
(b) *$B_S(u)$ is almost normally torsionfree.*
(c) *$u = x_1 x_2 \cdots x_{d-1} x_n$.*

In order to prove Corollary 3.3 and Corollary 3.4, we need the following lemmata.

**Lemma 3.5.** *Let $I$ be normally torsion free and let $w$ be a monomial such that* $\mathrm{Supp}(w) \cap \mathrm{Supp}(u) = \emptyset$ *for all $u \in G(I)$. Then $wI$ is normally torsion free.*

**Proof.** The assumption imply that

$$\mathrm{Ass}((wI)^k) = \mathrm{Ass}(I^k) \cup \mathrm{Ass}((w)^k) = \mathrm{Ass}(I^k) \cup \mathrm{Ass}((w)). \tag{2}$$

It follows that $\mathrm{Ass}((wI)^k) = \mathrm{Ass}(wI)$ for all $k \geq 1$ if and only if $\mathrm{Ass}(I^k) = \mathrm{Ass}(I)$ for all $k \geq 1$. □

**Lemma 3.6.** *Let $I$ be a monomial ideal. Then $I$ is almost normally torsionfree if and only if $I_{(j)}$ is normally torsionfree for all $j$.*

**Proof.** The proof follows from the fact that, similarly to ordinary localizations, one has

$$P \in \mathrm{Ass}_{S_{(j)}}((I_{(j)})^k) \text{ if and only if } PS \in \mathrm{Ass}_S(I^k) \text{ and } P \subset P_{\{j\}}, \tag{3}$$

see [16, Lemma 1.3]. □

**Proof of Corollary 3.3.** Note $B_S(u)_{(j)} = B_{S_{(j)}}(v)$, where $v$ is a monomial of degree 1. Therefore, $B_{S_{(j)}}(v)$ is normally torsionfree. By Lemma 3.6, it follows that $B_S(u)$ is almost normally torsionfree.

If $i_1 = 1$. Then Lemma 3.5 implies that $\mathrm{Ass}(B_S(u)^k) = (x_1) \cup (x_2, \ldots, x_n) = \mathrm{Min}(I)$. Therefore, $B_S(u)$ is normaly torsionfree. □

**Proof of Corollary 3.4.** (a) $\Rightarrow$ (b) is trivial. (b) $\Rightarrow$ (c) We prove the assertion by induction on $d$. Let $d = 3$ and $u = x_{i_1}x_{i_2}x_n$ with $i_1 < i_2 < n$. The assumption implies that $B_S(u)_{(1)} = B_{S_{(1)}}(v)$ is normally torsionfree, see Lemma 3.6. Here $v = x_{i_2x_n}$. Now Corollary 3.3 implies that $i_2 = 2$, and therefore and $u = x_1x_2x_n$.

Now let $d > 3$, and let $u = x_{i_1}x_{i_2}\cdots x_{i_{d-1}x_n}$ with $i_1 < i_2 < \ldots < i_{d-1} < n$. Then Theorem 3.1 implies that $B_S(u)_{(1)} = B_{S_{(1)}}(v)$, where $v = x_{i_2}\cdots x_{i_{d-1}}x_n$. By Lemma 3.6 it follows that $B_{S_{(1)}}(v)$ is normally torsionfree. Hence by induction hypothesis, we have $v = x_2x_3\cdots x_{d-1}x_n$. Therefore, $u = x_{i_1}x_2\cdots x_{d-1}x_{i_d}$. This implies that $i_1 = 1$, and proves (c).

(c) $\Rightarrow$ (a) Since $u = x_1x_2\cdots x_{d-1}x_n$, it follows that $B_S(u) = w(x_d,\ldots,x_n)$, where $w = x_1x_2\cdots x_{d-1}$. Therefore, Lemma 3.5 implies that $B_S(u)$ is normally torsionfree. □

Let $u \in S$ be a squarefree monomial. Then $B_S(u)$ does not have any embedded prime ideals, because $B_S(u)$ is a squarefree monomial ideal, and hence a radical ideal. In other words, $\mathrm{Ass}(B_S(u)) = \mathrm{Min}(B_S(u))$. Here, for any ideal $I \subset S$, $\mathrm{Min}(I)$ denotes the set of minimal prime ideals of $I$.

Let $B_S(u)^\vee$ denote the Alexander dual of $B_S(u)$. Then $v = x_{i_1}\cdots x_{i_k} \in G(B_S(u)^\vee)$, if and only if $(x_{i_1},\ldots,x_{i_k}) \in \mathrm{Min}(B_S(u))$, see for example [14, Theorem 1.4.6]. Therefore, the set $\mathrm{Min}_S(B_S(u))$ is determined, once we know the set $G(B_S(u)^\vee)$. In [12, Theorem 3.18], $G(B_S(u)^\vee)$ has been computed.

**Theorem 3.7** (Francisco, Mermin, Schweig). *Let $u \in S$, $u = x_{i_1}\cdots x_{i_d}$ be a monomial with $i_1 < i_2 < \ldots < i_d$. Then $B_S(u)^\vee$ is a squarefree strongly stable ideal with Borel generators $x_r\cdots x_{i_r}$ for $r = 1,\ldots,d$.*

This theorem has been generalized to $t$-spread principal Borel ideals, see [1, Theorem 1.2].

Let $I \subset S$ be any ideal. Recall that

$$\mathrm{height}(I) = \min\{\mathrm{height}(P)\colon\ P \in \mathrm{Ass}(I)\},$$

and

$$\mathrm{bigheight}(I) = \max\{\mathrm{height}(P)\colon\ P \in \mathrm{Ass}(I)\}.$$

As an immediate consequence of Theorem 3.7 we obtain

**Corollary 3.8.** *Let $u \in S$, $u = x_{i_1}\cdots x_{i_d}$ be a monomial with $i_1 < i_2 < \ldots < i_d$ and let $I = B_S(u)$. Then*

(a) $\mathrm{height}(I) = i_1$ *and* $\mathrm{bigheight}(I) = i_d - d + 1$.

(b) *The following conditions are equivalent:*

(i) $\text{height}(I) = \text{bigheight}(I)$.
(ii) $i_j = i_1 + j - 1$ *for* $j = 1, \ldots, d$.
(iii) $S/I$ *is Cohen-Macaulay.*

**Proof.** (i) $\iff$ (ii) We have

$$i_1 \le i_2 - 1 \le \ldots \le i_j - j + 1 \le i_{j+1} - (j+1) + 1 \le \ldots \le i_d - d + 1.$$

From (a) it follows that $i_1 = i_d - d + 1$. Therefore, $i_j - j + 1 = i_{j+1} - (j+1) + 1$ for all $j$ which implies that $i_{j+1} = i_j$ for all $j$. This yields the desired conclusion.

(iii) $\Rightarrow$ (i) Since $S/I$ is Cohen-Macaulay, the ideal $I$ is unmixed, see for example [6, Theorem 2.1.6]. Since $I$ is unmixed if and only if $\text{height}(I) = \text{bigheight}(I)$, the assertion follows.

(ii) $\Rightarrow$ (iii) By Theorem 3.7, $I^\vee = B_S(x_d \cdots x_{i_d})$ for ideals satisfying condition (ii). It is known that $B_S(x_d \cdots x_{i_d})$ has a linear resolution, see [1, Proposition 2.4]. Therefore, $S/I$ is Cohen-Macaulay, see for example [14, Theorem 8.1.9]. □

The following results [3, Theorem 2.1 and Corollary 2.3] is also important for this paper. We denote by $\mathfrak{m}$ the graded maximal ideal of $S = K[x_1, \ldots, x_n]$. For a monomial $v \neq 1$ we set $\min(v) = \min\{j : j \in \text{supp}(v)\}$ and $\max(v) = \max\{j : j \in \text{supp}(v)\}$. If $v = 1$, we set $\min(v) = \max(v) = 0$.

**Theorem 3.9** (Aslam). *Let* $u \in S = K[x_1, \ldots, x_n]$ *be a squarefree monomial of degree* $d$ *and let* $I = B_S(u)$. *Then* $\mathfrak{m} \in \text{Ass}(I^k)$ *for some* $k$ *if and only if* $\min(u) > 1$ *and* $\max(u) = n$. *If this is the case, then* $\mathfrak{m} \in \text{Ass}(I^k)$ *for all* $k \ge d$.

An alternate proof of this fact is given in Proposition 4.6.

By Brodmann ([4] and [5]), for any graded ideal $I \subset S = K[x_1, \ldots, x_n]$, there exists an integer $k_0$ such that $\text{Ass}(I^k) = \text{Ass}(I^{k_0})$ for all $k \ge k_0$. We set $\text{Ass}^\infty(I) = \text{Ass}(I^{k_0})$.

In our case, $I = B_S(u)$ with $u = x_{i_1} \cdots x_{i_d}$ and $i_1 < i_2 < \ldots < i_d$. Therefore, $I$ is a squarefree monomial ideal, and hence $\text{Ass}(I) = \text{Min}(I)$. Since $\text{Min}(I) \subset \text{Ass}^\infty(I)$, it follows in our case that $\text{Ass}(I) \subset \text{Ass}^\infty(I)$. In Corollary 3.4, we have seen that $\text{Ass}(I) = \text{Ass}^\infty(I)$ if and only if $u = x_1 x_2 \cdots x_{d-1} x_{i_d}$.

Let $T \subset [n]$. By using Theorem 3.1 and induction on $|T|$ it follows that there exists a (unique) monomial $u_T \subset S_T$ of degree $d - |T|$ such that $B_S(u)_T = B_{S_T}(u_T)$.

For example, if $u = x_2 x_5 x_6 x_7 x_9$ and $T = \{3, 5, 9\}$, then

$$u_T = u_{359} = (u_9)_{35} = ((x_2 x_5 x_6 x_7)_5)_3 = (x_2 x_6 x_7)_3 = x_2 x_7.$$

Note that the monomial prime ideal $P$ belongs to $\text{Ass}^\infty(I)$ if and only if $\mathfrak{m}_P \in \text{Ass}_{S(P)}(I(P)^k)$, for all $k \gg 0$. Here $\mathfrak{m}_P = PS(P)$ denotes the graded maximal ideal of $S(P)$. By Corollary 3.2 the monomial localization $I(P)$ of the squarefree principal monomial $I$ is again a squarefree principal monomial ideal.

A graded ideal $I \subset S$ is said to satisfy the *persistence property*, if $\mathrm{Ass}_S(I^k) \subset \mathrm{Ass}_S(I^{k+1})$ for all $k \geq 1$. Since squarefree principal monomial ideals satisfy the persistence property (see [1, Corollary 2.6]), we conclude that $P$ belongs to $\mathrm{Ass}^\infty(I)$ if and only if $\mathfrak{m}_P \in \mathrm{Ass}_{S(P)}(I(P)^k)$ for some $k$. From this observation together with Theorem 3.9 one obtains [3, Theorem 3.2].

**Theorem 3.10** (Aslam). *Let $T$ be any subset of $[n]$ and assume that $\min(u) > 1$. Then $P_T \in \mathrm{Ass}^\infty(B_S(u))$ if and only if $\min(T^c) < \min(u_T)$ and $\max(u_T) = \max(T^c)$.*

For the next result we need

**Lemma 3.11.** *Suppose $\min(u) > 1$. Then $\min(T^c) < \min(u_T)$ if and only if $u_T \neq 1$.*

**Proof.** It is enough to show that $\min(T^c) < \min(u_T)$ if $u_T \neq 1$. The other direction is trivial. We use induction on $|T|$. If $|T| = 0$, the assertion is trivial, since $\min(u) > 1$ and $T^c = [n]$. Now let $|T| > 0$. Note that $T \neq [n]$, because we assume that $u_T \neq 1$. This means that $T$ is a non-empty proper subset of $[n]$. Let $u = x_{i_1}x_{i_2}\cdots x_{i_d}$ with $i_1 < i_2 < \ldots < i_d$ and let $T = \{j_1, j_2, \ldots, j_s\}$ with $j_1 < j_2 < \ldots < j_s$. Let $r$ be the biggest number such that $j_i = i$ for $i = 1, \ldots, r-1$. Then $\min(T^c) = r$, and

$$u_T = (x_{i_r}\cdots x_{i_d})_{\{j_r,\ldots,j_s\}}$$

by Theorem 3.1. Therefore, $\min(u_T) \geq i_r \geq r+1$. Thus we see that $\min(T^c) = r < r+1 \leq \min(u_T)$, as desired. □

**Corollary 3.12.** *Let $u = x_{i_1}\cdots x_{i_d}$ be a monomial in $S = K[x_1, \ldots, x_n]$ with $i_1 < i_2 < \ldots < i_d$, and $T = \{j_1, \ldots, j_s\}$ with $j_1 < j_2 < \ldots < j_s$. Suppose that $i_j = j$ for $j = 1, \ldots, r$ and $i_{r+1} > r+1$. Then*

$$\mathrm{Ass}^\infty(B_S(u)) = \{(x_1), \ldots, (x_r)\} \cup \{P_T \colon [r] \subseteq T, u_T \neq 1, \max(T^c) = \max(u_T)\}.$$

*Moreover, $u_T \neq 1$ if and only if $s < d$ or $s \geq d$ and $j_l > i_l$ for some $l$.*

**Proof.** By assumption we have $u = x_1\cdots x_r u'$, where $u' = x_{i_{r+1}}\cdots x_{i_d}$ with $i_{r+1} > r+1$. Therefore, $B_S(u) = x_1\cdots x_r B_{S'}(u')S$, where $S' = K[x_{r+1}, \ldots, x_n]$. Hence (2) implies that $\mathrm{Ass}^\infty(B_S(u)) = \{(x_1), \ldots, (x_r)\} \cup \mathrm{Ass}^\infty(B_{S'}(u'))S$. Since $\min(u') > r+1$, we may apply Theorem 3.10 to $u'$ in $S'$. Therefore, the result follows from Theorem 3.10 and Lemma 3.11. □

**Corollary 3.13.** *Let $u \in S$, $u = x_{i_1} \cdots x_{i_d}$ be a monomial with $i_1 < i_2 < \ldots < i_d$ and let $I = B_S(u)$. Suppose that $i_j = j$ for $j = 1, \ldots, r$ and $i_{r+1} > r+1$. Then* $\text{bigheight}(I^l) = i_d - r$ *for $l \gg 0$.*

**Proof.** Since $l \gg 0$, it follows that $\text{bigheight}(I^l) = \max\{\text{height } P_T \colon P_T \in \text{Ass}^\infty(I)\}$. The ideal $P_T$ for $T = [r]$ has height $i_d - r$, and this the largest possible height among the ideals $P_T$. Since $u_T = x_{i_{r+1}} \cdots x_{i_d}$, Corollary 3.12 implies that $P_T \in \text{Ass}^\infty(I)$. □

**Example 3.14.** Let $u = x_1x_3x_5$ and $I = B_S(u)$. Then

$$\{T \colon [1] \subseteq T, u_T \neq 1, \max(T^c) = \max(u_T)\} = \{\{1\}, \{1,2\}, \{1,3\}, \{1,4,5\}\}.$$

Hence

$$\text{Ass}^\infty(I) = \{(x_1)\} \cup \{(x_2, x_3, x_4, x_5), (x_3, x_4, x_5), (x_2, x_4, x_5), (x_2, x_3)\}.$$

## 4. The socle of $S/B_S(u)^k$

In this section we determine the powers of $B_S(u)$ and their socle. We first recall a result of De Negri [7, Proposition 3.4], and present a proof it by using sortability. This makes the proof substantially shorter.

**Theorem 4.1.** *Let $u \in S$ be a squarefree monomial. Then $B_S(u)^k = B_S^k(u^k)$ for all $k \geq 1$. In particular, $B_S(u)^k$ is $k$-strongly stable.*

**Proof.** Let $u = x_{l_1} \cdots x_{l_d}$ with $l_1 \leq l_2 \leq \cdots \leq l_d$, and suppose that $v \in B_S^k(u^k)$, and let $v = x_{i_1} \cdots x_{i_{kd}}$ with $i_1 \leq i_2 \leq \cdots \leq i_{kd}$. Since $v \preceq_k u^k$ it follows that $i_{(r-1)k+1}, \ldots, i_{rk} \leq l_r$ for $r = 1, \ldots, d$, see Corollary 2.5.

For $j = 1, \ldots, k$, let $v_j = \prod_{r=1}^d x_{i(r-1)k+j}$. Then $v = v_1 \cdots v_k$ and $v_j \preceq u$. It remains that to be shown that each $v_j$ is squarefree. Suppose $v_j$ is not squarefree. Then there exists $r$ with $1 \leq r < d$ such that $x_{i_{(r-1)k+j}} = x_{i_{rk+j}}$. Hence $i_{(r-1)k+j} = i_{(r-1)k+j+1} = \cdots = i_{rk+j}$. This implies that $\nu_{i(r-1)k+j}(v) > k$, a contradiction.

Conversely, let $v \in B_S(u)^k$. By [1, Proposition 2.4], $G(B_S(u))$ is sortable. Therefore, $v = v_1v_2 \cdots v_k$ with $v_j \in B_S(u)$ and $(v_i, v_j)$ is sorted for all $i, j$ with $1 \leq i < j \leq k$. It follows from [9, (6.3)], that if $v_1 = x_{i_1} \cdots x_{i_d}$ with $i_1 < i_2 < \ldots < i_d$, $v_2 = x_{j_1} \cdots x_{j_d}$ with $j_1 < j_2 < \ldots < j_d$, $\ldots$, $v_k = x_{s_1} \cdots x_{s_d}$ with $s_1 < s_2 < \ldots < s_d$, then $i_1 \leq j_1 \leq \ldots \leq s_1 \leq i_2 \leq j_2 \leq \ldots \leq s_2 \leq \ldots \leq i_k \leq j_k \leq \ldots \leq s_k$. From this it follows that $v \in B_S^\infty(u^k)$. Since each $v_j$ is squarefree it follows that $v \in B_S^k(u^k)$. □

For any monomial ideal we denote by $\mathrm{Soc}(S/I)$ the (finite) set of monomial $v \in S$ with $v \notin I$ and $vx_j \in I$ for $j = 1, \ldots, n$. Note that $\operatorname{depth} S/I = 0$ if and only if $\mathrm{Soc}(S/I) \neq \emptyset$.

**Corollary 4.2.** *Let $u \in S$ be a squarefree monomial of degree $d$, and let $I = B_S(u)$. Suppose that* $\operatorname{depth} S/I^k = 0$*. Then* $\mathrm{Soc}(S/I^k)$ *is a $(k-1)$-stable set of monomials generated in degree $kd-1$.*

**Proof.** By [1, Proposition 2.4] the ideal $I^k$ has $kd$-linear resolution. Suppose now that $\operatorname{depth} S/I^k = 0$. Then the graded minimal free resolution of $S/I^k$ is of the form

$$0 \to F_n \to F_{n-1} \to \cdots \to F_1 \to F_0 \to S/I^k \to 0,$$

with $F_0 = S$ and $F_i = S(-kd-i+1)^{\beta_i}$ for $i = 1, \ldots, n$. So $F_n = S(-kd-n-1)^{\beta_n}$. This shows that all elements of $\mathrm{Soc}(S/I^k)$ have degree $(kd+n-1)-n = kd-1$. Indeed, $F_n/\mathfrak{m}F_n$ and the Koszul homology $H_n(x_1, \ldots, x_n; S/I^k)$ are isomorphic as the graded $K$-vector spaces. The generators of $H_n(x_1, \ldots, x_n; S/I^k)$ are $ve_1 \wedge \ldots \wedge e_n$ with $v \in \mathrm{Soc}(S/I^k)$. This gives us the above formula for the degree of the socle elements.

By [15, Corollary 1.2], $\nu_i(v) \leq k-1$ for all $v \in \mathrm{Soc}(S/I^k)$ and $i = 1, \ldots, n$. Thus it remains to be shown that if $v \in \mathrm{Soc}(S/I^k)$, $x_j|v$, $i < j$ and $\nu_i(x_iv) \leq k-1$, then $v_0 := x_i(v/x_j) \in \mathrm{Soc}(S/I^k)$, that is, $x_lv_0 \in I^k$ for $l = 1, \ldots, n$. Indeed, if $l = j$, then $x_lv_0 = x_iv \in I^k$. If $l \neq j$, then $x_lv_0 = x_i(x_lv)/x_j \in I^k$ because $x_lv \in I^k$ and $I^k$ is $k$-stable, see Theorem 4.1. □

Let $1 < d < n$ be integers and for $k \geq 2$ let $c_k, r_k$ be integers such that

$$kd - 1 = c_k(k-1) + r_k \quad \text{with} \quad 0 \leq r_k < k-1. \tag{4}$$

Then we define the monomial $w_k = x_1^{k-1} \cdots x_{c_k}^{k-1} x_{c_k+1}^r$. Note $w_k \in S$ if and only if $c_k \leq n$ when $r = 0$, and $c_k + 1 \leq n$ if $r \neq 0$.

**Corollary 4.3.** *Let $u \in S$ be a squarefree monomial of degree $d$, and let $I = B_S(u)$. Then* $\operatorname{depth} S/I^k = 0$ *if and only if $w_kx_n \preceq u^k$.*

**Proof.** If $\operatorname{depth} S/I^k = 0$, then $\mathrm{Soc}(S/I^k)$ is a non-empty $(k-1)$-stable set. Let $v \in \mathrm{Soc}(S/I^k)$. Then $\{w\colon\ w \preceq_{k-1} v\} \subseteq \mathrm{Soc}(S/I^k)$. Since $w_k \preceq_{k-1} v$ it follows that $w_k \in \mathrm{Soc}(S/I^k)$. Therefore, $w_kx_n \in I^k = B_S^k(u^k)$. This implies that $w_kx_n \preceq_k u^k$, and hence $w_kx_n \preceq u^k$, see Remark 2.3.

Conversely, suppose that $w_kx_n \preceq u^k$. Therefore, $x_iw_k = x_i(w_kx_n)/x_n \preceq u^k$. Since the exponents of $w_k$ are bounded by $k-1$, the exponents of $x_iw_k$ are bounded by $k$. Therefore, by Remark 2.3, $x_i w_k \preceq_k u^k$ for all $i$. This means that $x_i w_k \in I^k$ for all $i$. Hence $w_k \in \mathrm{Soc}(S/I^k)$, and so $\operatorname{depth} S/I^k = 0$. □

Note that $w_k x_n \preceq u^k$ only if $w_k \in S$.

**Remark 4.4.** Let $d < n$, $f_d(k) = \lceil (kd-1)/(k-1) \rceil$ and $w_k = x_1^{k-1} \cdots x_{c_k}^{k-1} x_{c_k+1}^r$. Then

(a) $f_d(k) = c_k$ if $r = 0$ and $f_d(k) = c_k + 1$ if $r \neq 0$. Therefore, $w_k \in S$ if and only if $f_d(k) \leq n$.
(b) The function $f_d(k)$ is a non-increasing function with $f_d(k) \leq n$ for $k \geq d$.
(c) Let $k_0 = \min\{k\colon\ f_d(k) \leq n\}$. Then $k_0 \leq d$.

**Proof.** (a) follows from the definition of $c_k$, see (4).

(b) It is obvious that $f_d(k)$ is a non-increasing function. Since $f_d(d) = d+1 \leq n$, we also have $f_d(k) \leq n$ for $k \geq d$.

(c) follows from (b). □

**Corollary 4.5.** *Let $k_0$ be defined as in Remark 4.5. Then* $\min\{k\colon\ \operatorname{depth} S/I^k = 0\} \geq k_0$. *In particular, this lower bound for* $\min\{k\colon\ \operatorname{depth} S/I^k = 0\}$ *depends only on $d$ and $n$.*

In general the inequality $\min\{k\colon\ \operatorname{depth} S/I^k = 0\} \geq k_0$ may be strict, as the following example shows: Let $u = x_2x_3x_4x_5x_6 \in S = K[x_1, \ldots, x_6]$. Then $d = 5$ and $n = 6$. Then $f_5(2) = 9$, $f_5(3) = 7$ and $f_5(4) = 6$. Therefore, $k_0 = 4$, but on the other hand 5 is the smallest number $k$ for which $\operatorname{depth} S/I^k = 0$.

We use this characterization of the socle elements to show

**Proposition 4.6.** *Let $I = B_S(x_{i_1}x_{i_2}\cdots x_{i_d}) \subset S = K[x_1, \ldots, x_n]$ with $1 < i_1 < i_2 < \ldots < i_d = n$. Then $\mathfrak{m} \in \mathrm{Ass}(I^k)$ for $k \geq d$.*

**Proof.** It is enough to show that $\mathfrak{m} \in \mathrm{Ass}(I^d)$, since $I$ satisfies the persistence property, see [1, Proposition 2.4].

Note that $w_d = x_1^{d-1} \cdots x_{d+1}^{d-1}$. We show that $w_d x_n \in I^d$. From Corollary 4.5 it then follows that $\operatorname{depth}(S/I^d) = 0$ which implies that $\operatorname{dstab}(I) \leq d$. To see that $w_d x_n \in I^d$, we must show that $w_d x_n \prec_d u^d$, where $u = x_{i_1} \cdots x_{i_d}$ with $1 < i_1 < i_2 < \ldots i_d = n$. Since $v = x_2 \cdots x_d x_n \prec_1 u$ it follows $v^d \prec_d u^d$. Therefore, it suffices to show that $w_d x_n \prec_d v^d$.

We write $w_d x_n = x_{k_1} \ldots x_{k_{d^2}}$ with $k_1 \leq k_2 \leq \ldots \leq k_{d^2}$ and $v^d = x_{l_1} \cdots x_{l_{d^2}}$ with $l_1 \leq l_2 \leq \ldots \leq l_{d^2}$. Since $\nu_r(w_d x_n), \nu_r(v^d) \leq d$ for all $r$, we must show that $k_r \leq l_r$

for $r = 1, \ldots, d^2$, see Corollary 2.5. Let $r$ be an integer with $1 \le r \le d^2$. We may assume that $r \le (d-1)d$, because $l_r = n$, for $(d-1)d < r \le d^2$. Let

$$r - 1 = qd + t_1 \text{ with } 0 \le t_1 \le d-1, \text{ and } r - 1 = q'(d-1) + t_2 \text{ with } 0 \le t_2 \le d-2.$$

Then $l_r = q + 2$ and $k_r = q' + 1$. Now $k_r \le l_r$ if and only if $q' - q \le 1$. Indeed, we have $qd + t_1 = q'(d-1) + t_2$. This implies that $d(q'-q) = t_1 - t_2 + q' < 2d$. Therefore, $q' - q < 2$, as desired. □

## 5. Comparison of $\operatorname{astab}(B_S(u))$ with $\operatorname{dstab}(B_S(u))$

As mentioned in Section 3, for any graded ideal $I \subset S = K[x_1, \ldots, x_n]$ there exists an integer $k_1$ such that $\operatorname{Ass}_S(I^k) = \operatorname{Ass}_S(I^{k_1})$ for all $k \ge k_1$. The smallest integer $k_1$ with this property is denoted by $\operatorname{astab}(I)$. Similarly there exists an integer $k_2$ such that $\operatorname{depth} S/I^k = \operatorname{depth} S/I^{k_2}$ for all $k \ge k_2$. The smallest integer $k_2$ with this property is denoted by $\operatorname{dstab}(I)$. The purpose of the section is to compute $\operatorname{dstab}(I)$ and $\operatorname{astab}(I)$ when $I$ is a squarefree principal Borel ideal.

Let $I = B_S(x_{i_1}x_{i_2}\cdots x_{i_d}) \subset S = K[x_1, \ldots, x_n]$ with $i_1 < i_2 < \ldots < i_d$. Assume that $i_d = m \le n$, and let $S' = K[x_1, \ldots, x_m]$ and $J = B_{S'}(x_{i_1}x_{i_2}\cdots x_{i_d})$. Then, we obviously one gets

**Lemma 5.1.** *With the assumptions and notation introduced, we have*

$$\operatorname{depth} S/I^k = \operatorname{depth} S'/J^k + n - m \quad \textit{for all} \quad k,$$

*and*

$$\operatorname{Ass}_S(I^k) = \{PS : P \in \operatorname{Ass}_{S'}(J^k)\} \quad \textit{for all} \quad k.$$

*In particular,* $\operatorname{dstab}(I) = \operatorname{dstab}(J)$ *and* $\operatorname{astab}(I) = \operatorname{astab}(J)$.

Therefore, for the rest of this section, if not otherwise stated, we may assume that $i_d = n$.

Next we show

**Lemma 5.2.** *Let $I = B_S(u) \subset S = K[x_1, \ldots, x_n]$, where $u = x_{i_1}x_{i_2}\cdots x_{i_d}$ with $1 = i_1 < i_2 < \ldots < i_d = n$. Let $J = B_{S'}(u') \subset S' = K[x_2, \ldots, x_{n-1}]$, where $u' = u/x_1$. Then for all $k$,*

$$\operatorname{Ass}_S(I^k) = \{PS\colon\ P \in \operatorname{Ass}_{S'}(J^k)\} \cup (x_1) \quad \textit{and} \quad \operatorname{depth} S/I^k = \operatorname{depth} S'/J^k + 1.$$

*In particular,* $\operatorname{dstab}(I) = \operatorname{dstab}(J)$ *and* $\operatorname{astab}(I) = \operatorname{astab}(J)$.

**Proof.** We observe that $I^k = x_1^k \tilde{J}^k$ for all $k$, where $\tilde{J} = JS$. Therefore, the statement about $\mathrm{Ass}_S(I^k)$ follows from (2). Since $(x_1^k) \cap \tilde{J}^k = x_1^k \tilde{J}^k$ we get the short exact sequence

$$0 \to S/I^k \to S/(x_1^k) \oplus S/\tilde{J}^k \to S/(x_1^k, \tilde{J}^k) \to 0.$$

By the depth lemma (see [8, Corollary 18.6]) we have

$$\begin{aligned} \operatorname{depth} S/I^k &\geq \min\{\operatorname{depth}(S/(x_1^k) \oplus S/\tilde{J}^k), \operatorname{depth} S/(x_1^k, \tilde{J}^k) + 1\} \\ &= \min\{\operatorname{depth} S/(x_1^k), \operatorname{depth} S/\tilde{J}^k, \operatorname{depth} S/(x_1^k, \tilde{J}^k) + 1\}. \end{aligned}$$

Since $x_1^k$ is regular on $S/\tilde{J}^k$, it follows that $\operatorname{depth} S/(x_1^k, \tilde{J}^k) = \operatorname{depth} S/\tilde{J}^k - 1$. Therefore, $\operatorname{depth} S/I^k \geq \operatorname{depth} S/\tilde{J}^k$.

By using again the depth lemma we also get

$$\operatorname{depth} S/\tilde{J}^k - 1 = \operatorname{depth} S/(x_1^k, \tilde{J}^k) \geq \min\{\operatorname{depth} S/(x_1^k), \operatorname{depth} S/\tilde{J}^k, \operatorname{depth} S/I^k - 1\}.$$

This implies that $\operatorname{depth} S/\tilde{J}^k \geq \operatorname{depth} S/I^k$, and hence $\operatorname{depth} S/I^k = \operatorname{depth} S/\tilde{J}^k$. Since $\tilde{J} = JS$, it follows that $\operatorname{depth} S/\tilde{J} = \operatorname{depth} S'/J$. This yields the desired conclusion. □

Because of Lemma 5.2 we may also assume for the rest of the section that $i_1 > 1$.

Combining Lemma 5.2 with Corollary 4.5 we obtain

**Proposition 5.3.** *Let $I = B_S(u) \subset S = K[x_1, \ldots, x_n]$ with $u = x_{i_1} x_{i_2} \cdots x_{i_d}$ and $1 < i_1 < i_2 < \ldots < i_d = n$. Then we have:*

(a) $\operatorname{dstab}(I) = \min\{k\colon\ \operatorname{depth} S/I^k = 0\}$.

(b) *Let $k_0$ be defined as in Corollary 4.5. Then*

$$\operatorname{dstab}(I) = \min\{k \geq k_0\colon\ w_k x_n \in I^k\}.$$

(c) $k_0 \leq \operatorname{dstab}(I) \leq \operatorname{astab}(I) \leq d$.

The following result tells us how dstab behaves under monomial localization.

**Theorem 5.4.** *Let $I = B_S(u) \subset S = K[x_1, \ldots, x_n]$ with $u = x_{i_1} x_{i_2} \cdots x_{i_d}$ and $1 < i_1 < i_2 < \ldots < i_d = n$. Then*

$$\operatorname{dstab}(I_{(a)}) \leq \operatorname{dstab}(I) \quad \textit{for} \quad a = 1, \ldots, n.$$

**Proof.** We first consider the case that $a \leq i_{d-1}$. Then $I_{(a)} = B_S(u)_{(a)} = B_{S_(a)}(u_a)$ and $\min(u_a) > \min(T^c)$ and $\max(u_a) = \max(T^c) = n$. Then Proposition 4.6 implies that $\operatorname{depth} S/I^k = 0$ for some $k$ and $\operatorname{depth} S_{(a)}/I_{(a)}^l = 0$ for some $l$. Let $k$ and $l$ be the smallest numbers with this property. Then $k = \operatorname{dstab}(I)$ and $l = \operatorname{dstab}(I_{(a)})$, and we want to show that $l \leq k$. For this it is enough to show that $\operatorname{depth} S_{(a)}/I_{(a)}^k =$

0. Since depth $S/I^k = 0$, Corollary 4.5 implies that $w_k x_n \in I^k$. By Theorem 4.1, $I^k = B_S^k(u_k)$. Therefore, $w_k x_n \preceq_k u^k$. Let $w_k x_n = x_{t_1(k)} x_{t_2(k)} \cdots x_{t_{kd}(k)}$ with $t_1(k) \le \ldots \le t_{kd}(k)$ and $u^k = x_{l_1(k)} x_{l_2(k)} \cdots x_{l_{kd}(k)}$ with $l_1(k) \le \cdots \le l_{kd}(k)$. By Corollary 2.5, $w_k x_n \preceq u^k$ if and only if $t_s(k) \le l_s(k)$ for $s = 1, \ldots, kd$. The numbers $l_s(k)$ and $t_s(k)$ can be computed.

With the notation introduced we have

$$l_s(k) = i_j, \quad \text{where} \quad j = \lceil s/k \rceil,$$

and

$$t_s(k) = \begin{cases} \lceil s/(k-1) \rceil, & \text{if } s \le (k-1)c; \\ c+1, & \text{if } (k-1)c < s < kd; \\ n, & \text{if } s = kd. \end{cases}$$

Let $b$ be the unique number with $i_{b-1} < a \le i_b$. Then $u_a = u/i_b \in S_{(a)}$ is of degree $d-1$, see Theorem 3.1. Let $w'_k \in S_{(a)}$ be the socle test element for $I^k_{(a)}$ in $S_{(a)}$. We have depth $S_{(a)}/I^k_{(a)} = 0$, once we have shown that $w'_k x_m \preceq_k u_a^k$.

Let

$$w'_k x_n = x_{t'_1(k)} x_{t'_2 k)} \cdots x_{t'_{k(d-1)}(k)} \quad \text{with} \quad t'_1(k) \le \ldots \le t'_{k(d-1)}(k),$$

and

$$u_a^k = x_{l'_1(k)} x_{l'_2(k)} \cdots x_{l'_{k(d-1)}(k)} \quad \text{with} \quad l'_1(k) \le \cdots \le l'_{k(d-1)}(k).$$

It remains to be shown that $t'_s(k) \le l'_s(k)$ for $s = 1, \ldots, k(d-1)$.

Note that

$$l'_s(k) = \begin{cases} l_s(k), & \text{if } s \le k(b-1); \\ l_{s+k}(k), & \text{if } k(b-1) < s \le k(d-1). \end{cases}$$

and

$$t'_s(k) = \begin{cases} t_s(k), & \text{if } s \le (k-1)(a-1); \\ t_{s+k-1}(k), & \text{if } (k-1)(a-1) < s \le k(d-1). \end{cases}$$

Note that $b \le a$, because $b - 1 \le i_{b-1} < a$. We consider different cases.

Case 1: $s \le (k-1)(a-1)$: Then, since $b \le a$, we have

$$\begin{aligned} t'_s(k) &= t_s(k) \le l_s(k) = \begin{cases} l'_s(k), & \text{if } s \le k(b-1); \\ l'_{s-k}(k), & \text{if } k(b-1) < s \le k(d-1), \end{cases} \\ &\le l'_s(k), \end{aligned}$$

because $l'_s(k)$ is a non-decreasing function.

Case 2: $s > (k-1)(a-1)$: Then

$$t'_s(k) = t_{s+k-1}(k) \le l_{s+k-1}(k) = \begin{cases} l'_{s+k-1}(k), & \text{if } s \le k(b-1); \\ l'_{s-1}(k), & \text{if } k(b-1) < s \le k(d-1), \end{cases}$$

Therefore, if $s > k(b-1)$, then $t'_s(k) \le l'_s(k)$.

Suppose now that $(k-1)(a-1) < s < k(b-1)$. Then $t_s(b) = \lceil s/(k-1) \rceil \geq a$ and $l_s(k) = i_j$, where $j = \lceil s/k \rceil$. Therefore, $l_s(k) = i_c$ with $c \leq b-1$. Hence,

$$a \leq t_s(k) \leq l_s(k) = i_c \leq i_{b-1},$$

a contradiction. So this case cannot happen.

It remains to treat the case that $a > i_{d-1}$. Then $u_a = x_{i_1} \cdots x_{i_{d-1}}$, and $I_{(a)} = B_{S_{(a)}}(u_a)$. Let $J = B_{S'}(u_a)$, where $S' = k[x_1, \ldots, x_{i_{d-1}}]$. Then, by Lemma 5.1, $\operatorname{dstab}(J) = \operatorname{dstab}(I_{(a)})$, and furthermore by Proposition 5.3 we have $\operatorname{dstab}(J) = \min\{k\colon \ \operatorname{depth} S'/J^k = 0\}$. Therefore, we must show that $\operatorname{depth} S'/J^k = 0$ if $\operatorname{depth} S/I^k = 0$. Since $\operatorname{depth} S/I^k = 0$ it follows $w_k x_n \preceq_k u^k$. Let $w_k = \prod_{s=1}^{kd-1} x_{t_s(k)}$. Then the socle test element for $J^k$ is $w'_k = \prod_{s=1}^{k(d-1)-1} x_{t_s(k)}$. Then it is clear that $w'_k x_{i_{d-1}} \preceq_k u_a^k$. This shows that $\operatorname{depth} S'/J^k = 0$, as desired. □

As the main result of the section we show

**Theorem 5.5.** *Let $S = K[x_1, \ldots, x_n]$ be the polynomial ring in $n$ variables, let $u \in S$ be a squarefree monomial ideal and let $I = B_S(u)$. Then* $\operatorname{astab}(I) = \operatorname{dstab}(I)$.

**Proof.** Let $u = x_{i_1} \cdots x_{i_d}$ with $i_1 < i_2 < \cdots < i_d$. By the discussion at the beginning of this section and by Lemma 5.2 we may assume that $1 < i_1$ and $i_d = n$. Let $k = \operatorname{dstab}(I)$. By Proposition 5.3, we have $\mathfrak{m} \in \operatorname{Ass}_S(I^k)$. Now let $P \in \operatorname{Ass}_S^\infty(I)$ with $P \neq \mathfrak{m}$. Then there exists $j$ and $P' \in \operatorname{Ass}_{S_{(j)}}^\infty(I_{(j)})$ with $P = P'S$. By induction we may assume that $\operatorname{astab}(I_{(j)}) = \operatorname{dstab}(I_{(j)})$. Thus, if $k' = \operatorname{dstab}(I_{(j)})$, then $P' \in \operatorname{Ass}_{S_{(j)}}(I_{(j)}^{k'})$. Therefore, $P \in \operatorname{Ass}_S(I^{k'})$, by (3). Theorem 5.4 implies that $k' \leq k$. Since squarefree principal monomial ideals satisfy the persistence property, we conclude that $P \in \operatorname{Ass}_S(I^k)$, as desired. This shows that $\operatorname{astab}(I) \leq \operatorname{dstab}(I)$. The other inequality is shown in Proposition 5.3. □

## 6. On the depth of $S/B_S(u)^k$

We provide some partial results regarding $\operatorname{depth} S/I^k$ for $I = B_S(u)$. Since all powers have a linear resolution it follows that $\operatorname{depth} S/I^{k+1} \leq \operatorname{depth} S/I^k$ for all $k \geq 1$, as mentioned in the proof of Proposition 5.3. Actually all powers of $I$ have linear quotients for a suitable order, as shown in [1, Proposition 2.4]. Here we show

**Proposition 6.1.** *Let $I = B_S(u)$. Then for all $k$, the ideal $I^k$ has linear quotients with respect to the lexicographic order induced by $x_1 > x_2 > \cdots > x_n$.*

**Proof.** Let $v, w \in I^k$ with $w > v$ with respect to the lexicographic order. Let $v = x_{j_1} x_{j_2} \cdots x_{j_{kd}}$ with $j_1 \leq j_2 \leq \ldots \leq j_{kd}$, and $w = x_{l_1} x_{l_2} \cdots x_{l_{kd}}$ with $l_1 \leq l_2 \leq \ldots \leq l_{kd}$. Since $w > v$ there exists an integer $r$ such that $l_s = j_s$ for $s = 1, \ldots, r-1$

and $j_r > l_r$. Then $x_{l_r}$ divides $w/gcd(w,v)$. Let $w' = x_{l_r}v/x_{j_r}$. Since $l_r < j_r$ it follows that $w' > v$ and $w'/\gcd(w',v) = x_{l_r}$ which divides $w/gcd(w,v)$. It remains to be shown that $w' \in I^k$. To see that we use Theorem 4.1 which says $I^k = B_S^k(u^k)$. Since $v \in B_S^k(u^k)$ it follows that $v \prec_k u^k$. We now show that $w' \prec_k v$. Then it follows $w' \prec_k u^k$, and hence belongs to $I^k$. Indeed, since $x_{j_r}$ in $v$ is replaced by $x_{l_r}$ with $l_r < j_r$ to obtain $w'$ it follows that $w' \prec v$. Let $w' = x_{f_1}x_{f_2}\cdots x_{f_{kd}}$ with $f_1 \leq f_2 \leq \ldots \leq f_{kd}$. Then $f_s = j_s$ for $s \neq r$ and $f_r = l_r$. To have $w' \prec_k v$ we must show that there is no $t$ such that $f_t = f_{t+1} = \cdots = f_{t+k}$, or equivalently there is no $t$ such that $f_t = f_{t+k}$. If $t+k < r$ or $t > r$ we have $f_t = j_t$ and $f_{t+k} = j_{t+k}$. Therefore, $f_t \neq f_{t+k}$ because the exponents of the monomial $v$ are bounded by $k$. Now let $t \leq r \leq t+k$. Suppose first that $r < t+k$. Then $r+1 \leq t+k$, and $f_{r+1} = j_{r+1} \geq j_r > l_r = f_r$. So not all $f_s$ are the same for $s$ with $t \leq s \leq t+k$. Finally, assume that $r = t+k$ and $f_{r-k} = \cdots = f_r$. For $s < r$ we have $f_s = j_s = l_s$ and $f_r = l_r$. Thus our assumption implies that $l_{r-k} = \cdots = l_r$, a contradiction because $w \in B_S^k(u^k)$. □

In all examples considered, the depth function $f(k) = \operatorname{depth} S/I^k$ is strictly decreasing until it becomes stable. In some special case we show that this is indeed the case.

We first observe

**Proposition 6.2.** *Let $I = B_S(u)$ with $u = x_{i_1}x_{i_2}\cdots x_{i_d}$ and $i_1 < i_2 < \cdots < i_d = n$. Then* $\operatorname{depth} S/I = d-1$.

**Proof.** We have to show that $\operatorname{proj\,dim} I = n-d$. Let $\tau$ be the inverse of the spreading operator $\sigma$. By definition, if $v = x_{k_1}x_{k_2}\cdots x_{k_d}$ with $k_1 < k_2 < \cdots < k_d$, then $\tau(v) = \prod_{j=1}^d x_{k_j-(j-1)}$, and when $I$ is a squarefree monomial ideal with $G(I) = \{v_1,\ldots,v_m\}$, one sets $I^\tau$ to be the ideal with $G(I^\tau) = \{\tau(v_1),\ldots,\tau(v_m)\}$.

By [10, Proposition 2.1] we have $B_S(u)^\tau = B_S^\infty(\tau(u))$, where $B_S^\infty(\tau(u))$ is the principal Borel ideal with Borel generator $\tau(u)$. It follows from [10, Theorem 1.11] that $\operatorname{proj\,dim} B_S(u) = \operatorname{proj\,dim} B_S^\infty(\tau(u))$. Note that all generators of $B_S^\infty(\tau(u))$ belong to $T = K[x_1,\ldots,x_{n-d+1}]$. Let $J = G(B_S^\infty(\tau(u)))T$. Then $B_S^\infty(\tau(u)) = JS$. It follows that

$$\operatorname{depth} S/I = \operatorname{depth} S/B_S^\infty(\tau(u)) = \operatorname{depth} T/J + d - 1.$$

It remains to be shown that $\operatorname{depth} T/J = 0$. It is well-known and easy to prove that

$$J = \prod_{j=1}^d (x_1,\ldots,x_{i_j-(j-1)}).$$

Since $i_d = n$ it follows that $J = J_0\mathfrak{n}$, where $J_0 = \prod_{j=1}^{d-1}(x_1, \ldots, x_{i_j-(j-1)})$ and $\mathfrak{n}$ is the maximal ideal of $T$. This shows that any minimal generator of $J_0$ defines a socle element of $T/J$. In particular, $\operatorname{depth} T/J = 0$. □

The next result tells us when $\operatorname{depth} S/I^2 = 0$ for $I = B_S(u)$.

**Proposition 6.3.** *Let $u = x_{i_1}x_{i_2}\cdots x_{i_d} \subset S$ be a monomial with $1 < i_1 < i_2 < \ldots < i_d = n$, and let $I = B_S(u)$. The following conditions are equivalent:*

(a) $\operatorname{depth} S/I^2 = 0$.

(b) $x_1 \cdots x_{2d-1} \in \operatorname{Soc}(S/I^2)$.

(c) $i_j \geq 2j$ *for* $j = 1, \ldots, d-1$.

**Proof.** (a) $\Longleftrightarrow$ (b) follows from Corollary 4.3.

(b) $\Longleftrightarrow$ (c) $w_2 = x_1 \cdots x_{2d-1} \in \operatorname{Soc}(S/I^2)$, if and only if $x_1 \cdots x_{2d-1}x_n \in I^2 = B_S(u^2)$, and this is the case if and only if

$$x_1 \cdots x_{2d-1}x_n \prec u^2 = x_{i_1}x_{i_1}x_{i_2}x_{i_2} \cdots x_{i_{d-1}}x_{i_{d-1}}x_nx_n.$$

By Corollary 2.5 this is the case if and only if $i_j \geq 2j$ for $j = 1, \ldots, d-1$, as desired. □

**Corollary 6.4.** *Let $I = B_S(x_ix_n)$. Then* $\operatorname{depth} S/I = 1$, *and for $k \geq 2$ we have*

$$\operatorname{depth} S/I^k = \begin{cases} 1, & \text{if } i = 1, \\ 0, & \text{if } i > 1. \end{cases}$$

**Proof.** The fact $\operatorname{depth}(S/I) = 1$ follows from Proposition 6.2. In order to compute $\operatorname{depth}(S/I^k)$, we first suppose that $i = 1$. Then Lemma 5.2 implies that $\operatorname{depth} S/I^k = \operatorname{depth} S'/B_{S'}(x_{n-1})^k + 1$. Since $B_{S'}(x_{n-1}) = (x_1, \ldots, x_{n-1})$, it follows that $\operatorname{depth} S/I^k = 1$ for all $k$. Finally, for $i > 1$ and $k > 1$, the result follows from Proposition 6.3. □

Next we consider the case $d = 3$.

**Proposition 6.5.** *Let $I = B_S(x_ix_jx_n)$ with $i < j < n$.*

(a) *If $i = 1$, then* $\operatorname{depth} S/I = 2$, *and for $k \geq 2$ we have*

$$\operatorname{depth} S/I^k = \begin{cases} 2, & \text{if } j = 2, \\ 1, & \text{if } j > 2. \end{cases}$$

(b) *If $i > 1$, then* $\operatorname{depth} S/I = 2$, *and* $\operatorname{depth} S/I^k = 0$ *for $k \geq 3$. Moreover,*

$$\operatorname{depth} S/I^2 = \begin{cases} 1, & \text{if } j = 3, \\ 0, & \text{if } j > 3. \end{cases}$$

**Proof.** (a) follows from Lemma 5.2 and Proposition 6.4.

(b) Proposition 6.2 shows that $\operatorname{depth} S/I = 2$. Moreover, $\operatorname{depth} S/I^k = 0$ for $k \geq 3$, see Proposition 5.3. Thus it remains to consider the case $k = 2$. If $j > 3$, then $\operatorname{depth} S/I^2 = 0$. Finally assume that $j = 3$. Then $x_i x_j x_n = x_2 x_3 x_n$. Then $\operatorname{depth} S/I^2 > 0$, by Proposition 6.3. On the other hand, $\operatorname{depth} S/I^2 < \operatorname{depth} S/I = 2$, by the next Theorem 6.6. Therefore, $\operatorname{depth} S/I^2 = 1$, as desired. □

The following result supports what we expect, namely that depth $S/I^k <$ depth $S/I^{k-1}$ if $k \leq \operatorname{astab}(I)$.

**Theorem 6.6.** *Let $I = B_S(u)$. Then the following conditions are equivalent:*

(a) $u = x_1 x_2 \cdots x_{d-1} x_n$.
(b) $\operatorname{depth} S/I = \operatorname{depth} S/I^k$ *for all $k$.*
(c) $\operatorname{depth} S/I = \operatorname{depth} S/I^k$ *for some $k \geq 2$.*
(d) $\operatorname{depth} S/I = \operatorname{depth} S/I^2$.

*Moreover, if one of these equivalent conditions fails, then* $\operatorname{depth} S/I^2 < \operatorname{depth} S/I$.

For the proof of Theorem 6.6 we use

**Lemma 6.7.** *Let $I = B_S(u)$, where $u = x_{i_1} \cdots x_{i_d}$ with $1 < i_1 < i_2 < \ldots < i_d = n$. Let $v \in I^2$ and let $J = (w \in G(I^2)\colon\ v <_{\text{lex}} w)$. Assume that $x_n$ divides $v$. Then*

$$\{x_i\colon\ i < n \text{ and } x_i^2 \nmid v\} \subseteq J : v.$$

**Proof.** Let $i$ be such that $i < n$ and $x_i^2 \nmid v$, and let $w = v x_i / x_n$. Then $\nu_j(w) \leq 2$ for all $j$ and $w \prec v$. Therefore, $w \in I^2$. Since $v <_{\text{lex}} w$, it follows that $w \in J$. Moreover, $v x_i = w x_n$, and hence $x_i \in J : v$. □

**Proof of Theorem 6.6.** (a) $\Rightarrow$ (b) Note that $I = x_1 \cdots x_{d-1}(x_d, \ldots, x_n)$. Therefore, $I^k \cong (x_d, \ldots, x_n)^k$, and hence $\operatorname{depth} S/I^k = \operatorname{depth} I^k - 1 = \operatorname{depth}(x_d, \ldots, x_n)^k - 1 = \operatorname{depth} S/(x_d, \ldots, x_n)^k$. Since $\operatorname{depth} K[x_d, \ldots, x_n]/(x_d, \ldots, x_n)^k = 0$, it follows that

$$\operatorname{depth} S/I^k = \operatorname{depth} S/(x_d, \ldots, x_n)^k = d - 1 = \operatorname{depth} S/I.$$

(b) $\Rightarrow$ (c) is trivial.

(c) $\Rightarrow$ (d) Since $\operatorname{depth} S/I \geq \operatorname{depth} S/I^2 \geq \operatorname{depth} S/I^k$, it is obvious that (c) implies (d).

(d) $\Rightarrow$ (a) We show that if $u \neq x_1 \cdots x_{d-1} x_n$, then $\operatorname{depth} S/I^2 < \operatorname{depth} S/I$. Since $u \neq x_1 \cdots x_{d-1} x_n$ it follows that $u = x_1 x_2 \cdots x_i v$ with $i < d - 1$ and $v = x_{j_1} x_{j_2} \cdots x_n$ with $j_1 < j_2 < \cdots < n$ and $j_1 > i + 1 \geq 1$. Then $I = B_S(u) = x_1 \cdots x_i J$, where $J = B_S(v)$. From this it follows that $\operatorname{depth} S/I^2 < \operatorname{depth} S/I$ if and only if depth $S/J^2 <$ depth $S/J$. Therefore, we may assume that $u = x_{i_1} \cdots x_{i_d}$ with $1 < i_1 < i_2 < \cdots < i_d = n$.

We claim that there exits $v \in G(I^2)$ such that $x_n$ divides $v$ and such that $|\mathcal{S}| \geq n - d + 1$, where $\mathcal{S} = \{x_i\colon\ i < n \text{ and } x_i^2 \nmid v\}|$. By Lemma 6.7, the claim implies that $\operatorname{proj\,dim} I^2 \geq n - d + 1$, and hence depth $S/I^2 \leq d - 2 <$ depth $S/I$.

We write $u^2 = x_{l_1} \cdots x_{l_{2d}}$ with $l_1 \leq l_2 \leq \cdots \leq l_{2d}$, and let $v = x_{t_1} x_{t_2} \cdots x_{t_{2d}}$ with $t_1 \leq t_2 \leq \ldots \leq t_{2d}$.

For the proof of the claim we consider two cases.

Case 1: $i_j = i_1 + j - 1$ for $j = 1, \ldots, d$. Then we let $t_j = j$ for $j = 1, \ldots, i_1$ and $t_j = l_j$ for $j = i_1 + 1, \ldots, 2d$. Then $|\mathcal{S}| \geq i_1 = n - d + 1$, because $n = i_1 + d - 1$.

Case 2: There exists $k$ such that $i_{k+1} > i_k + 1$. Then for all $j$ we let $t_j = l_j - 1$ if $j$ is odd, and let $t_j = l_j$ if $j$ is even. Then $v$ is divided by as many squares as we have integers $j$ with $i_{j+1} = i_j + 1$. Therefore, $v$ has at most $d - 1$ squares as factors. This implies that $|\mathcal{S}| \geq n - d + 1$, as desired. □

## References


[1] C. Andrei, V. Ene and B. Lajmiri, *Powers of t-spread principal Borel ideals*, Arch. Math. (Basel), 112(6) (2019), 587-597.

[2] A. Aramova, J. Herzog and T. Hibi, *Squarefree lexsegment ideals*, Math. Z., 228(2) (1998), 353-378.

[3] A. Aslam, *The stable set of associated prime ideals of a squarefree principal Borel ideal*, Bull. Math. Soc. Sci. Math. Roumanie (N.S.), 57(105) (2014), 243-252.

[4] M. Brodmann, *Asymptotic stability of* $\operatorname{Ass}(M/I^n M)$, Proc. Amer. Math. Soc., 74(1) (1979), 16-18.

[5] M. Brodmann, *The asymptotic nature of the analytic spread*, Math. Proc. Cambridge Philos. Soc., 86 (1979), 35-39.

[6] W. Bruns and J. Herzog, Cohen-Macaulay Rings, Cambridge Studies in Advanced Mathematics, 39, Cambridge University Press, Cambridge, 1993.

[7] E. De Negri, *Toric rings generated by special stable sets of monomials*, Math. Nachr., 203(1) (1999), 31-45.

[8] D. Eisenbud, Commutative Algebra: with a view toward algebraic geometry, Graduate Texts in Mathematics, 150, Springer-Verlag, New York, 1995.

[9] V. Ene and J. Herzog, Gröbner Bases in Commutative Algebra, Graduate Studies in Mathematics, 130, American Mathematical Society, Providence, RI, 2012.

[10] V. Ene, J. Herzog and A. Asloob Qureshi, *t-spread strongly stable ideals*, arXiv:1805.02368 [math.AC].

[11] C. A. Francisco, *Minimal graded Betti numbers and stable ideals*, Comm. Algebra, 31(10) (2003), 4971-4987.

[12] C. A. Francisco, J. Mermin and J. Schweig, *Borel generators*, J. Algebra, 332(1) (2011), 522-542.

[13] J. Herzog and T. Hibi, *The depth of powers of an ideal*, J. Algebra, 291(2) (2005), 534-550.

[14] J. Herzog and T. Hibi, Monomial Ideals, Graduate Texts in Mathematics, 260, Springer-Verlag London, Ltd., London, 2011.

[15] J. Herzog and T. Hibi, *Bounding the socles of powers of squarefree monomial ideals*, Commutative Algebra and Noncommutative Algebraic Geometry, Vol. II, Math. Sci. Res. Inst. Publ., 68, Cambridge Univ. Press, New York, (2015), 223-229.

[16] J. Herzog, A. Rauf and M. Vladoiu, *The stable set of associated prime ideals of a polymatroidal ideal*, J. Algebraic Combin., 37(2) (2013), 289-312.

[17] G. Kalai, *Algebraic shifting*, in: Computational Commutative Algebra and Combinatorics, (Osaka, 1999), Adv. Stud. Pure Math., 33, Math. Soc. Japan, Tokyo, (2002), 121-163.

[18] I. Peeva and M. Stillman, *The minimal free resolution of a Borel ideal*, Expo. Math., 26(3) (2008), 237-247.

**Jürgen Herzog** (Corresponding Author)
Universität Duisburg-Essen
Fachbereich Mathematik
Campus Essen
45117, Essen, Germany
e-mail: juergen.herzog@uni-essen.de

**Bahareh Lajmiri** and **Farhad Rahmati**
Department of Mathematics and Computer Science
Amirkabir University of Technology
424 Hafez Ave, Tehran, Iran
e-mails: bahareh.lajmiri@aut.ac.ir (B. Lajmiri)
frahmati@aut.ac.ir (F. Rahmati)

# 13

# ON FINITE DIMENSIONAL ALGEBRAS WHICH ARE SUMS OF TWO SUBALGEBRAS

M. Tamer Koşan and Jan Žemlička

Communicated by Abdullah Harmancı

ABSTRACT. In this paper, we give a general method of the construction of a 3-dimensional associative algebra $R$ over an arbitrary field $F$ that is a sum of two subalgebras $R_1$ and $R_2$ (i.e. $R = R_1 + R_2$).

**Keywords:** Sum of rings, sum of algebras

## 1. Introduction

Let $R$ be an associative ring. Writing $R = R_1 + R_2$, we mean that $R_1$ and $R_2$ are subrings of $R$ and, for every $r \in R$, there exist $r_1 \in R_1$ and $r_2 \in R_2$ such that $r = r_1 + r_2$.

Many authors studied problems and relationships among properties of $R_1$, $R_2$ and $R$ (see [4], [5], [7], [8], [10], [11], [12], [13]). For instance, the question "is $R$ nil provided that $R_1^2 = 0$ and $R_2$ is nil?" is equivalent to the famous Köthe's nil ideal problem (see [14], [15] and [16]). Let us mention three other open problems in this area.

Suppose that $R_1$ and $R_2$ are rings satisfying a polynomial identity. Does $R$ satisfies a polynomial identity ([1])? The answer to this problem is known in several particular cases (cf. [3], [7], [8], [9], [10], [11], [12], [13]), and a full positive answer in [12].

It is known [14] that if both $R_1$ and $R_2$ are nilpotent rings, then so is $R$. It is also known that there exists a function $f(n, m)$ such that if $R_1^n = 0$ and $R_2^m = 0$, then $R^{f(n,m)} = 0$. However the best such a function is still unknown. It is conjectured that $f(m, n) = mn$.

For the last problem, suppose that $R_1$ is nil and $R_2$ is reduced (it has no nonzero nilpotent elements). Is $R_1$ an ideal of $R$? The answer to this question is known to be "yes" in many particular cases [13], but in general, is still unknown.

We remark that a general method of constructing rings which are sums of two subrings was invited by Kelarev [6]. As we mentioned before, all these problems, on their substantial parts, concern due structure of rings which are sums of two subrings. This is particularly clear for the question 3. Note that it also concerns questions first and second. Namely, if $R$ is a semiprime ring which satisfies a polynomial identity, then the ring of Ore extension of $R$ is isomorphic to a direct sum of matrix rings over division rings. Thus, when studying whether of $R = R_1 + R_2$, and each $R_i$ satisfies a polynomial identity implies $R$ satisfies such an identity in the case of semiprime $R$ to study the structure of subrings of matrix rings of two subrings. Here, we also note that, Bokut [2] proved that every algebra over a field can be embedded into a simple algebra which is a sum of three nilpotent subalgebras. In view of these two facts, we will a general method of constructing 2 and 3 dimensional algebras over a field $F$ which are sums of to subalgebras with respect to the isomorphisms. This general method may give us to describe semiprime rings (even finite dimensional algebras) as a sum of two subrings.

## 2. Some general results

Suppose that $F$ is a field and $R$ is a finite dimensional (non-unitary) associative $F$-algebra. Put $A := F \times R$ and define binary operation $+$ and $\cdot$ on $A$ by the rule

$$(f_1, r_1) + (f_2, r_2) = (f_1 + f_2, r_1 + r_2)$$

and

$$(f_1, r_1) \cdot (f_2, r_2) = (f_1 \cdot f_2, f_1 r_2 + f_2 r_1 + r_1 r_2),$$

where $f_1, f_2 \in F$ and $r_1, r_2 \in R$. It is easy to say that $A$ is an $F$-vector space and we may identify elements $r \in R$ with elements $(0, r)$ in $A$.

**Proposition 2.1.** *$A$ is a finite dimensional $F$-algebra with a unit and $R$ is a two sided ideal of $A$ such that $\dim_F(A/R) = 1$. Moreover, $A/R \cong F$.*

**Proof.** Since $A$ is a vector space over $F$, it is easy to see that

$$\begin{aligned} f(a, b) &= (f, 0) \cdot (a, b) \\ &= (fa, fb) \\ &= (a, b) \cdot (f, 0), \end{aligned}$$

which shows that $(1, 0)$ is a unit.

For each $(f_i, r_i) \in A$, we get

$$\begin{aligned}(f_1,r_1)[(f_2,r_2)(f_3,r_3)] &= (f_1,r_1)(f_2f_3, f_2r_3+f_3r_2+r_2r_3)\\ &= (f_1f_2f_3, f_1f_2r_3+f_1f_3r_2+f_1r_2r_3+f_2f_3r_1+f_2r_1r_3\\ &\quad +f_3r_1r_2+r_1r_2r_3)\\ &= (f_1f_2, f_1r_2+f_2r_1+r_1r_2)(f_3,r_3)\\ &= [(f_1,r_1)(f_2,r_2)](f_3,r_3).\end{aligned}$$

Similarly

$$\begin{aligned}(f_1,r_1)[(f_2+f_3, r_2+r_3)] &= (f_1f_2+f_1f_3, f_1r_2+f_1r_3+f_2r_1+f_3r_1+r_1r_2+r_1r_3)\\ &= (f_1,r_1)(f_2,r_2)+(f_1,r_1)(f_3,r_3)\end{aligned}$$

and

$$[(f_2+f_3, r_2+r_3)](f_1,r_1) = (f_2,r_2)(f_1,r_1)+(f_3,r_3)(f_1,r_1).$$

Clearly, $R$ is $F$-subspace of $A$ such that $\dim_F(A) = \dim_F(R)+1$ and

$$\begin{aligned}(f,r)(0,s) &= (0, fs+rs)\\ (0,s)(f,r) &= (0, sf+sr)\end{aligned}$$

for every $(f,r) \in A$ and $s \in R$. Hence $R$ is a two-sided ideal of $A$. Finally, $A/R \cong F$. □

For an algebra $R$, the symbol $R^*$ denotes $R$ with an identity adjoined.

**Theorem 2.2.** *If $R$ does not contain a unit, then it is an ideal of finite dimensional $F$-algebra $A$ such that $\dim_F(A) = \dim_F(R)+1$, i.e. there exists a complete orthogonal set of primitive idempotents $e_0, \ldots, e_m$ such that $e_0 \in A \setminus R$, $e_1, \ldots, e_m \in R$, $R = A(1-e_0)A$ and $\dim_F(e_0A/e_0J(A)) = 1$.*

**Proof.** The first part follows from Proposition 2.1. The rest of the claim follows from [17, Proposition VIII.4.1], since every orthogonal set of idempotents of $R + J(A)/J(A)$ can be lifted to an orthogonal set of idempotents of $R$. □

**Remark 2.3.** Fix $A$ from Theorem 2.2 if $R$ does not contain a unit and put $A = R$ otherwise. We know that there exists a complete orthogonal set of primitive idempotents $e_0, \ldots, e_m$ of $A$ such that $e_1, \ldots, e_m \in R$ and $\dim_F(e_0A/e_0J(A)) = 1$ if $A \neq R$. Then $m < \dim_F R$.

**Proof.** Obviously, $m \leq \dim_F R \leq \dim_F A$ and, moreover, $m \neq \dim_F R$ if $R = A$. Let $R \neq A$. If $e_0A$ is simple, then $1 - e_0 \in R$ is a unit of $R$, a contradiction. Thus $\dim_F(e_0A) > 1$, hence $m < \dim_F R$. □

## 3. Two structure theorems

The following observation is easy.

**Lemma 3.1.** *If* $0 \neq R_1$ *and* $0 \neq R_2$ *are proper subalgebras of the algebra* $R$ *over a field* $F$ *and* $\dim_F R = 2$*, then* $\dim_F R_1 = 1 = \dim_F R_2$.

Assume that $\dim_F R = 2$ such that $R$ is a sum of subalgebras $R_1$ and $R_2$. By Lemma 3.1, we may assume that $\dim_F R_1 = 1 = \dim_F R_2$. Let $R_1 = Fe$ and $R_2 = Ff$, where $e^2 = e \in R$ and $f^2 = f \in R$, or $e^2 = 0$ or $f^2 = 0$ if $R_1 \cong F^0$ or $R_2 \cong F^0$, where $F^0$ denotes $F$ with zero multiplication. Then we have the following cases on $R_1$ (or $R_2$).

* If $R_1^2 = 0$, then $R_1 \cong F^0$. Hence we have either $R_2 \cong F^0$ or $R_2 \cong F$.

** If $R_1^2 \neq 0$, then $R_1 \cong F$. Hence we have either $R_2 \cong F^0$ or $R_2 \cong F$.

**Case 1.** Let $R_1^2 = 0$, $R_1 \cong F^0$ and $R_2 \cong F^0$. For $\alpha \in R_1$ and $\beta \in R_2$, write

$$ef = \alpha e + \beta f. \tag{1}$$

Note that, in this case, $e^2 = 0 = f^2$. Multiplying the equation (1) by $f$ on the right, we get $\alpha ef = 0$ which implies $\alpha = 0$ or $ef = 0$. If $\alpha = 0$, we get $ef = \beta f$. Thus $\beta ef = 0$. It follows $ef = 0$. Similarly, we obtain that $fe = 0$. Now it is enough to take $R = F^0 \times F^0$, $e = (1, 0)$ and $f = (0, 1)$.

**Case 2.** Let $R_1^2 \neq 0$, $R_1 \cong F$ and $R_2 \cong F$. Then, for $\alpha \in R_1$ and $\beta \in R_2$, we get again the equation (1). Multiplying the equation (1) by $e$ on the left, we get

$$ef = \alpha e + \beta ef. \tag{2}$$

Now, $0 = \beta(ef - f)$ implies either $\beta = 0$ or $ef = f$. If $\beta = 0$, then, by equation (1), $ef = \alpha e$. Since $\alpha e = ef = ef^2 = \alpha ef = \alpha^2 e$, we get either $\alpha = 1$ or $\alpha = 0$. By equation (1), if $\beta = 0$ and $\alpha = 0$, then we have $ef = 0$ and so $fe = 0$, if $\beta = 0$ and $\alpha = 1$ then $ef = e$. Therefore $ef \in \{0, e, f\}$.

By similar reasoning as above we can obtain that $fe \in \{0, e, f\}$. Assume that $ef = 0$. Multiplying the equation (3) by $f$ on the right, we have $\beta_1 f = 0$. So $\beta_1 = 0$. Consequently $0 = efe = \alpha_1 e$ which implies $\alpha_1 = 0$. Hence $fe = 0$. We can take $R = F \times F$ and $e = (1, 0)$, $f = (0, 1)$.

Let $ef = f = fe$. So $e$ is an identity element of $R$. Therefore $e - f$ and $f$ are orthogonal idempotents of $R$. Consequently $R = F \times F$ and we can take $e = (1, 1)$ and $f = (0, 1)$. The case where $ef = e = fe$ can be treated analogously.

In the case where $ef = f$ and $fe = e$, consider $R = \begin{pmatrix} F & F \\ 0 & 0 \end{pmatrix}$, where $e = \begin{pmatrix} 1 & 0 \\ 0 & 0 \end{pmatrix}$ and $f = \begin{pmatrix} 1 & 1 \\ 0 & 0 \end{pmatrix}$.

Analogously, if $ef = e$ and $fe = f$, let $R = \begin{pmatrix} F & 0 \\ F & 0 \end{pmatrix}$, where $e = \begin{pmatrix} 1 & 0 \\ 0 & 0 \end{pmatrix}$ and $f = \begin{pmatrix} 1 & 0 \\ 1 & 0 \end{pmatrix}$.

**Case 3.** Let $R_1^2 \neq 0$, $R_1 \cong F$ and $R_2 \cong F^0$. Multiplying the equation (1) by $f$ on the right, we get $0 = \alpha ef$ which implies either $\alpha = 0$ or $ef = 0$.

If $\alpha = 0$, then $ef = \beta f$ by equation (1). Now $ef = e^2 f = eef = e\beta f = \beta ef = \beta^2 f$ which implies either $\beta = 0$ or $\beta = 1$. Clearly, if $\beta = 0$ then $ef = 0$ and if $\beta = 1$, then $ef = f$.

Finally, multiplying the equation (3) by $f$ on the right, we get $\alpha_1 ef = fef$. So $\alpha_1 e = 0$ because $ef = f$ or $ef = 0$. Hence $fe = \beta_1 f = \beta_1^2 f$. It follows that $\beta_1 = 1$ or $\beta_1 = 0$. Consequently, $fe = 0$ or $fe = f$.

In the case where $ef = fe = 0$, we take $R = F \times F^0$, where $e = (1, 0)$ and $f = (0, 1)$. Note that if $R = \begin{pmatrix} F & F \\ 0 & 0 \end{pmatrix}$, $e = \begin{pmatrix} 1 & 0 \\ 0 & 0 \end{pmatrix}$ and $f = \begin{pmatrix} 0 & 0 \\ 1 & 0 \end{pmatrix}$, then $ef = 0$ and $fe = f$.

Finally, let $R = (F^0)^*$, $e = (1, 0)$ and $f = (0, 1)$. Clearly, $ef = f = fe$.

As a result of above calculations, we have the following observation.

**Theorem 3.2.** *Assume that* $\dim_F R = 2$ *and* $R = R_1 + R_2$ *is a sum of proper subalgebras* $R_1$ *and* $R_2$. *Then we have the following cases.*

(I) *If* $R_1 \cong F^0$ *and* $R_2 \cong F^0$, *then* $R = F^0 \times F^0$.

(II) *If* $R_1 \cong F$ *and* $R_2 \cong F$, *then one of the following cases holds true:*

  (1) $R = F \times F$,

  (2) $R = \begin{pmatrix} F & F \\ 0 & 0 \end{pmatrix}$,

  (3) $R = \begin{pmatrix} F & 0 \\ F & 0 \end{pmatrix}$.

(III) *If* $R_1 \cong F$ *and* $R_2 \cong F^0$, *then one of the following cases holds true:*

  (1) $R = F \times F^0$,

  (2) $R = \begin{pmatrix} F & F \\ 0 & 0 \end{pmatrix}$,

(3) $R = \begin{pmatrix} F & 0 \\ F & 0 \end{pmatrix}$,

(4) $R = (F^0)^*$.

In the next Theorem, we will use the following.

**Remark 3.3.** Let $M$ be an $A$-$A$-bimodule, denote $\mu(M)$ a matrix $\mu(M) = (d_{ij})$ where $d_{ij} = \dim_F(e_i M e_j)$. Denote by $S$ a ring of polynomials in two noncommuting variables $x_1$, $x_2$ over $F$. If $\mathbf{A} = (a_{kl})$ is a matrix where $a_{kl} \in F$, $1 \leq k, l \leq 2$, then define

$$S(\mathbf{A}) = S/\langle a_{22}x_1^2 - a_{11}x_2^2, a_{21}x_1x_2 - a_{12}x_2x_1, a_{22}x_1x_2 - a_{12}x_2^2, a_{22}x_2x_1 - c_{21}x_2^2, x_1^3, x_2^3\rangle,$$

where $a_{11}, a_{12}, a_{21}, a_{22} \neq 0$. Moreover,

$$S(\mathbf{A}) = S/\langle x_1^2, a_{21}x_1x_2 - a_{12}x_2x_1, a_{22}x_1x_2 - a_{12}x_2^2, a_{22}x_2x_1 - a_{21}x_2^2, x_2^3\rangle$$

if $a_{11} = 0$,

$$S(\mathbf{A}) = S/\langle x_1^2, x_1x_2, a_{22}x_2x_1 - a_{21}x_2^2, x_2^3\rangle$$

if $a_{11} = a_{12} = 0$,

$$S(\mathbf{A}) = S/\langle x_1^2, x_1x_2, x_2x_1, x_2^3\rangle$$

if $a_{11} = a_{12} = a_{21} = 0$. Analogously we define $S(\mathbf{A})$ in the case $a_{11} = c_{22} = 0$ or $a_{11} = a_{12} = a_{22} = 0$.

Assume that $\dim_F R = 3$ such that $R$ is a sum of subalgebras order of $R_1$ and $R_2$.

**Theorem 3.4.** *Assume that* $\dim_F R = 3$ *such that* $R$ *is a sum of subalgebras order of* $R_1$ *and* $R_2$.

(I) *Let* $m = 0$ *and* $R = A$. *Then one of the following cases holds true:*

(a) *If* $J(R) = 0$, *then* $R \cong K$, *where* $K$ *is a field of* $F$*-dimension* $3$ *and either* $R_1 = R$ *or* $R_2 = R$.

(b) *If* $R/J(R) \cong F$ *and* $J^2(R) = 0$, *then* $R \cong F[x_1, x_2]/\langle x_1^2, x_1x_2, x_2^2\rangle$ *and*

(1) $R_1 = R$ *and* $R_2$ *is an arbitrary subalgebra of* $R$.

(2) $R_1 \cong F_0 \times F_0$ *and* $R_2 \cong F$.

(3) $R_1 \cong F[x]/\langle x^2\rangle$ *and* $R_2 \cong F_0$.

(4) $R_1 \cong F[x]/\langle x^2\rangle$ *and* $R_2 \cong F_0 \times F_0$.

(5) $R_1 \cong R_2 \cong F[x]/\langle x^2\rangle$.

(c) *If* $R/J(R) \cong F$ *and* $J^2(R) \neq 0$, *then* $R \cong F[x]/\langle x^3\rangle$ *and*

(6) $R_1 = R$ *and* $R_2$ *is an arbitrary subalgebra of* $R$,

(7) $R_1 \cong xF[x]/\langle x^3\rangle$ *and* $R_2 \cong F$,

(8) $R_1 \cong xF[x]/\langle x^3\rangle$, $R_2 \cong F[x]/\langle x^2\rangle$, *where* $F[x]/\langle x^2\rangle \cong F + x^2F \subset R$.

*Furthermore, all the cases in* (1) *are commutative.*

(II) *Let* $m = 1$ *and* $R = A$. *Then one of the following cases holds true:*

(a) *If* $J(R) = 0$, *then* $R \cong F \times G$ *where* $G$ *is a field of* $F$*-dimension* 2 *and*

(1) $R_1 = R$ *and* $R_2$ *is an arbitrary subalgebra of* $R$,

(2) $R_1 \cong G$ *and* $R_2 \cong F$,

(3) $R_1 \cong G$ *and* $R_2 \cong F \times F$.

(b) *If* $J(R) \neq 0$, *then* $R \cong \begin{pmatrix} F & F \\ 0 & F \end{pmatrix}$ *and we have one of the following cases.*

(4) $R_1 = R$ *and* $R_2$ *is an arbitrary subalgebra of* $R$,

(5) $R_1 \cong F^2$ *and* $R_2 \cong F_0$,

(6) $R_1 \cong \begin{pmatrix} F & F \\ 0 & 0 \end{pmatrix}$ *and* $R_2 \cong F$,

(7) $R_1 \cong \begin{pmatrix} 0 & F \\ 0 & F \end{pmatrix}$ *and* $R_2 \cong F$,

(8) $R_1 \cong F^2$ *and* $R_2 \cong \begin{pmatrix} 0 & F \\ 0 & F \end{pmatrix}$,

(9) $R_1 \cong F^2$ *and* $R_2 \cong \begin{pmatrix} F & F \\ 0 & 0 \end{pmatrix}$,

(10) $R_1 \cong \begin{pmatrix} F & F \\ 0 & 0 \end{pmatrix}$ *and* $R_2 \cong \begin{pmatrix} 0 & F \\ 0 & F \end{pmatrix}$.

(III) *Let* $m = 2$ *and* $R = A$. *Then* $R \cong F \times F \times F$ *whenever* $m = 2$ *and so we have one of the following cases.*

(a) $R_1 = R$ *and* $R_2$ *is an arbitrary (semisimple) subalgebra of* $R$,

(b) $R_1 \cong F \times F$ *and* $R_2 \cong F$,

(c) $R_1 \cong R_2 \cong F \times F$.

(IV) *Let* $m = 0$ *and* $R \neq A$. *If* $J^2(R) = 0$, *then* $R \cong F_0^3$, *and so* $R_1 \cong F_0^k$ *and* $R_2 \cong F_0^l$ *for* $k, l \leq 3$. *We have one of the following cases.*

(a) *If* $J^2(R) = 0$, *then* $R \cong F_0^3$, *and so* $R_1 \cong F_0^k$ *and* $R_2 \cong F_0^l$ *for* $k, l \leq 3$.

(b) *If* $J^2(R) \neq 0$, $J^3(R) = 0$ *and* $\dim_F(J(R)/J^2(R)) > \dim_F(J^2(R)/J^3(R))$, *then* $R \cong \{x_1, x_2\}S(\mathbf{A})$ *and*

(1) $R_1 = R$ *and* $R_2$ *is an arbitrary subalgebra of* $R$,

(2) $R_1 \cong xF[x]/\langle x^3\rangle$ *and* $R_2 \cong F_0$,

(3) $R_1 \cong F_0^2$ *and* $R_2 \cong F_0$,

(4) $R_1 \cong R_2 \cong xF[x]/\langle x^3\rangle$,

(5) $R_1 \cong xF[x]/\langle x^3\rangle$ *and* $R_2 \cong F_0^2$,

(6) $R_1 \cong F_0^2$ *and* $R_2 \cong F_0^2$.

(c) *If* $J^3(R) \neq 0$, *then* $R = R_1 \cong xF[x]/\langle x^4\rangle$ *and* $R_2$ *is an arbitrary (chain) subalgebra of* $R$.

(V) *Let* $m = 1$ *and* $R \neq A$. *Then* $A/J(R) \cong F \times G$ *where* $G$ *is a field and* $\dim_F(G) \leq 2$. *Let* $G \cong F$. *We have one of the following cases.*

(a) *If* $J^2(R) = 0$ *and* $\mu(J(R)) = \begin{pmatrix} 1 & 0 \\ 0 & 1 \end{pmatrix}$, *then* $R \cong F_0 \times F[x]/\langle x^2\rangle$ *and*

(1) $R_1 = R$ *and* $R_2$ *is an arbitrary subalgebra of* $R$,

(2) $R_1 \cong F[x]/\langle x^2\rangle$ *and* $R_2 \cong F_0$,

(3) $R_1 \cong F_0 \times F$ *and* $R_2 \cong F_0$,

(4) $R_1 \cong F[x]/\langle x^2\rangle$ *and* $R_2 \cong F_0^2$,

(5) $R_1 \cong F[x]/\langle x^2\rangle$ *and* $R_2 \cong F_0 \times F$.

(b) *If* $J^2(R) = 0$ *and* $\mu(J(R)) = \begin{pmatrix} 0 & 1 \\ 0 & 1 \end{pmatrix}$, *then* $R \cong \left\{ \begin{pmatrix} 0 & 0 & b \\ 0 & a & c \\ 0 & 0 & a \end{pmatrix} \mid a, b, c \in F \right\}$

*and*

(6) $R_1 = R$ *and* $R_2$ *is an arbitrary subalgebra of* $R$,

(7) $R_1 = J(R) \cong F_0^2$ *and* $R_2 \cong F$,

(8) $R_1 = \begin{pmatrix} F & 0 \\ F & 0 \end{pmatrix}$ *and* $R_2 \cong F_0$,

(9) $R_1 \cong F[x]/\langle x^2\rangle$ *and* $R_2 \cong F_0$,

(10) $R_1 = J(R) \cong F_0^2$ *and* $R_2 \cong F[x]/\langle x^2\rangle$,

(11) $R_1 = J(R) \cong F_0^2$ *and* $R_2 \cong \begin{pmatrix} F & 0 \\ F & 0 \end{pmatrix}$,

(12) $R_1 \cong \begin{pmatrix} F & 0 \\ F & 0 \end{pmatrix}$ *and* $R_2 \cong F[x]/\langle x^2\rangle$.

(c) *If* $J^2(R) = 0$ *and* $\mu(J(R)) = \begin{pmatrix} 0 & 0 \\ 1 & 1 \end{pmatrix}$, *then* $R \cong \left\{ \begin{pmatrix} 0 & 0 & 0 \\ 0 & a & 0 \\ b & c & a \end{pmatrix} \mid a, b, c \in F \right\}$.

*In fact, this case is antiisomorphic to the case (b) (i.e. it has the structure of* $R^{op}$*).*

(d) *If* $J^2(R) = 0$ *and* $\mu(J(R)) = \begin{pmatrix} 1 & 0 \\ 1 & 0 \end{pmatrix}$, *then* $R \cong \begin{pmatrix} F & 0 & F \\ 0 & 0 & F \\ 0 & 0 & 0 \end{pmatrix}$ *and*

(13) $R_1 = R$ *and* $R_2$ *is an arbitrary subalgebra of* $R$,

(14) $R_1 \cong F_0^2$ *and* $R_2 \cong F$,

(15) $R_1 = \begin{pmatrix} F & 0 \\ F & 0 \end{pmatrix}$ *and* $R_2 \cong F_0$,

(16) $R_1 \cong F \times F_0$ *and* $R_2 \cong F_0$,

(17) $R_1 = J(R) \cong F_0^2$ *and* $R_2 \cong F \times F_0$,

(18) $R_1 = J(R) \cong F_0^2$ *and* $R_2 \cong \begin{pmatrix} F & 0 \\ F & 0 \end{pmatrix}$,

(19) $R_1 \cong \begin{pmatrix} F & 0 \\ F & 0 \end{pmatrix}$ *and* $R_2 \cong F \times F_0$.

(e) *If* $J^2(R) = 0$ *and* $\mu(J(R)) = \begin{pmatrix} 1 & 1 \\ 0 & 0 \end{pmatrix}$, *then* $R \cong \begin{pmatrix} F & 0 & 0 \\ 0 & 0 & 0 \\ F & F & 0 \end{pmatrix}$.

(f) *If* $J^2(R) = 0$ *and* $\mu(J(R)) = \begin{pmatrix} 0 & 1 \\ 1 & 0 \end{pmatrix}$, *then* $R \cong \left\{ \begin{pmatrix} a & b & 0 \\ 0 & 0 & c \\ 0 & 0 & a \end{pmatrix} \mid a, b, c \in F \right\}$

*and*

(20) $R_1 = R$ *and* $R_2$ *is an arbitrary subalgebra of* $R$,

(21) $R_1 \cong F_0^2$ *and* $R_2 \cong F$,

(22) $R_1 = \begin{pmatrix} F & 0 \\ F & 0 \end{pmatrix}$ *and* $R_2 \cong F_0$,

(23) $R_1 \cong \begin{pmatrix} F & F \\ 0 & 0 \end{pmatrix}$ *and* $R_2 \cong F_0$,

(24) $R_1 = J(R) \cong F_0^2$ *and* $R_2 \cong \begin{pmatrix} F & F \\ 0 & 0 \end{pmatrix}$,

(25) $R_1 = J(R) \cong F_0^2$ *and* $R_2 \cong \begin{pmatrix} F & 0 \\ F & 0 \end{pmatrix}$,

(26) $R_1 \cong \begin{pmatrix} F & 0 \\ F & 0 \end{pmatrix}$ *and* $R_2 \cong \begin{pmatrix} F & F \\ 0 & 0 \end{pmatrix}$.

(g) *If* $\dim_F G = 2$, *then* $J(R)$ *is not a* $G$*-module, hence* $R \cong G \times F_0$ *and*

(27) $R_1 = R$ *and* $R_2$ *is an arbitrary subalgebra of* $R$,

(28) $R_1 \cong G$ *and* $R_2 \cong F_0$,

(29) $R_1 \cong G$ *and* $R_2 \cong F \times F_0$.

(h) *If* $J^2(R) \neq 0$, *then* $R \cong F \times (xF[x]/\langle x^3 \rangle)$ *and*

(30) $R_1 = R$ *and* $R_2$ *is an arbitrary subalgebra of* $R$,

(31) $R_1 \cong xF[x]/\langle x^3 \rangle$ *and* $R_2 \cong F$,

(32) $R_1 \cong xF[x]/\langle x^3 \rangle$ *and* $R_2 \cong F \times F_0$.

(VI) *Let* $m = 2$ *and* $R \neq A$. *Then* $A/J(R) \cong F^3$ *and we have one of the following cases.*

(a) $R \cong F^2 \times F_0$ *and*

(1) $R_1 = R$ *and* $R_2$ *is an arbitrary subalgebra of* $R$,

(2) $R_1 \cong F^2$ *and* $R_2 \cong F_0$,

(3) $R_1 \cong F \times F_0$ *and* $R_2 \cong F$,

(4) $R_1 \cong F^2$ *and* $R_2 \cong F \times F_0$.

(b) $R \cong F \times \begin{pmatrix} F & 0 \\ F & 0 \end{pmatrix}$ *and*

(5) $R_1 = R$ *and* $R_2$ *is an arbitrary subalgebra of* $R$,

(6) $R_1 \cong F^2$ *and* $R_2 \cong F_0$,

(7) $R_1 \cong \begin{pmatrix} F & 0 \\ F & 0 \end{pmatrix}$ *and* $R_2 \cong F$,

(8) $R_1 \cong F \times F_0$ *and* $R_2 \cong F$,

(9) $R_1 \cong F^2$ *and* $R_2 \cong F \times F_0$,

(10) $R_1 \cong F^2$ *and* $R_2 \cong \begin{pmatrix} F & 0 \\ F & 0 \end{pmatrix}$,

(11) $R_1 \cong \begin{pmatrix} F & 0 \\ F & 0 \end{pmatrix}$ *and* $R_2 \cong F \times F_0$.

(c) $R \cong F \times \begin{pmatrix} F & F \\ 0 & 0 \end{pmatrix}$ *and*

(12) $R_1 = R$ *and* $R_2$ *is an arbitrary subalgebra of* $R$,

(13) $R_1 \cong F^2$ *and* $R_2 \cong F_0$,

(14) $R_1 \cong \begin{pmatrix} F & F \\ 0 & 0 \end{pmatrix}$ *and* $R_2 \cong F$,

(15) $R_1 \cong F \times F_0$ *and* $R_2 \cong F$,

(16) $R_1 \cong F^2$ *and* $R_2 \cong F \times F_0$,

(17) $R_1 \cong F^2$ *and* $R_2 \cong \begin{pmatrix} F & F \\ 0 & 0 \end{pmatrix}$,

(18) $R_1 \cong \begin{pmatrix} F & F \\ 0 & 0 \end{pmatrix}$ *and* $R_2 \cong F \times F_0$.

**Proof.** We can suppose without loss of generality that $\dim_F(R_1) \geq \dim_F(R_2)$.
**(I)** Let $m = 0$ and $R = A$. Then $m = 0$, $R$ is a local algebra and either $R/J(R) \cong F$ or $J(R) = 0$ and $R$ is a field. Note that the case $\dim_F(R/J(R)) = 2$ is excluded, since $J(R)/J^2(R)$ should be an $R/J(R)$-vector space and hence $2 \mid \dim_F(R)$.

If $J(R) = 0$, then $R$ is a field and its subfields of $F$-dimension is either 1 or 3. Hence either $R_1 = R$ or $R_2 = R$.

If $R/J(R) \cong F$ and $J^2(R) = 0$, then $R \cong F[x_1, x_2]/\langle x_1^2, x_1x_2, x_2^2\rangle$. Hence we have one of the possibilities (1)-(5).

If $R/J(R) \cong F$ and $J^2(R) \neq 0$, then $J^3(R) = 0$ and there exists a basis $1, j, j^2$ of $R_F$, where $j \in J(R)$ and $j^2 \in J^2(R)$. Hence $R \cong F[x]/\langle x^3\rangle$. Now we have one of the possibilities (6)-(8).

**(II)** If $m = 1$, $\dim_F(R/J(R)) \geq 2$ and so $\dim_F(J(R)) \leq 1$.

If $J(R) = 0$, then clearly, $R \cong F \times G$, where $G$ is a field of $F$-dimension 2 and we have one of the possibilities (1)-(3).

If $J(R) \neq 0$, then $R/J(R) \cong F \times F$ and $\dim_F(J(R)) = 1$. As $J(R)$ is an $F \times F$-module, we get that $R \cong \begin{pmatrix} F & F \\ 0 & F \end{pmatrix}$. Note that $R \cong \begin{pmatrix} F & 0 \\ F & F \end{pmatrix}$. Hence we have one of the possibilities (4)-(10).

**(III)** This is clear.

**(IV)** We assume that $m = 0$. Then $R = J(R)$ for a local algebra $A$ of the dimension 4. If $J^2(R) = 0$, then $R \cong F_0^3$. So (a) is clear. If $J^2(R) \neq 0$ and $J^3(R) = 0$, then we get $\dim_F(J(R)/J^2(R)) = 2$ and $\dim_F(J^2(R)/J^3(R)) = 1$. Then there exists a basis $\{j_1, j_2, j_3\}$ of $R$ and elements $a_{kl} \in F$, $k, l \in \{1, 2\}$ such that $j_1, j_2 \in J(R) \setminus J^2(R)$, $j_3 \in J^2(R)$ and $j_kj_l = a_{kl}j_3$ for all $k, l \in \{1, 2\}$. Now it is easy to see that $A \cong S((a_{kl}))$ and so we have one of the possibilities (1)-(6).

If $J^3(R) \neq 0$, then $J^4(R) = 0$ and $\dim_F(J(R)/J^2(R)) = \dim_F(J^2(R)/J^3(R)) = \dim_F(J^3(R)) = 1$ and there is a base $\{j, j^2, j^3\}$ of $R$. Hence $R = R_1 \cong xF[x]/\langle x^4\rangle$ and $R_2$ is an arbitrary (chain) subalgebra of $R$.

**(V)** Let $m = 1$ and $R \neq A$. Then either $e_0J(R) \neq 0$ or $J(R)e_0 \neq 0$ and $A/J(R) \cong F \times G$ where $G$ is a field and $\dim_F(G) \leq 2$. Let $G \cong F$. First, moreover, suppose that $J^2(R) = 0$ and we will discuss cases of $\mu(J(R))$, denote $\{j_1, j_2\}$ a base of $J(R)$ such that $j_k \in e_{a_k}J(R)e_{b_k}$ for a suitable $a_k$ and $b_k$. Note that either the first row or the first column is nonzero and $j_kj_l = 0$. Hence it remains to express multiplications $e_1j_k$ and $j_ke_1$, which follows immediately from the occurrence of $j_k$ in $e_{a_k}Je_{b_k}$.

If $\mu(J(R)) = \begin{pmatrix} 1 & 0 \\ 0 & 1 \end{pmatrix}$, then $j_1 \in e_0J(R)e_0$, $j_2 \in e_1J(R)e_1$, $e_1j_1 = j_1e_1 = 0$ and $e_1j_2 = j_2e_1 = j_2$. Thus $R \cong F_0 \times F[x]/\langle x^2\rangle$ and one of the possibilities (1)-(5) holds true.

If $\mu(J(R)) = \begin{pmatrix} 0 & 1 \\ 0 & 1 \end{pmatrix}$, then $j_1 \in e_0 J(R) e_1$, $j_2 \in e_1 J(R) e_1$, $e_1 j_1 = 0$, $j_1 e_1 = j_1$ and $e_1 j_2 = j_2 e_1 = j_2$. Hence $R \cong \left\{ \begin{pmatrix} 0 & 0 & b \\ 0 & a & c \\ 0 & 0 & a \end{pmatrix} \mid a, b, c \in F \right\}$. Hence one of the possibilities (6)-(12) holds true.

If If $\mu(J(R)) = \begin{pmatrix} 1 & 0 \\ 1 & 0 \end{pmatrix}$, then $j_1 \in e_0 J(R) e_0$, $j_2 \in e_1 J(R) e_0$, $e_1 j_1 = e_1 j_2 = j_1 e_1 = 0$ and $j_2 e_1 = j_2$. Thus $R \cong \begin{pmatrix} F & 0 & F \\ 0 & 0 & F \\ 0 & 0 & 0 \end{pmatrix}$ and one of the possibilities (13)-(19) holds true

If $\mu(J(R)) = \begin{pmatrix} 1 & 1 \\ 0 & 0 \end{pmatrix}$, then $j_1 \in e_0 J(R) e_0$ and $j_2 \in e_0 J(R) e_1$. Hence $e_1 j_1 = j_2 e_1 = j_1 e_1 = 0$ and $e_1 j_2 = j_2$, and so $R \cong \begin{pmatrix} F & 0 & 0 \\ 0 & 0 & 0 \\ F & F & 0 \end{pmatrix}$. Similarly as in (c), we need not to describe $R_1$ and $R_2$, since the case is antiisomorphic to the case (d).

If $\mu(J(R)) = \begin{pmatrix} 0 & 1 \\ 1 & 0 \end{pmatrix}$, then $j_1 \in e_0 J(R) e_1$, $j_2 \in e_1 J(R) e_0$, $e_1 j_1 = j_2 e_1 = 0$, $j_1 e_1 = j_1$ and $e_1 j_2 = j_2$, which implies that $R \cong \left\{ \begin{pmatrix} a & b & 0 \\ 0 & 0 & c \\ 0 & 0 & a \end{pmatrix} \mid a, b, c \in F \right\}$. Hence one of the possibilities (20)-(26) holds true.

The the possibilities (27)-(29) are clear.

Finally assume that $J^2(R) \neq 0$. Then $\dim_F(J(R)/J^2(R)) = \dim_F(J^2(R)) = 1$ and there exist a base $j_1, j_2$ of $J(R)$ such that $j_1^2 = j_2 \in J^2(R)$. Note that $j_1 \in e_0 J(R) e_0$ since $j_1^2 \neq 0$ and $e_0 J(R) \neq 0$. Hence we get $R \cong F \times (xF[x]/\langle x^3 \rangle)$ and the possibilities (30)-(22) are clear.

**(VI)** Let $m = 2$ and $R \neq A$. Clearly, $A/J(R) \cong F^3$ and either $e_0 J(R) \neq 0$ or $J(R) e_0 \neq 0$. As a similar calculation in (V), we get that:

If $e_0 J(R) \neq 0$ then $R \cong F^2 \times F_0$ and one of the possibilities (1)-(4) holds.

If either $e_0 J(R) e_1 \neq 0$ or $e_0 J(R) e_2 \neq 0$, then $R \cong F \times \begin{pmatrix} F & 0 \\ F & 0 \end{pmatrix}$ and one of the possibilities (5)-(11) holds.

If either $e_1 J(R) e_0 \neq 0$ or $e_2 J(R) e_0 \neq 0$, then $R \cong F \times \begin{pmatrix} F & F \\ 0 & 0 \end{pmatrix}$ and one of the possibilities (12)-(18) holds. □

We conclude the paper with the following problems for further studies.

**Problem 1.** Describe all finite dimensional algebras over a field which are sums of two subalgebras.

**Problem 2.** Describe all finite dimensional algebras over a field which are sums of two nilpotent subalgebras.

**Problem 3.** Describe all finite dimensional algebras which are sums of two matrix algebras over a field.

**Acknowledgements:** First author wishes to thank Prof. Edmund Puczylowski for his valuable discussions during the preparation of this paper. We warmly thank the referee for carefully reading and commenting a first version of this article.

## References

[1] K. I. Beidar and A. V. Mikhalev, *Generalized polynomial identities and rings that are sums of two subrings*, Algebra i Logika, 34(1) (1995), 3-11.

[2] L. A. Bokut, *Imbeddings into simple associative algebras*, Algebra i Logika, 15(2) (1976), 117-142.

[3] B. Felzenszwalb, A. Giambruno and G. Leal, *On rings which are sums of two PI-subrings: a combinatorial approach*, Pacific J. Math., 209(1) (2003), 17-30.

[4] O. H. Kegel, *Zur Nilpotenz gewisser assoziativer Ringe*, Math. Ann., 149 (1962/63), 258-260.

[5] O. H. Kegel, *On rings that are sums of two subrings*, J. Algebra, 1 (1964), 103-109.

[6] A. V. Kelarev, *A sum of two locally nilpotent rings may be not nil*, Arch. Math. (Basel), 60 (1993), 431-435.

[7] M. Kepczyk, *Note on algebras which are sums of two PI subalgebras*, J. Algebra Appl., 14 (2015), 1550149 (10 pp).

[8] M. Kepczyk, *A note on algebras that are sums of two subalgebras*, Canad. Math. Bull., 59 (2016), 340-345.

[9] M. Kepczyk, *A ring which is a sum of two PI subrings is always a PI ring*, Israel J. Math., 221(1) (2017), 481-487.

[10] M. Kepczyk and E. R. Puczylowski, *On radicals of rings which are sums of two subrings*, Arc. Math. (Basel), 66(1) (1996), 8-12.

[11] M. Kepczyk and E. R. Puczylowski, *Rings which are sums of two subrings*, Ring Theory (Miskolc, 1996), J. Pure App. Algebra, 133(1-2) (1998), 151-162.

[12] M. Kepczyk and E. R. Puczylowski, *Rings which are sums of two subrings satisfying polynomial identities*, Comm. Algebra, 29(5) (2001), 2059-2065.

[13] M. Kepczyk and E. R. Puczylowski, *On the structure of rings which are sums of two subrings*, Arc. Math. (Basel), 83(5) (2004), 429-436.

[14] G. Köthe, *Die Struktur der Ringe, deren Restklassenring nach dem Radikal vollstanding irreduzibel ist.*, Math. Z., 32 (1930), 161-186.

[15] A. Smoktunowicz, *On some results related to Köthe's conjecture*, Serdica Math. J., 27 (2001), 159-170.

[16] A. Smoktunowicz, *A simple nil ring exists*, Comm. Algebra, 30(1) (2002), 27-59.

[17] B. Stenström, Rings of Quotients: Die Grundlehren der Mathematischen Wissenschaften, Band 217, An introduction to methods of ring theory, Springer-Verlag, New York-Heidelberg, 1975.

**M. Tamer Koşan** (Corresponding Author)
Department of Mathematics
Faculty of Sciences
Gazi University
Ankara, Turkey
e-mail: mtamerkosan@gazi.edu.tr, tkosan@gmail.com

**Jan Žemlička**
Department of Algebra
Charles University in Prague
Faculty of Mathematics and Physics
Sokolovská 83, 186 75 Praha 8, Czech Republic
e-mail: zemlicka@karlin.mff.cuni.cz

# 14

# ON QUASI COMULTIPLICATION MODULES OVER PULLBACK RINGS

S. Ebrahimi Atani and F. Esmaeili Khalil Saraei

Communicated by Alberto Facchini

ABSTRACT. We classify all indecomposable quasi comultiplication modules over pullback of two Dedekind domains. We extend the definitions and the results of comultiplication modules over pullback rings to a more general quasi comultiplication modules case.

**Keywords**: Pullback, separated module, quasi comultiplication module, Dedekind domain, pure-injective module

## 1. Introduction

One of the aims of the modern representation theory is to solve classification problems for subcategories of modules over a unitary ring $R$. We make the general point that over most rings it is impossible to classify all modules: even algebras of tame representation type typically are "wild" when their infinitely generated representations are considered. The reader is referred to [3], [24], [25, Chapter 1 and 6] and [26] for a detailed discussion of classification problems, their representation types (finite, tame, or wild), and useful computational reduction procedures. Pure-injective modules seem to form one of the classes of modules which arise in practice and where there is hope of some kind of classification. Pure-injective modules play a central role in the model theory of modules. Let $R_i$ be a local Dedekind domain, $\bar{R}$ be a common field and let $v_i : R_i \to \bar{R}$ be a homomorphism of $R_i$ onto $\bar{R}$ for both $i = 1, 2$. Denote the pullback $R = \{(r_1, r_2) \in R_1 \oplus R_2 : v_1(r_1) = v_2(r_2)\}$ by $(R_1 \xrightarrow{v_1} \bar{R} \xleftarrow{v_2} R_2)$, where $\bar{R} = R_1/J(R_1) = R_2/J(R_2)$. Then $R$ is a ring under coordinate-wise multiplication. Denote the kernel of $v_i$, $i = 1, 2$, by $P_i$. Then $\mathrm{Ker}(R \to \bar{R}) = P = P_1 \times P_2$, $R/P \cong \bar{R} \cong R_1/P_1 \cong R_2/P_2$, and $P_1P_2 = P_2P_1 = 0$ (so $R$ is not a domain). Furthermore, there is an exact sequence $0 \to P_i \to R \to R_j \to 0$ of $R$-modules (see [21]), for $i \neq j$. For such a pullback ring $R$, indecomposable pure-injective modules with finite-dimensional top (for any module $M$ we define its top as $M/rad(M)$) over $R$ have already been classified by the first author [5]. Also, the classification of an arbitrary indecomposable pure-injective module over the $\bar{R}$-algebra $\bar{R}[x,y : xy = 0]_{(x,y)}$ which is the pullback $(\bar{R}[x]_{(x)} \to \bar{R} \leftarrow \bar{R}[y]_{(y)})$ (see [2, Section 6]) appears to be a very difficult problem. Therefore the classification of subclass of pure-injective modules over a pullback of two local Dedekind domains over a common factor field is very important. One point of this paper is to introduce a subclass of pure-injective modules over such rings. Indeed, this article includes the classification of all indecomposable quasi comultiplication modules over $\bar{R}[x,y : xy = 0]_{(x,y)}$.

Modules over pullback rings have been studied by several authors (see for example [4], [8], [9], [11], [12], [14], [15], [18], [19], [23] and [28]). Notably, there is the monumental work of Levy [22], resulting in the classification of all finitely generated indecomposable modules over Dedekind-like rings. Common to all these classification is the reduction to a "matrix problem" over a division ring (see [25, Section 17.9] for background on matrix problems and their applications).

In the present paper we introduce a new class of $R$-modules, called quasi comultiplication modules (see Definition 2.1), and we study them in detail from the classification point of view. We are mainly interested in the case where $R$ is either a Dedekind domain or a pullback ring of two local Dedekind domains. The classification is divided into two stages: the description of all indecomposable separated quasi comultiplication $R$-modules and then, using this list of separated quasi comultiplication modules, we show that the only non-zero indecomposable quasi comultiplication non-separated $R$-module, up to isomorphism, is $E(R/P)$, the $R$-injective hull of $R/P$. For the sake of completeness, we state some definitions and notations used throughout. In this paper all rings are commutative with identity and all modules are unitary.

**Definition 1.1.** An $R$-module $S$ is defined to be separated if there exist $R_i$-modules $S_i$, $i = 1,2$, such that $S$ is a submodule of $S_1 \oplus S_2$ (the latter is made into an $R$-module by setting $(r_1,r_2)(s_1,s_2) = (r_1s_1, r_2s_2)$).

Equivalently, $S$ is separated if it is a pullback of an $R_1$-module and an $R_2$-module and then, using the same notation for pullbacks of modules as for rings, $S = (S/P_2S \to S/PS \leftarrow S/P_1S)$ [21, Corollary 3.3] and $S \subseteq (S/P_2S) \oplus (S/P_1S)$. Also, we show $S$ is separated if and only if $P_1S \cap P_2S = 0$ [21, Lemma 2.9].

If $R$ is a pullback ring, then every $R$-module is an epimorphic image of a separated $R$-module, indeed every $R$-module has a "minimal" such representation: a separated representation of an $R$-module $M$ is an epimorphism $\varphi = (S \xrightarrow{f} S' \to M)$ of $R$-modules where $S$ is separated and, if $\varphi$ admits a factorization $\varphi : S \xrightarrow{f} S' \to M$

with $S'$ separated, then $f$ is one-to-one. The module $K = \mathrm{Ker}(\varphi)$ is an $\bar{R}$-module, since $\bar{R} = R/P$ and $PK = 0$ [21, Proposition 2.3]. An exact sequence $0 \to K \to S \to M \to 0$ of $R$-modules with $S$ separated and $K$ an $\bar{R}$-module is a separated representation of $M$ if and only if $P_iS \cap K = 0$ for each $i$ and $K \subseteq PS$ [21, Proposition 2.3]. Every module $M$ has a separated representation, which is unique up to isomorphism [21, Theorem 2.8]. Moreover, $R$-homomorphisms lift to a separated representation, preserving epimorphisms and monomorphisms [20, Theorem 2.6].

Now, in the following definition, we have collected several notions, which we use.

**Definition 1.2.** (a) If $R$ is a ring and $N$ is a submodule of an $R$-module $M$, then the ideal $\{r \in R : rM \subseteq N\}$ is denoted by $(N :_R M)$. The ideal $(0 :_R M)$ is the annihilator of $M$.

(b) A proper ideal $I$ of $R$ is said to be quasi-prime if for each pair of ideals $A$ and $B$ of $R$, $A \cap B \subseteq I$ yields either $A \subseteq I$ or $B \subseteq I$ (see [16]). A proper submodule $N$ of an $R$-module $M$ is called quasi-prime if $(N :_R M)$ is a quasi-prime ideal of $R$. The set of all quasi-prime submodules of $M$ is denoted by $qSpec_R(M)$ (see [1]). Every maximal submodule of an $R$-module $M$ is prime and every prime submodule of $M$ is a quasi-prime submodule. Therefore $Max_R(M) \subseteq Spec_R(M) \subseteq qSpec_R(M)$ by [1, Remark 2.3].

(c) An $R$-module $M$ is defined to be a comultiplication module if for each submodule $N$ of $M$, $N = (0 :_M I)$, for some ideal $I$ of $R$. In this case we can take $I = \mathrm{Ann}(N)$.

(d) A submodule $N$ of an $R$-module $M$ is called pure submodule if any finite system of equations over $N$ which is solvable in $M$ is also solvable in $N$. A submodule $N$ of an $R$-module $M$ is called relatively divisible (or an $RD$-submodule) in $M$ if $rN = N \cap rM$ for all $r \in R$.

(e) A module $M$ is pure-injective if it has the injective property relative to all pure exact sequences.

(f) A ring $R$ is called to be a serial ring, if the set of all ideals of $R$ is linearly ordered.

**Remark 1.3.** (1) An $R$-module $M$ is pure-injective if and only if it is algebraically compact (see [17] and [27]).

(2) Let $R$ be a Dedekind domain, $M$ an $R$-module and $N$ a submodule of $M$. Then $N$ is pure in $M$ if and only if $IN = N \cap IM$ for each ideal $I$ of $R$. Moreover, $N$ is pure in $M$ if and only if $N$ is an $RD$-submodule of $M$ [27].

## 2. Some properties of quasi comultiplication modules

In this section, we collect some basic properties concerning quasi comultiplication modules. We begin with the key definition of this paper.

**Definition 2.1.** Let $R$ be a commutative ring and $M$ be an $R$-module. Then $M$ is defined to be a quasi comultiplication module if $\mathrm{qSpec}(M) = \emptyset$ or for every quasi-prime submodule $N$ of $M$, $N = (0 :_M I)$ for some ideal $I$ of $R$.

One can easily show that if $M$ is a quasi comultiplication module, then $N = (0 :_M \mathrm{ann}(N))$ for every quasi-prime submodule $N$ of $M$. It is easy to see that the class of quasi comultiplication modules contains the class of comultiplication modules defined in [7].

**Lemma 2.2.** *Let $R$ be a commutative ring and $M$ be an $R$-module. If $I, J$ are proper ideals of $R$ and $I \subseteq (0 :_R M)$, then the following hold:*

(1) *If $I \subseteq J$, then $J$ is a quasi-prime ideal of $R$ if and only if $J/I$ is a quasi-prime ideal of $R/I$.*

(2) *If $N$ is a proper submodule of $M$, then $N$ is a quasi-prime $R$-submodule of $M$ if and only if $N$ is a quasi-prime $R/I$-submodule of $M$.*

(3) *$M$ is a quasi comultiplication $R$-module if and only if $M$ is a quasi comultiplication $R/I$-module.*

**Proof.** (1) The proof is straightforward.
(2) It is easy to see that $(N :_R M)/I = (N :_{R/I} M)$. So the result follows from Part(1).
(3) One can show that $(0 :_R N)/I = (0 :_{R/I} N)$. So the result follows from Part (2). □

**Lemma 2.3.** *Let $R$ be a commutative ring and $M$ be an $R$-module. If $N \subseteq L$ are submodules of $M$, then the following hold:*

(1) *$L$ is a quasi-prime submodule of $M$ if and only if $L/N$ is a quasi-prime submodule of $M/N$.*

(2) *If $M$ is a quasi comultiplication $R$-module and $N$ is a pure submodule of $M$, then $M/N$ is a quasi comultiplication $R$-module.*

(3) *If $M$ is a quasi comultiplication $R$-module, then every direct summand of $M$ is a quasi comultiplication $R$-module.*

**Proof.** (1) The proof is straightforward, since $(L :_R M) = (L/N :_R M/N)$.
(2) Let $M$ be a quasi comultiplication module and let $L/N$ be a quasi-prime submodule of $M/N$. Then by (1), $L$ is a quasi-prime submodule of $M$, so $L = (0 :_M I)$

for some ideal $I$ of $R$. Now we show that $L/N = (0 :_{M/N} I)$. Since $L = (0 :_M I)$, so $I(L/N) = (IL + N)/N = 0_{M/N}$. Hence $L/N \subseteq (0 :_{M/N} I)$. Assume that $m + N \in (0 :_{M/N} I)$. Then $a(m+N) = am + N = 0_{M/N}$ for every $a \in I$. Therefore $am \in N$ and then $Im \subseteq N \cap IM = IN = 0$, since $IN \subseteq IL = 0$. Then $m \in L$, so $m + N \in L/N$ and we have the equality.

(3) The result follows from (2), since direct summands are pure submodules. □

**Proposition 2.4.** *Let $R$ be a local Dedekind domain and $M$ be an $R$-module. Then the following hold:*

(1) *Every proper submodule of $M$ is quasi-prime.*
(2) *$M$ is a quasi comultiplication $R$-module if and only if $M$ is a comultiplication $R$-module.*
(3) *If $M$ is a quasi comultiplication $R$-module, then $M$ is indecomposable.*

**Proof.** (1) Since every local Dedekind domain is a serial ring, the proof follows from [1, Lemma 2.4].

(2) Follows from (1).

(3) Let $M$ be a quasi comultiplication $R$-module such that $M = N \oplus K$ with $N \neq 0$ and $K \neq 0$. By (1), there are positive integers $m, n$, with $m < n$, such that $M = (0 :_M P^n) + (0 :_M P^m) = (0 :_M P^n)$ and this contradicts $N \cap K = 0$. Thus either $N = 0$ or $K = 0$, as required. □

**Theorem 2.5.** *Let $R$ be a local Dedekind domain with a unique maximal ideal $P = Rp$. Then the quasi comultiplication modules over $R$ are:*

(1) *$R/P^n$, $n \geq 1$;*
(2) *$E(R/P)$, the injective hull of $R/P$.*

**Proof.** The result follows from Proposition 2.4 and [7, Theorem 2.5]. □

## 3. The separated quasi comultiplication modules

Throughout this paper we shall assume unless otherwise stated, that

$$R = (R_1 \xrightarrow{v_1} \bar{R} \xleftarrow{v_2} R_2) \tag{1}$$

is the pullback of two local Dedekind domains $R_1, R_2$ with maximal ideals $P_1, P_2$ generated respectively by $p_1, p_2$, $P$ denotes $P_1 \oplus P_2$ and $R_1/P_1 \cong R_2/P_2 \cong R/P \cong \bar{R}$ is a field. In particular, $R$ is a commutative Noetherian local ring with unique maximal ideal $P$. The other prime ideals of $R$ are easily seen to be $P_1$ (that is $P_1 \oplus 0$) and $P_2$ (that is $0 \oplus P_2$).

**Theorem 3.1.** *Let $R$ be the pullback ring as described in (1), $M$ be an $R$-module and $N$ be a proper submodule of $M$. Then $N$ is a quasi-prime submodule of $M$ if and only if either $(P_1 \oplus 0)M \subseteq N$ or $(0 \oplus P_2)M \subseteq N$.*

**Proof.** Let $N$ be a quasi-prime submodule of $M$. Since $(P_1 \oplus 0) \cap (0 \oplus P_2) = 0 \subseteq (N :_R M)$, we have that $P_1 \oplus 0 \subseteq (N :_R M)$ or $0 \oplus P_2 \subseteq (N :_R M)$. Therefore $(P_1 \oplus 0)M \subseteq N$ or $(0 \oplus P_2)M \subseteq N$. Conversely, suppose that $N$ is a proper submodule of $M$ and $(P_1 \oplus 0)M \subseteq N$. Since $R_2$ is a local Dedekind domain, so $N/(P_1 \oplus 0)M$ is a quasi-prime $R_2$-submodule of $M/(P_1 \oplus 0)M$ by Proposition 2.4. So $N/(P_1 \oplus 0)M$ is a quasi-prime $R$-submodule of $M/(P_1 \oplus 0)M$ by Lemma 2.2. Then $N$ is a quasi-prime $R$-submodule of $M$ by Lemma 2.3. So $(N :_R M)/(P_1 \oplus 0)$ is a quasi-prime ideal, then $(N :_R M)$ is a quasi-prime ideal of $R$ by Lemma 2.2 and so $N$ is a quasi-prime submodule of $M$. □

**Remark 3.2.** Let $R$ be the pullback ring as described in (1), and let $T$ be an $R$-submodule of a separated module $S = (S_1 \xrightarrow{f_1} \bar{S} \xleftarrow{f_2} S_2)$, with projection maps $\pi_i : S \to S_i$. Set

$$T_1 = \{t_1 \in S_1 : (t_1, t_2) \in T \text{ for some } t_2 \in S_2\},$$

$$T_2 = \{t_2 \in S_2 : (t_1, t_2) \in T \text{ for some } t_1 \in S_1\}.$$

Then for each $i$, $i = 1, 2$, $T_i$ is an $R_i$-submodule of $S_i$ and $T \leq T_1 \oplus T_2$. Moreover, we can define a mapping $\pi'_1 = \pi_1|T : T \to T_1$ by sending $(t_1, t_2)$ to $t_1$; hence $T_1 \cong T/((0 \oplus Ker(f_2)) \cap T) \cong T/(T \cap P_2S) \cong (T + P_2S)/P_2S \subseteq S/P_2S$. So we may assume that $T_1$ is a submodule of $S_1$. Similarly, we may assume that $T_2$ is a submodule of $S_2$ (note that $\text{Ker}(f_1) = P_1S_1$ and $\text{Ker}(f_2) = P_2S_2$).

**Proposition 3.3.** *Let $R$ be the pullback ring as described in (1) and $S$ be a non-zero separated $R$-module with $\bar{S} = 0$. Then $(0 :_R S) \in \{P_1 \oplus 0, 0 \oplus P_2, 0\}$.*

**Proof.** It is clear that $S = PS$ since $\bar{S} = 0$. First suppose that $(0 :_R S) = P_1^n \oplus P_2^m$ for some positive integers $n$ and $m$. Now we consider the various possibilities for $m$ and $n$.

**Case 1.** If $n > 1$ and $m > 1$, then $(P_1^{n-1} \oplus P_2^{m-1})S = (P_1^{n-1} \oplus P_2^{m-1})PS = (P_1^n \oplus P_2^m)S = 0$. So $P_1^{n-1} \oplus P_2^{m-1} \subseteq (0 :_R S)$ which is a contradiction.

**Case 2.** If $n = 1$ and $m > 1$, then $(P_1 \oplus P_2^{m-1})S = (P_1 \oplus P_2^{m-1})PS = (P_1^2 \oplus P_2^m)S = 0$ since $P_1^2 \oplus P_2^m \subseteq P_1 \oplus P_2^m = (0 :_R S)$. So $P_1 \oplus P_2^{m-1} \subseteq (0 :_R S)$ which is a contradiction.

**Case 3.** If $n > 1$ and $m = 1$, then the proof is similar to Case 2.

**Case 4.** If $n = 1$ and $m = 1$, then $(0 :_R S) = P$. So $S = PS = 0$, which is a

contradiction. Now suppose that $(0 :_R S) = P_1^n \oplus 0$ for some positive integer $n > 1$. So $(P_1^{n-1} \oplus 0)S = (P_1^{n-1} \oplus 0)PS = (P_1^n \oplus 0)S = 0$. So $P_1^{n-1} \oplus 0 \subseteq (0 :_R S)$ which is a contradiction.
The case $(0 :_R S) = 0 \oplus P_2^m$ for some positive integer $m > 1$ is similar. □

Now, we find the separated quasi comultiplication modules over the pullback ring $R$. We begin with the following proposition.

**Proposition 3.4.** *Let $S$ be any separated quasi comultiplication module over the pullback ring as described in (1). Then the following hold:*

(1) *If $(0 :_R S) = 0$, then $\bar{S} = 0$.*
(2) *If $\bar{S} \neq 0$, then $(0 :_R S) \neq P_1^n \oplus 0$ and $(0 :_R S) \neq 0 \oplus P_2^n$ for every positive integer $n$.*
(3) *If $\bar{S} \neq 0$ and $(0 :_R S) = P_1^n \oplus P_2^m$ for some positive integers $n, m$, then either $m = 1$ or $n = 1$.*

**Proof.** (1) Suppose $\bar{S} \neq 0$. Then $PS$ is a quasi-prime submodule of $S$ by Theorem 3.1. Let $(r_1, r_2) \in (0 :_R PS)$. Then $(r_1, r_2)(p_1, p_2)S \subseteq (r_1, r_2)PS = 0$, so $r_1 p_1 = 0$ and $r_2 p_2 = 0$; hence $r_1 = 0$ and $r_2 = 0$, since $R_i$ is an integral domain for $i = 1, 2$. Therefore, $(0 :_R PS) = 0$. Then $S$ quasi comultiplication gives $PS = (0 :_S (0 :_R PS)) = (0 :_S 0) = S$, which is a contradiction.
(2) Let $(0 :_R S) = P_1^n \oplus 0$. If $(0 \oplus P_2)S = 0$, then $0 \oplus P_2 \subseteq P_1^n \oplus 0$, which is a contradiction. So $(0 \oplus P_2)S \neq 0$ and $(0 :_R (0 \oplus P_2)S) \neq R$. Then $(0 :_R (0 \oplus P_2)S) = P_1 \oplus 0$ by [10, Proposition 3.8]. Moreover, by Theorem 3.1, $(0 \oplus P_2)S$ is a quasi-prime submodule of $S$, so $(0 \oplus P_2)S = (0 :_S P_1 \oplus 0)$ since, $S$ is quasi comultiplication. We may assume that $n > 1$. Since $(P_1 \oplus 0)(P_1^{n-1} \oplus P_2)S = 0$, we must have $(P_1^{n-1} \oplus P_2)S \subseteq (0 :_S P_1 \oplus 0) = (0 \oplus P_2)S$. Let $s_1 \in S_1$. Then there is an element $s_2 \in S_2$ such that $(s_1, s_2) \in S$. Hence $(p_1^{n-1}, p_2)(s_1, s_2) \in (0 \oplus P_2)S$; hence $p_1^{n-1} s_1 = 0$ and so $P_1^{n-1} S_1 = 0$. Therefore, $P_1^{n-1} \subseteq (0 :_{R_1} S_1) = P_1^n$ by [10, Proposition 3.6], which is a contradiction. Thus $(0 :_R S) \neq P_1^n \oplus 0$ for every positive integer $n$. Similarly, $(0 :_R S) \neq 0 \oplus P_2^n$ for every positive integer $n$.
(3) Suppose not. We may assume that $n > 1$ and $m > 1$. Clearly, $0 \neq (P_1 \oplus 0)S \subseteq PS \neq S$, $0 \neq (0 \oplus P_2)S \subseteq PS \neq S$, and they are quasi-prime submodules of $S$ by Theorem 3.1. Since $S$ is a quasi comultiplication $R$-module, we must have $(P_1 \oplus 0)S = (0 :_S (0 :_R (P_1 \oplus 0)S)) = (0 :_S (P_1^{n-1} \oplus P_2))$ and $(0 \oplus P_2)S = (0 :_S (P_1 \oplus P_2^{m-1}))$ by [10, Lemma 3.4]. Let $s_1 \in S_1$. There exists $s_2 \in S_2$ such that $(s_1, s_2) \in S$. It follows that $p_1^n s_1 = 0$ and $p_2^m s_2 = 0$ by [10, Proposition 3.6]. Therefore, $(p_1, p_2^{m-1})(p_1^{n-1} s_1, p_2 s_2) = 0$, so $(p_1^{n-1} s_1, p_2 s_2) \in (0 :_S P_1 \oplus P_2^{m-1}) =$

$(0 \oplus P_2)S$; hence $p_1^{n-1}s_1 = 0$. By a similar way, we get $p_1 s_1 = 0$. Therefore, $P_1 S_1 \cong (P_1 \oplus 0)S = 0$, which is a contradiction. □

**Theorem 3.5.** *Let $R$ be the pullback ring as described in (1), and let $S = (S_1 \to \bar{S} \leftarrow S_2)$ be a separated $R$-module. Then $S$ is a quasi comultiplication $R$-module if and only if each $S_i$ is a quasi comultiplication $R_i$-module, for both $i = 1, 2$.*

**Proof.** Let $S$ be a quasi comultiplication $R$-module. If $\mathrm{qSpec}(S) = \emptyset$, then $(P_1 \oplus 0)S = (0 \oplus P_2)S = S$ by Theorem 3.1. Then for each $i = 1, 2$, $S_i = 0$ is a quasi comultiplication $R_i$-module by [21, Corollary 3.3]. So, we may assume that $\mathrm{qSpec}(S) \neq \emptyset$. If $\bar{S} = 0$. Then by [5, Lemma 2.7], $S = S_1 \oplus S_2$; hence for each $i = 1, 2$; $S_i$ is a quasi comultiplication $R$-module by Lemma 2.2. Therefore for each $i = 1, 2$; $S_i$ is a quasi comultiplication $R_i$-module by Lemma 2.2. So, we may assume that $\bar{S} \neq 0$.
Let $L$ (resp. $L'$) be a quasi-prime submodule of $S_1$ (resp $S_2$). Then there exists a separated submodule $T = (T/P_2S = T_1 \xrightarrow{g_1} \bar{T} = T/PT \xleftarrow{g_2} T_2 = T/P_1T)$ (resp. $T' = (T'/P_2T' = T_1' \xrightarrow{g_1'} \bar{T}' = T'/PT' \xleftarrow{g_2'} T_2' = T'/P_1S))$ of $S$, where $g_i$ (resp. $g_i'$) is the restriction of $f_i$ over $T_i$ (resp. $T_i'$), $i = 1, 2$ such that $L = T_1$ (resp. $L' = T_2'$). Since $(0 \oplus P_2)S \subseteq T$ $((P_1 \oplus 0)S \subseteq T')$; hence $T$ (resp. $T'$) is a proper quasi-prime $R$-submodule of $S$ by Theorem 3.1. We split the proof into two cases for $(0 :_R S)$ by Proposition 3.4.
**Case 1.** $(0 :_R S) = P_1 \oplus P_2^m$ for some positive integer $m$. If $m = 1$, then $(0 :_R S) = P_1 \oplus P_2 = P$. Hence $PT \subseteq PS = 0$ and so $P \subseteq (0 :_R T)$. Then we have $(0 :_R T) = P$. Thus $S$ is quasi comultiplication implies that $T = (0 :_S P) = S$, which is a contradiction. So we may assume that $m > 1$. By [10, Proposition 3.6], $(0 :_{R_1} S_1) = P_1$ and $(0 :_{R_2} S_2) = P_2^m$. Since $(P_1 \oplus 0)S \cong P_1 S_1 = 0$ and $(0 \oplus P_2)S \subseteq T$, we get $PS \subseteq T \subseteq S$, so $(0 :_R S) \subseteq (0 :_R T) \subseteq (0 :_R PS)$; thus $P_1 \oplus P_2^m \subseteq (0 :_R T) \subseteq P_1 \oplus P_2^{m-1}$ by [10, Proposition 3.7]. Therefore, either $(0 :_R T) = P_1 \oplus P_2^m$ or $(0 :_R T) = P_1 \oplus P_2^{m-1}$. Since $S$ is quasi comultiplication, we have either $T = (0 :_S P_1 \oplus P_2^m) = S$ or $T = (0 :_S P_1 \oplus P_2^{m-1}) = PS$; hence $T = PS$ and $T_1 = (PS)/PS = 0$. Then $L = T_1 = (0 :_{S_1} R_1)$ gives $S_1$ is quasi comultiplication. Now we will prove that $S_2$ is a quasi comultiplication $R_2$-module. By hypothesis, $T' = (0 :_S P_1^s \oplus P_2^t)$ for some positive integers $s, t$. We show that $T_2' = (0 :_{S_2} P_2^m)$. Since the inclusion $T_2' \subseteq (0 :_{S_2} P_2^m)$ is clear, we will prove the reverse inclusion. Let $s_2 \in (0 :_{S_2} P_2^m)$. Then $P_2^m s_2 = 0$ and there exists $s_1 \in S_1$ such that $(s_1, s_2) \in S$, so $(P_1^s \oplus P_2^t)(s_1, s_2) = 0$; hence $(s_1, s_2) \in T'$. Therefore, $s_2 \in T_2'$, and so we have the equality.

**Case 2.** $(0 : S) = P_1^m \oplus P_2$ for some positive integer $m$. The proof is similar to that in Case 1.

Conversely, assume that $S_i$ is a quasi comultiplication $R_i$-module for each $i$, $i = 1, 2$ and let $T$ be a proper quasi-prime submodule of $S$. We consider the various possibilities for $(0 :_R T)$.

**Case 1.** If $(0 :_R T) = 0$, then $(0 :_{R_1} T_1) = 0$ and $(0 :_{R_2} T_2) = 0$. So $T_1 = S_1$ and $T_2 = S_2$ implies that $T = S$ which is a contradiction.

**Case 2.** If $(0 :_R T) = P_1^n \oplus P_2^m$ for some positive integer $n$ and $m$, then $T_1 = (0 :_{S_1} P_1^n)$ and $T_2 = (0 :_{S_2} P_2^m)$ by Proposition 2.4 and [10, Proposition 3.6]. So $T = (0 :_S P_1^n \oplus P_2^m)$.

**Case 3.** If $(0 :_R T) = P_1^n \oplus 0$ for some positive integer $n$, then $T_1 = (0 :_{S_1} P_1^n)$, $T_2 = S_2$ by Proposition 2.4 and [10, Proposition 3.6]. So it is easy to see that $T = (0 :_S P_1^n \oplus 0)$.

The case $(0 :_R T) = 0 \oplus P_2^m$ is similar. So $S$ is a quasi comultiplication $R$-module. □

**Proposition 3.6.** *Let $R$ be the pullback ring as described in (1), and let $S = (S_1 \xrightarrow{f_1} \bar{S} \xleftarrow{f_2} S_2)$ be a separated quasi comultiplication $R$-module with $\bar{S} \neq 0$. Then $S$ is an indecomposable $R$-module.*

**Proof.** Let $S$ be a separated quasi comultiplication module. Then, for both $i$, $i = 1, 2$, $S_i$ is a quasi comultiplication $R_i$-module by Theorem 3.5. Therefore for both $i$, $i = 1, 2$, $S_i$ is an indecomposable $R_i$-module by Proposition 2.4, and so, $S$ is an indecomposable $R$-module by [5, Lemma 2.7]. □

**Theorem 3.7.** *Let $R$ be the pullback ring as described in (1). Then the indecomposable separated quasi comultiplication modules over $R$ are:*

(I) $S = (E(R_1/P_1) \to 0 \leftarrow 0)$ *and* $S = (0 \to 0 \leftarrow E(R_2/P_2)$, *where* $E(R_i/P_i)$ *is the* $R_i$*-injective hull of* $R_i/P_i$ *for* $i = 1, 2$;

(II) $S = (R_1/P_1^n \to \bar{R} \leftarrow R_2/P_2^m)$.

**Proof.** By [5, Lemma 2.8], these modules are indecomposable and by Theorems 2.5 and 3.5 they are quasi comultiplication.

Now, let $S$ be an indecomposable separated quasi comultiplication $R$-module. First, suppose that $S = PS$. Then by [5, Lemma 2.7 (i)], $S = S_1$ or $S_2$ and so, $S$ is an indecomposable quasi comultiplication $R_i$-module for some $i$ and, since $PS = S$, is type (I) in the list. So we may assume that $S \neq PS$. By Theorem 3.5, $S_i$ is a quasi comultiplication $R_i$-module, for each $i = 1, 2$. Hence, by the structure of quasi comultiplication modules over a local Dedekind domain (see Theorem 2.5), we must have $S_i = E(R_i/P_i)$ or $R_i/P_i^n$ $(n \geq 1)$. Since $S$ is indecomposable and $S/PS \neq 0$, it follows that for each $i = 1, 2$, $S_i$ is torsion and it is not divisible $R_i$-module. Then there are positive integers $m, n$ and $k$ such that $P_1^m S_1 = 0$, $P_2^k S_2 = 0$ and $P^n S = 0$. For $t \in S$, let $o(t)$ denote the least positive integer $l$ such that $P^l t = 0$. Now choose $t \in S_1 \cup S_2$ with $\bar{t} \neq 0$ and such that $o(t)$ is maximal. There exists a $t = (t_1, t_2)$ such that $o(t) = n$, $o(t_1) = m$ and $o(t_2) = k$. Then $R_i t_i$ is pure in $S_i$ for $i = 1, 2$ (see [5, Theorem 2.9]). Therefore, $R_1 t_1 \cong R_1/P_1^m$ (resp. $R_2 t_2 \cong R_2/P_2^k$) is a direct summand of $S_1$ (resp. $S_2$), since for each $i$, $R_i t_i$ is pure-injective. Hence $S_1 = R_1 t_1 \cong R_1/P_1^m$ since, $S_1$ is indecomposable. Similarly, $S_2 = R_2 t_2 \cong R_2/P_2^k$. Let $\bar{M}$ be the $\bar{R}$-subspace of $\bar{S}$ generated by $\bar{t}$. Then $\bar{M} \cong \bar{R}$. Let $M = (R_1 t_1 = M_1 \rightarrow \bar{M} \leftarrow M_2 = R_2 t_2)$. Then $M$ is an $R$-submodule of $S$ which is quasi comultiplication by Theorem 3.5 and is a direct summand of $S$; this implies that $S = M$, and $S$ is as in (II) in the list (see [5, Theorem 2.9]). □

We refer to modules of type (I) in Theorem 3.7 as $P_1$-Prüfer and $P_2$-Prüfer respectively.

**Theorem 3.8.** *Let $R$ be the pullback ring as described in (1) and let $S$ be a separated quasi comultiplication $R$-module. Then $S$ has finite-dimensional top.*

**Proof.** Apply Theorem 3.7 (note that $S = U \oplus X$, where $dim_{\bar{R}}(U/PU) \leq 1$ and $X/PX = 0$). □

## 4. The non-separated quasi comultiplication modules

We continue to use the notation already established, so $R$ is the pullback ring as described in (1). In this section, we will determine all the indecomposable non-separated quasi comultiplication $R$-modules over $R$. It turns out that each can be obtained by amalgamating finitely many indecomposable separated quasi comultiplication modules.

We begin by the following lemma.

**Lemma 4.1.** *Let $R$ be a pullback ring as described in (1) and $M$, $S$ be two $R$-modules. Let $\varphi : S \longrightarrow M$ be an epimorphism.*

(1) *If $N$ is a submodule of $M$, then $(N :_R M) = (\varphi^{-1}(N) :_R S)$.*
(2) *If $T$ is a proper submodule of $S$, then $(0 :_R T) = (0 :_R \varphi(T))$.*
(3) *If either $(T :_R S) = P_1 \oplus 0$ or $(T :_R S) = 0 \oplus P_2$, then*

$$(T :_R S) = (\varphi(T) :_R M).$$

(4) *If $N$ is a quasi-prime submodule of $M$, then $\varphi^{-1}(N)$ is a quasi-prime submodule of $S$.*

(5) *If $T$ is a quasi-prime submodule of $S$, then $\varphi(T)$ is a quasi-prime submodule of $M$.*

**Proof.** (1) Suppose $(r_1, r_2) \in (N :_R M)$ and $(s_1, s_2) \in S$. Then $\varphi(s_1, s_2) \in M$ and so $\varphi(r_1 s_1, r_2 s_2) = (r_1, r_2)\varphi(s_1, s_2) \in (r_1, r_2)M \subseteq N$. Thus $(r_1, r_2)(s_1, s_2) = (r_1 s_1, r_2 s_2) \in \varphi^{-1}(N)$. Hence $(r_1, r_2)S \subseteq \varphi^{-1}(N)$ and we have $(r_1, r_2) \in (\varphi^{-1}(N) :_R S)$. For the reverse inclusion, suppose that $(r'_1, r'_2) \in (\varphi^{-1}(N) :_R S)$ and let $m \in M$. Then $\varphi(s'_1, s'_2) = m$ for some $(s'_1, s'_2) \in S$. Thus $(r'_1, r'_2)(s'_1, s'_2) \in \varphi^{-1}(N)$. Hence $(r'_1, r'_2)m = (r'_1, r'_2)\varphi(s'_1, s'_2) = \varphi(r'_1 s'_1, r'_2 s'_2) \in \varphi(\varphi^{-1}(N)) \subseteq N$, and we have the equality.
(2) First suppose that $(0 :_R T) = 0$ and $(r_1, r_2) \in (0 :_R \varphi(T))$. Therefore $\varphi((r_1, r_2)T) = (r_1, r_2)\varphi(T) = 0$, hence $(r_1, r_2)T \subseteq K$. So $(r_1, r_2)(p_1, p_2)T = 0$ since $PK = 0$ by [21, Proposition 2.4 ]. Thus $(r_1 p_1, r_2 p_2) \in (0 :_R T)$ implies that $r_1 = 0$ and $r_2 = 0$ since $R_i$ is a domain for each $i = 1, 2$. So we get $(0 :_R \varphi(T)) = 0$ and we have the equality. It is clear that $(0 :_R T) \subseteq (0 :_R \varphi(T))$. Now, we consider the possibilities for $(0 :_R \varphi(T))$:
**Case 1.** If $(0 :_R \varphi(T)) = P_1^n \oplus 0$ for some positive integer $n$, then $\varphi((P_1^n \oplus 0)T) = (P_1^n \oplus 0)\varphi(T) = 0$. Hence $(P_1^n \oplus 0)T \subseteq K \cap (P_1 \oplus 0)S = 0$ by [21, Proposition 2.3] and so, $P_1^n \oplus 0 \subseteq (0 :_R T)$ as required.
**Case 2.** If $(0 :_R \varphi(T)) = 0 \oplus P_2^m$ for some positive integer $m$, the proof is similar to Case 1.
**Case 3.** If $(0 :_R \varphi(T)) = P_1^n \oplus P_2^m$ for some positive integer $n$, $m$, then by an argument like in Case 1, we get $P_1^n \oplus 0 \subseteq (0 :_R T)$. Similarly, by Case 2, $0 \oplus P_2^m \subseteq (0 :_R \varphi(T))$ implies that $0 \oplus P_2^m \subseteq (0 :_R T)$. Thus $(0 :_R \varphi(T)) = P_1^n \oplus P_2^m \subseteq (0 :_R T)$ and we have the equality.
(3) Let $(T :_R S) = P_1 \oplus 0$. It is clear that $(P_1 \oplus 0)M \subseteq \varphi(T)$. Now, suppose that $(r_1, r_2) \in (\varphi(T) :_R M)$. It suffices to show that $r_2 = 0$. Let $(s_1, s_2) \in S$. Then $(r_1, r_2)\varphi(s_1, s_2) \in (r_1, r_2)M \subseteq \varphi(T)$. So $\varphi(r_1 s_1, r_2 s_2) = \varphi(t_1, t_2)$ for some $(t_1, t_2) \in T$. Hence $(r_1, r_2)(s_1, s_2) = (t_1, t_2) + (k_1, k_2)$ for some $(k_1, k_2) \in K$. This implies that $(r_1, r_2)S \subseteq T + K$. So $(0, p_2)(r_1, r_2)S \subseteq T$ since $(o, p_2)K \subseteq K \cap (0 \oplus P_2)S = 0$. Hence $(0, p_2 r_2) \in (T :_R S) = P_1 \oplus 0$. Then $p_2 r_2 = 0$ implies that $r_2 = 0$, since $R_2$ is a domain. Therefore $(\varphi(T) :_R M) = P_1 \oplus 0$. The case $(T :_R S) = 0 \oplus P_2$ is similar.
(4) Clear by Theorem 3.1 and Case (1).
(5) Let $T$ be a quasi-prime submodule of $S$. Then by Theorem 3.1, we may assume that $(P_1 \oplus 0)S \subseteq T$. We show that $(P_1 \oplus 0)M \subseteq \varphi(T)$. Assume that $m \in M$, then there exists $s \in S$ where $\varphi(s) = m$. Therefore $(p_1, 0)m = (p_1, 0)\varphi(s) = \varphi((p_1, 0)s) \in \varphi(T)$ for every $m \in M$. Hence $(P_1 \oplus 0)M \subseteq \varphi(T)$. Similarly, if $(0 \oplus P_2)S \subseteq T$, then $(0 \oplus P_2)M \subseteq \varphi(T)$. □

**Proposition 4.2.** *Let $R$ be the pullback ring as described in (1) and let $M$ be any $R$-module. Let $0 \to K \xrightarrow{i} S \xrightarrow{\varphi} M \to 0$ be a separated representation of $M$. Then the following hold:*

(1) $\text{qSpec}_R(S) = \emptyset$ *if and only if* $\text{qSpec}_R(M) = \emptyset$.

(2) $(0 :_R S) = 0$ *if and only if* $(0 :_R M) = 0$.

(3) *If* $(0 :_R S) = 0$, *then* $\text{qSpec}_R(S) \neq \emptyset$.

(4) *If* $(0 :_R M) = 0$, *then* $\text{qSpec}_R(M) \neq \emptyset$.

(5) *If either* $P_1 \oplus 0 \subseteq (0 :_R S)$ *or* $0 \oplus P_2 \subseteq (0 :_R S)$, *then $M$ is separated.*

**Proof.** (1) Follows from Lemma 4.1.
(2) Clear by [13, Proposision 5.2].
(3) Assume that $(0 :_R S) = 0$. If $(P_1 \oplus 0)S = S$, then $(0 \oplus P_2)S = (0 \oplus P_2)(P_1 \oplus 0)S = 0$ and so $0 \oplus P_2 \subseteq (0 :_R S)$ which is a contradiction. So $(P_1 \oplus 0)S$ is a quasi-prime submodule of $S$ by Theorem 3.1. Similarly, $(0 \oplus P_2)S \in \text{qSpec}_R(S)$.
(4) If $(0 :_R M) = 0$, then $(0 :_R S) = 0$ by Part (2). Thus $\text{qSpec}_R(S) \neq \emptyset$ by Part (3). Hence the result follows from Part (1).
(5) Let $P_1 \oplus 0 \subseteq (0 :_R S)$. Then $P_1 \oplus 0 \subseteq (0 :_R M)$ since $(0 :_R S) \subseteq (0 :_R M)$. Then $(P_1 \oplus 0)M = 0$ and $M$ is a separated $R$-module by [21, Lemma 2.9 ]. The case $0 \oplus P_2 \subseteq (0 :_R S)$ is similar. □

**Theorem 4.3.** *Let $R$ be the pullback ring as described in (1) and let $M$ be any non-separated $R$-module. Let $0 \to K \to S \to M \to 0$ be a separated representation of $M$. Then $S$ is a quasi comultiplication $R$-module if and only if $M$ is a quasi comultiplication $R$-module.*

**Proof.** Let $S$ be a quasi comultiplication $R$-module. By Proposition 4.2, we may assume that $\text{qSpec}(S) \neq \emptyset$. First suppose that $\bar{S} \neq 0$. Then either $(0 :_R S) = P_1 \oplus P_2^n$ or $(0 :_R S) = P_1^n \oplus P_2$ for some positive integer $n$ by Proposition 3.4. So either $P_1 \oplus 0 \subseteq (0 :_R S)$ or $0 \oplus P_2 \subseteq (0 :_R S)$; hence $M$ is a separated $R$-module by Proposition 4.2 which is a contradiction. Now, assume that $\bar{S} = 0$, if either $(0 :_R S) = P_1 \oplus 0$ or $(0 :_R S) = 0 \oplus P_2$, then $M$ is a separated $R$-module by Proposition 4.2 which is a contradiction. So we may assume that $(0 :_R S) = 0$ by Proposition 3.3. Let $N$ be a quasi-prime submodule of $M$. Then $\varphi^{-1}(N)$ is a quasi-prime submodule of $S$ by Lemma 4.1. So $\varphi^{-1}(N) = (0 :_S (0 :_R \varphi^{-1}(N)))$. We show that $N = (0 :_M (0 :_R \varphi^{-1}(N)))$. If $n \in N$, then $\varphi(s) = n$ for some $s \in S$. Hence $s = \varphi^{-1}(n) \in \varphi^{-1}(N)$. Let $(r_1, r_2) \in (0 :_R \varphi^{-1}(N))$. Then $(r_1, r_2)s = 0$. Then we have $(r_1, r_2)n = (r_1, r_2)\varphi(s) = \varphi((r_1, r_2)s) = 0$ and $n \in (0 :_M (0 :_R \varphi^{-1}(N)))$. Therefore $N \subseteq (0 :_M (0 :_R \varphi^{-1}(N)))$. Now assume that $m \in (0 :_M (0 :_R \varphi^{-1}(N)))$. By Theorem 3.1, we can assume that $(P_1 \oplus 0)M \subseteq N$

and so by Lemma 4.1, $(P_1 \oplus 0)S \subseteq \varphi^{-1}(N)$. Since $(0 :_R S) = 0$, it is easy to see that $(0 :_R (P_1 \oplus 0)S) = 0 \oplus P_2$. Since $(P_1 \oplus 0)S \subseteq \varphi^{-1}(N)$, we have $(0 :_R \varphi^{-1}(N)) \subseteq (0 :_R (P_1 \oplus 0)S) = 0 \oplus P_2$. Therefore $(0 :_R \varphi^{-1}(N)) = 0 \oplus P_2^k$ for some positive integer $k$. So $\varphi^{-1}(N) = (0 :_S 0 \oplus P_2^k)$ and $m \in (0 :_M 0 \oplus P_2^k)$. There exists $t = (t_1, t_2) \in S$ such that $\varphi(t) = \varphi(t_1, t_2) = m$. Thus $\varphi(0, p_2^k t_2) = \varphi((0, p_2^k)(t_1, t_2)) = (0, p_2^k)m = 0$. Therefore $(0, p_2^k t_2) \in K \cap (0 \oplus P_2)S = 0$ and so $p_2^k t_2 = 0$ by [21, Proposition 2.3]. Hence $t = (t_1, t_2) \in (0 :_S (0 \oplus P_2^k)) = \varphi^{-1}(N)$ and so $m = \varphi(t) \in N$. Then we have the equality.

Conversely, let $M$ be a quasi comultiplication $R$-module. By Proposition 4.2, we may assume that $\mathrm{qSpec}(S) \neq \emptyset$. Let $T$ be a non-zero quasi-prime submodule of $S$. Then $K \subseteq T$ by [7, Proposition 4.3 ], and so $T/K$ is a quasi-prime submodule of $S/K$ by Lemma 2.2. By an argument like that in [7, Theorem 4.4], $S$ is a quasi comultiplication $R$-module. □

We are ready to determine all indecomposable non-separated quasi comultiplication $R$-modules.

**Proposition 4.4.** *Let $R$ be the pullback ring as described in (1). Then the injective hull $E(R/P)$ of $R/P$ is a non-separated quasi comultiplication $R$-module.*

**Proof.** By [7, Proposition 4.2], $E(R/P)$ is a non-separated comultiplication $R$-module. So it is a non-separated quasi comultiplication $R$-module. □

**Proposition 4.5.** *Let $R$ be the pullback ring as described in (1) and let $M$ be a non-zero indecomposable non-separated quasi comultiplication $R$-module. Let $0 \to K \to S \to M \to 0$ be a separated representation of $M$. Then $\bar{S} = 0$.*

**Proof.** Assume to the contrary, $\bar{S} \neq 0$. By Theorem 4.3, $S$ is quasi comultiplication; hence $S$ is type (II) of Theorem 3.7 which are indecomposable. Then $P^m S = 0$ for some $m$, and so $P^m M = 0$. If $m = 1$, then $(P_1 \oplus 0)M \subseteq PM = 0$; so $(P_1 \oplus 0)M \cap (0 \oplus P_2)M = 0$ that is a contradiction. So suppose that $m \geq 2$. Let $k$ be the least positive integer such that $P^k M = 0$ (so $P^{k-1}M \neq 0$). It follows that $\varphi(P^k x) = 0$ for all $x \in S$; so $\varphi(P_1^k x) = \varphi(P_2^k x) = 0$. Since by [19, Proposition 2.3], $\varphi$ is one-to-one on $P_i S$ for each $i$, we get $P_1^k x = P_2^k x = 0$. Thus $P^k x = 0$ and hence $P^k S = 0$. Set $N = P^{k-1}M$. Then $0 \to K \to \varphi^{-1}(N) = P^{k-1}S \to N \to 0$ is a separated representation of $N$ by [6, Lemma 3.1]; hence $K \subseteq PP^{k-1}S = P^k S = 0$ which is a contradiction since $M$ is non-separated. Thus $\bar{S} = 0$. □

Before embarking on the proof of the next result let us explain its idea. Let $M$ be any $R$-module and let $0 \to K \xrightarrow{i} S \xrightarrow{\varphi} M \to 0$ be a separated representation of $M$. If $M$ is a quasi comultiplication non-separated $R$-module, then $S$ is quasi-comultiplication by Theorem 4.3. In this case, $S = S_1 \oplus S_2$, where $S_i$ is of type (I) (in statement of Theorem 3.7). In any separated representation $0 \to K \xrightarrow{i} S \xrightarrow{\varphi} M \to 0$ the kernel of the map $\varphi$ to $M$ is annihilated by $P$, hence is contained in the socle of the separated module $S$. Thus $M$ is obtained by amalgamation in the socles of the various direct summands of $S$.

Let $M$ be any non-zero indecomposable non-separated quasi comultiplication $R$-module and let $0 \to K \xrightarrow{i} S \xrightarrow{\varphi} M \to 0$ be a separated representation of $M$. By Proposition 4.4, the modules finite length do not occur among the direct summands of $S$ and $S = S_1 \oplus S_2$, where $S_i$ is of type (I) (in statement of Theorem 3.7). If there are two modules of type (I), then their generators cannot both be annihilated by the same $P_i$. This contradicts there being two copies of the $P_1$-Prüfer or two copies of the $P_2$-Prüfer. So $S_1$ is $P_1$-Prüfer and $S_2$ is $P_2$-Prüfer. It is clear that the module obtained this amalgamation is, indeed, $E(R/P)$, the $R$-injective hull of $R/P$ which is an indecomposable quasi-comultiplication non-separated $R$-module by Proposition 4.4 (also see [5, p. 4053]). Therefore we have the following theorem:

**Theorem 4.6.** *Let $R = (R_1 \to \bar{R} \leftarrow R_2)$ be the pullback of two discrete valuation domains $R_1, R_2$ with common factor field $\bar{R}$. Then the only non-zero indecomposable quasi comultiplication non-separated $R$-module, up to isomorphism, is $E(R/P)$, the $R$-injective hull of $R/P$.*

**Corollary 4.7.** *Let $R$ be the pullback ring as described in Theorem 4.6. Then every non-zero-indecomposable non-separated quasi comultiplication $R$-module is pure-injective.*

**Proof.** Apply [5, Theorem 3.5] and Theorem 4.6. □

**Acknowledgement.** The authors are deeply grateful to the referee for a careful reading of this manuscript and his valuable suggestions.

## References

[1] A. Abbasi and D. Hassanzadeh-Lelekaami, *Quasi-prime submodules and developed Zariski topology*, Algebra Colloq., 19 (2012), 1089-1108.

[2] D. M. Arnold and R. C. Laubenbacher, *Finitely generated modules over pullback rings*, J. Algebra, 184 (1996), 304-332.

[3] I. Assem, D. Simson and A. Skowroński, Elements of the Representation Theory of Associative Algebras, Vol.1, Techniques of Representation Theory, London Mathematical Society Student Texts, 65, Cambridge University Press, Cambridge, 2006.

[4] H. Bass, *On the ubiquity of Gorenstein rings*, Math. Z., 82 (1963), 8-28.

[5] S. Ebrahimi Atani, *On pure-injective modules over pullback rings*, Comm. Algebra, 28 (2000), 4037-4069.

[6] S. Ebrahimi Atani, *On secondary modules over pullback rings*, Comm. Algebra, 30 (2002), 2675-2685.

[7] R. Ebrahimi Atani and S. Ebrahimi Atani, *Comultiplication modules over a pullback of Dedekind domains*, Czechoslovak Math. J., 59 (2009), 1103-1114.

[8] R. Ebrahimi Atani and S. Ebrahimi Atani, *Weak comultiplication modules over a pullback of commutative local Dedekind domains*, Algebra Discrete Math., 1 (2009), 1-13.

[9] R. Ebrahimi Atani and S. Ebrahimi Atani, *On primarily multiplication modules over pullback rings*, Algebra Discrete Math., 11 (2011), 1-17.

[10] S. Ebrahimi Atani and F. Esmaeili Khalil Saraei, *Indecomposable primarily comultiplication modules over a pullback of two Dedekind domains*, Colloq. Math., 120 (2010), 23-42.

[11] S. Ebrahimi Atani and F. Esmaeili Khalil Saraei, *On P-torsion modules over a pullback of two Dedekind domains*, Georgian Math. J., 19 (2012), 473-488.

[12] S. Ebrahimi Atani, S. Dolati Pish Hesari, M. Khoramdel and M. Sedghi Shanbeh Bazari, *Absorbing comultiplication modules over a pullback ring*, Int. Electron. J. Algebra, 24 (2018), 31-49.

[13] F. Esmaeili Khalil Saraei, *On pseudo-prime multiplication modules over pullback rings*, Colloq. Math., 143 (2016), 63-78.

[14] A. Facchini, *Fiber products and Morita duality for commutative rings*, Rend. Sem. Mat. Univ. Padova, 67 (1982), 143-159.

[15] A. Facchini and P. Vamos, *Injective modules over pullbacks*, J. London Math. Soc., 31 (1985), 425-438.

[16] W. J. Heinzer, L. J. Ratlif, Jr. and D. E. Rush, *Strongly irreducible ideals of a commutative ring*, J. Pure Appl. Algebra, 166 (2002), 267-275.

[17] R. Kielpinski, *On $\Gamma$-pure injective modules*, Bull. Acad. Polon. Sci. Ser. Sci. Math. Astronom. Phys., 15 (1967), 127-131.

[18] V. V. Kiričenko, *Classification of the pairs of mutually annihilating operators in a graded space and representations of a dyad of generalized uniserial algebras, In: "Rings and Linear Groups"*, Zap. Naučn. Sem. Leningrad. Otdel. Mat. Inst. Steklov. (LOMI), 75 (1978), 91-109, 196-197.

[19] L. Klingler, *Integral representations of groups of square-free order*, J. Algebra, 129 (1990), 26-74.

[20] L. S. Levy, *Mixed modules over ZG, G cyclic of prime order, and over related Dedekind pullbacks*, J. Algebra, 71 (1981), 62-114.

[21] L. S. Levy, *Modules over pullbacks and subdirect sums*, J. Algebra, 71 (1981), 50-61.

[22] L. S. Levy, *Modules over Dedekind-like rings*, J. Algebra, 93 (1985), 1-116.

[23] L. A. Nazarova and A. V. Roiter, *Finitely generated modules over a dyad of two local Dedekind rings, and finite groups which possess an abelian normal divisor of index p*, Izv. Acad. Nauk SSSR Ser. Mat., 33 (1969), 65-89.

[24] M. Prest, Model Theory and Modules, London Mathematical Society Lecture Note Series, 130, Cambridge University Press, Cambridge, 1988.

[25] D. Simson, Linear Representations of Partially Ordered Sets and Vector Space Categories, Algebra, Logic and Applications, 4, Gordon and Breach Science Publishers, Switzerland-Australia, 1992.

[26] D. Simson and A. Skowroński, Elements of the Representation Theory of Associative Algebras, Representation-Infinite Tilted Algebras, Vol. 3, London Mathematical Society Student Texts, 72, Cambridge University Press, Cambridge, 2007.

[27] R. B. Warfield, Jr., *Purity and algebraic compactness for modules*, Pacific J. Math., 28 (1969), 699-719.

[28] A. N. Wiseman, *Projective modules over pullback rings*, Math. Proc. Cambridge Philos. Soc., 97 (1985), 399-406.

**S. Ebrahimi Atani**
Faculty of Mathematical Sciences
University of Guilan
P.O. Box 1914
Rasht, Iran
e-mail: ebrahimi@guilan.ac.ir

**F. Esmaeili Khalil Saraei** (Corresponding Author)
Fouman Faculty of Engineering
College of Engineering
University of Tehran
P.O. Box 43515-1155, Fouman
43516-66456, Guilan, Iran
e-mail: f.esmaeili.kh@ut.ac.ir

# Permissions

All chapters in this book were first published in IEJA, by ULAKBİM Dergi Sistemleri; hereby published with permission under the Creative Commons Attribution License or equivalent. Every chapter published in this book has been scrutinized by our experts. Their significance has been extensively debated. The topics covered herein carry significant findings which will fuel the growth of the discipline. They may even be implemented as practical applications or may be referred to as a beginning point for another development.

The contributors of this book come from diverse backgrounds, making this book a truly international effort. This book will bring forth new frontiers with its revolutionizing research information and detailed analysis of the nascent developments around the world.

We would like to thank all the contributing authors for lending their expertise to make the book truly unique. They have played a crucial role in the development of this book. Without their invaluable contributions this book wouldn't have been possible. They have made vital efforts to compile up to date information on the varied aspects of this subject to make this book a valuable addition to the collection of many professionals and students.

This book was conceptualized with the vision of imparting up-to-date information and advanced data in this field. To ensure the same, a matchless editorial board was set up. Every individual on the board went through rigorous rounds of assessment to prove their worth. After which they invested a large part of their time researching and compiling the most relevant data for our readers.

The editorial board has been involved in producing this book since its inception. They have spent rigorous hours researching and exploring the diverse topics which have resulted in the successful publishing of this book. They have passed on their knowledge of decades through this book. To expedite this challenging task, the publisher supported the team at every step. A small team of assistant editors was also appointed to further simplify the editing procedure and attain best results for the readers.

Apart from the editorial board, the designing team has also invested a significant amount of their time in understanding the subject and creating the most relevant covers. They scrutinized every image to scout for the most suitable representation of the subject and create an appropriate cover for the book.

The publishing team has been an ardent support to the editorial, designing and production team. Their endless efforts to recruit the best for this project, has resulted in the accomplishment of this book. They are a veteran in the field of academics and their pool of knowledge is as vast as their experience in printing. Their expertise and guidance has proved useful at every step. Their uncompromising quality standards have made this book an exceptional effort. Their encouragement from time to time has been an inspiration for everyone.

The publisher and the editorial board hope that this book will prove to be a valuable piece of knowledge for researchers, students, practitioners and scholars across the globe.

# List of Contributors


**Xigou Zhang**
College of Mathematics and Information Science Jiangxi Normal University Nanchang, Jiangxi 330022, China

**Xing Gao**
School of Mathematics and Statistics Key Laboratory of Applied Mathematics and Complex Systems Lanzhou University Lanzhou, Gansu 730000, China

**Li Guo**
Department of Mathematics and Computer Science Rutgers University – Newark Newark, NJ 07102, USA

**Pramod Kanwar**
Department of Mathematics, Ohio University-Zanesville, Zanesville, Ohio, USA

**Meenu Khatkar and R. K. Sharma**
Department of Mathematics, Indian Institute of Technology Delhi, New Delhi, 110016, India

**Lihong Dong and Shuan Xue**
College of Mathematics and Information Science, Henan Normal University, 453007 Xinxiang, Henan, China

**Robert Wisbauer**
Department of Mathematics University of Düsseldorf Germany

**Pradip Girase**
Department of Mathematics, K. K. M. College, Manwath Dist-Parbhani(M.S.)-431505, India

**Vandeo Borkar**
Department of Mathematics, Yeshwant Mahavidyalaya Nanded(M.S.)-431602, India

**Narayan Phadatare**
Department of Mathematics, Savitribai Phule Pune University, Pune (M.S.), India

**Gurninder S. Sandhu**
Department of Mathematics, Punjabi University, 147002 Patiala, India
Multani Mal Modi College, 147001 Patiala, India

**Deepak Kumar**
Department of Mathematics, Punjabi University, 147002 Patiala, India

**Hyun Seung Choi**
Department of Mathematics, Glendale Community College, Glendale, California 91208, U.S.A

**Andrew Walker**
Department of Mathematics, Midland University, Fremont, Nebraska, 68025, U.S.A

**Dario Spirito**
Dipartimento di Matematica e Fisica Universitá degli Studi "Roma Tre", Roma, Italy

**N. J. Groenewald**
Department of Mathematics, Faculty of Science, Nelson Mandela University, Port Elizabetyh, South Africa

**Bac T. Nguyen**
Nguyen Tat Thanh University, 300 A Nguyen Tat Thanh street, Ho Chi Minh city, Vietnam and Dept. of Basic Science Thainguyen Univ. of Economics and Business Administration, Thainguyen Vietnam

**Alberto Facchini**
Dipartimento di Matematica Pura e Applicata, Universitá di Padova, 35121 Padova, Italy

**Serap Şahinkaya**
Department of Mathematics, Faculty of Science, Gebze Technical University, 41400 Gebze, Kocaeli, Turkey

**Allen Herman and Gurmail Singh**
Department of Mathematics and Statistics, University of Regina, Regina, Canada, S4S 0A2

**Jürgen Herzog**
Universiti at Duisburg-Essen, Fachbereich Mathematik, Campus Essen, 45117, Essen, Germany

**Bahareh Lajmiri and Farhad Rahmati**
Department of Mathematics and Computer Science, Amirkabir University of Technology, 424 Hafez Ave, Tehran, Iran

**M. Tamer Koşan**
Department of Mathematics, Faculty of Sciences, Gazi University, Ankara, Turkey

**Jan Žemlička**
Department of Algebra, Charles University in Prague, Faculty of Mathematics and Physics, Sokolovská 83, 186 75 Praha 8, Czech Republic

**S. Ebrahimi Atani**
Faculty of Mathematical Sciences, University of Guilan, Rasht, Iran

**F. Esmaeili Khalil Saraei**
Fouman Faculty of Engineering, College of Engineering, University of Tehran, Fouman, 43516-66456, Guilan, Iran

# Index

## N



## P



## Q



## S



## T



## U



## V